Name Reactions

Springer
Berlin
Heidelberg
New York
Barcelona
Hong Kong
London
Milan
Paris
Tokyo

Jie Jack Li

Name Reactions

A Collection
of Detailed Reaction Mechanisms

Springer

Jie Jack Li, Ph. D

Pfizer Global Research and Development
Chemistry Department
2800 Plymouth Road
48105 Ann Arbor, MI
USA
e-mail: Jack.Li@Pfizer.com

ISBN 3-540-43024-5 Springer-Verlag Berlin Heidelberg New York

CIP data applied for

Die Deutsche Bibliothek - CIP-Einheitsaufnahme
Li, Jie Jack: Name reactions : a collection of detailed reaction mechanisms / Jie Jack Li. -
Berlin ; Heidelberg ; New York ; Barcelona ; Hongkong ; London ; Milan ; Paris ; Tokyo : Springer, 2002
ISBN 3-540-43024-5

Springer-Verlag Berlin Heidelberg New York
a member of BertelsmannSpringer Science+Business Media GmbH

http://www.springer.de

Printed in Germany

Typesetting: Dataconversion by author
Cover-design: design & production, Heidelberg
Printed on acid-free paper SPIN: 10901243 2 / 3020 hu - 5 4 3 2 1

To Vivien

Preface

What's in a name? That which we call a rose by any other name would smell as sweet.[a] On the other hand, *name reactions* in organic chemistry and the corresponding mechanisms are nevertheless fascinating for their far-reaching utility as well as their insight into organic reactions. Furthermore, understanding their mechanisms greatly enhances our ability to solve more complex chemical problems. As a matter of fact, some name reactions are the direct result of better understanding of the mechanisms, as exemplified by the Barton–McCombie reaction.[b] In addition, our knowledge of how reactions work can shed light on side reactions and by-products, or when a reaction does not give the "desired" product, the mechanism may provide clues to where the reaction has gone awry.

I started collecting named and unnamed organic reactions and their mechanisms while I was a graduate student. It occurred to me that many of my fellow practitioners are doing exactly the same, and that these efforts could be made easier through a monograph tabulating interesting and useful mechanisms of name reactions. To this end, I have updated my collection with many *contemporary* name reactions and added more recent references, especially up-to-date review articles. In reflecting the advent of asymmetric synthesis, relevant name reactions in this field have been included to the repertoire. Since the step-by-step mechanisms delineated within are mostly self-explanatory, detailed verbal explanations are not offered, although some important jargons entailing the types of transformations are highlighted. For some reactions, short descriptions are given as mnemonics rather than accurate definitions. With regard to the references, the first one is generally the original article, whereas the rest are articles and review articles. Readers interested in in-depth coverage of name reactions are encouraged to follow up with the references as well as the following five books covering the relevant topic:

1. Mundy, B. R.; Ellerd, M. G. *Name Reactions and Reagents in Organic Synthesis* John–Wiley & Sons, New York, **1988**.
2. Hassner, A.; Stumer, C. *Organic Synthesis Based on Named and Unnamed Reactions* Pergamon, **1994**.
3. Laue, L.; Plagens, A. *Named Organic Reactions* John–Wiley & Sons, New York, **1999**.
4. "*Organic Name Reactions*" section, *The Merck Index* (13th edition), **2001**.
5. Smith, M. B.; March, J. "*Advanced Organic Chemistry*" (5th edition), Wiley, New York, **2001**.

I would like to express my grateful thanks to Dr. Brian J. Myers of Wayne Sate University, Profs. Jeffrey N. Johnston of Indiana University and Christian M. Rojas of Bernard College, who read the manuscript and offered many invaluable comments and suggestions. Special thanks are due to Profs. Gordon W. Gribble of Dartmouth College, Louis S. Hegedus of Colorado State University, and Thomas R. Hoye of University of Minnesota for their critique of the drafts. In addi tion, I am very much indebted to Nadia M. Ahmad, John (Jack) Hodges, Michael D. Kaufman, Peter L Toogood, and Kim E. Werner for proofreading the manuscript. Any remaining errors are, of course, solely my own. I am also grateful to Ms. Ann Smith of Merck & Co., Inc. for her helpful communications and discussions. Last but not the least, I wish to thank my wife, Sherry Chun-hua Cai, for her understanding and support throughout the entire project.

Jack Li
Ann Arbor, Michigan
November, 2001

References

a. William Shakespeare, "*Romeo and Juliet*" Act II, Scene ii, **1594–1595**.
b. Derek H. R. Barton, "*Some Recollections of Gap Jumping*" American Chemical Society, Washington, DC, **1991**.

Table of Contents

Abbreviations and Acronyms

Ac	acetyl
AIBN	2,2'-azobisisobutyronitrile
Alpine-borane®	*B*-isopinocamphenyl-9-borabicyclo[3.3.1]-nonane
B:	generic base
9-BBN	9-borabicyclo[3.3.1]nonane
BINAP	2,2'-bis(diphenylphosphino)-1,1'-binaphthyl
Boc	*tert*-butyloxycarbonyl
t-Bu	*tert*-butyl
Cbz	benzyloxycarbonyl
m-CPBA	*m*-chloroperoxybenzoic acid
CuTC	copper thiophene-2-carboxylate
DABCO	1,4-diazabicyclo[2.2.2]octane
dba	dibenzylideneacetone
DBU	1,8-diazabicyclo[5.4.0]undec-7-ene
DCC	1,3-dicyclohexylcarbodiimide
DDQ	2,3-dichloro-5,6-dicyano-1,4-benzoquinone
DEAD	diethyl azodicarboxylate
Δ	solvent heated under reflux
$(DHQ)_2$-PHAL	1,4-bis(9-*O*-dihydroquinine)-phthalazine
$(DHQD)_2$-PHAL	1,4-bis(9-*O*-dihydroquinidine)-phthalazine
DIBAL	diisobutylaluminum hydride
DMA	*N,N*-dimethylacetamide
DMAP	*N,N*-dimethylaminopyridine
DME	1,2-dimethoxyethane
DMF	dimethylformamide
DMS	dimethylsulfide
DMSO	dimethylsulfoxide
DMSY	dimethylsulfoxonium methylide
DMT	dimethoxytrityl
dppb	1,4-bis(diphenylphosphino)butane
dppe	1,2-bis(diphenylphosphino)ethane
dppf	1,1'-bis(diphenylphosphino)ferrocene
dppp	1,3-bis(diphenylphosphino)propane
E1	unimolecular elimination
E2	bimolecular elimination
E1cb	2-step, base-induced β-elimination *via* carbanion
Eq	equivalent
HMPA	hexamethylphosphoric triamide
Imd	imidazole
LAH	lithium aluminum hydride
LDA	lithium diisopropylamide
LHMDS	lithium hexamethyldisilazane
LTMP	lithium 2,2,6,6-tetramethylpiperidine

Mes	mestyl
MVK	methyl vinyl ketone
NBS	*N*-bromosuccinimide
NCS	*N*-chlorosuccinimide
NIS	*N*-iodosuccinimide
NMP	1-methyl-2-pyrrolidinone
Nu	nucleophile
PCC	pyridinium chlorochromate
PDC	pyridinium dichromate
SET	single electron transfer
S_NAr	nucleophilic substitution on an aromatic ring
S_N1	unimolecular nucleophilic substitution
S_N2	bimolecular nucleophilic substitution
TBAF	tetrabutylammonium fluoride
TBDMS	*tert*-butyldimethylsilyl
TBS	*tert*-butyldimethylsilyl
Tf	trifluoromethanesulfonyl (triflyl)
TFA	trifluoroacetic acid
TFAA	trifluoroacetic anhydride
TFP	tri-*o*-furylphosphine
THF	tetrahydrofuran
TIPS	triisopropylsilyl
TMEDA	*N,N,N',N'*-tetramethylethylenediamine
TMP	tetramethylpiperidine
TMS	trimethylsilyl
Tol	toluene or tolyl
Tol-BINAP	2,2'-bis(di-*p*-tolylphosphino)-1,1'-binaphthyl
Ts	tosylate

Abnormal Claisen rearrangement

Further rearrangement of the normal Claisen rearrangement product with the β-carbon becoming attached to the ring.

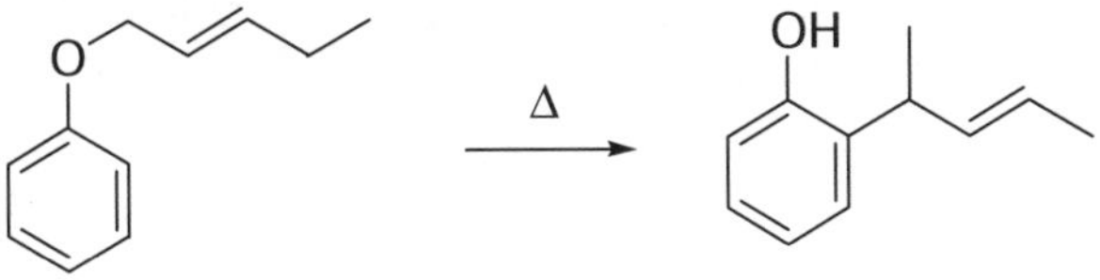

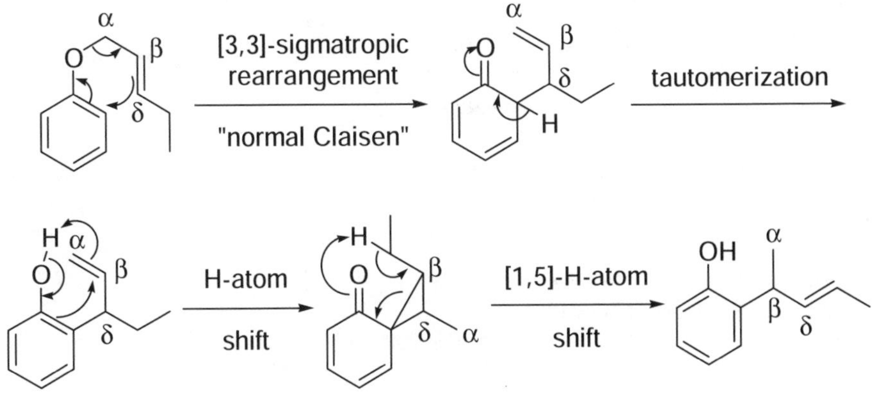

References

1. Hansen, H.-J. In *Mechanisms of Molecular Migrations,* vol. 3, Thyagarajan, B. S. ed. Wiley-Interscience: New York, **1971**, pp 177–200.
2. Shah, R. R.; Trivedi, K. N. *Curr. Sci.* **1975**, *44*, 226.
3. Kilenyi, S. N.; Mahaux, J. M.; Van Durme, E. *J. Org. Chem.* **1991**, *56*, 2591.
4. Nakamura, S.; Ishihara, K.; Yamamoto, H. *J. Am. Chem. Soc.* **2000**, *122*, 8131.
5. Schobert, R.; Siegfried, S.; Gordon, G.; Mulholland, D.; Nieuwenhuyzen, M. *Tetrahedron Lett.* **2001**, *42*, 4561.

Alder ene reaction

ene

reaction

References

1. Alder, K.; Pascher, F.; Schmitz, A. *Ber.* **1943**, *76*, 27.
2. Oppolzer, W. *Angew. Chem.* **1984**, *96*, 840.
3. Johnson, J. S.; Evans, D. A. *Acc. Chem. Res.* **2000**, *33*, 325.

Allan–Robinson reaction

Synthesis of flavones or isoflavones.

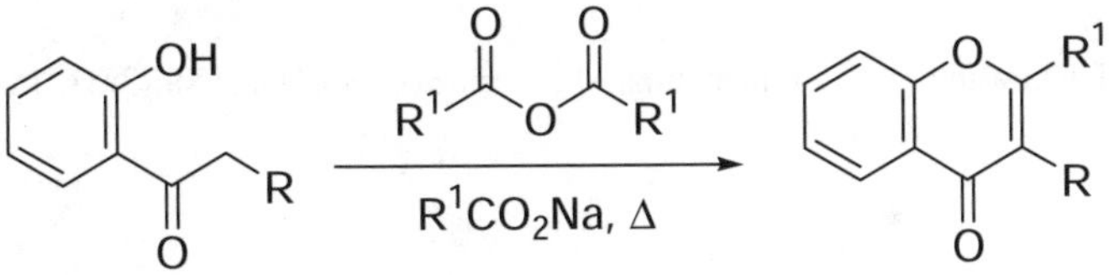

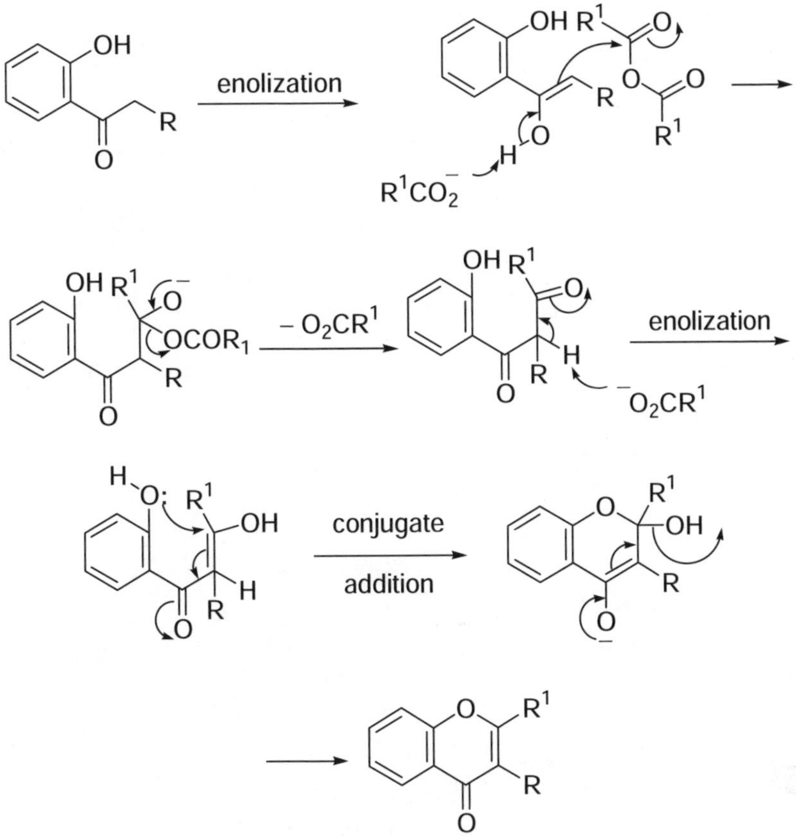

References

1. Allan, J.; Robinson, R. *J. Chem. Soc.* **1924**, *125*, 2192.
2. Szell, T.; Dozsai, L.; Zarandy, M.; Menyharth, K. *Tetrahedron* **1969**, *25*, 715.
3. Wagner, H.; Maurer, I.; Farkas, L.; Strelisky, J. *ibid.* **1977**, *33*, 1405.
4. Dutta, P. K.; Bagchi, D.; Pakrashi, S. C. *Indian J. Chem., Sect. B* **1982**, *21B*, 1037.

5. Patwardhan, S. A.; Gupta, A. S. *J. Chem. Res., (S)* **1984**, 395.
6. Horie, T.; Tsukayama, M.; Kawamura, Y.; Seno, M. *J. Org. Chem.* **1987**, *52*, 4702.
7. Horie, T.; Tsukayama, M.; Kawamura, Y.; Yamamoto, S. *Chem. Pharm. Bull.* **1987**, *35*, 4465.
8. Horie, T.; Kawamura, Y.; Tsukayama, M.; Yoshizaki, S. *ibid.* **1989**, *37*, 1216.

Alper carbonylation

$Pd(PPh_3)_4$, CO

$Pd(Ph_3)_4 \rightleftharpoons Pd(PPh_3)_3 + PPh_3$

$Pd(PPh_3)_3 + CO \longrightarrow Pd(PPh_3)_3CO$

:$Pd(PPh_3)_3CO$ — oxidative addition → PPh_3 + Ph, N, $Pd(II)(PPh_3)_2CO$

dimerization → Ph, N, N, $OC(Ph_3P)_2(II)Pd$ — PPh_3 →

Ph, N, N, PPh_3, :$Pd(PPh_3)_2CO$ — $- PPh_3$ →

Ph, N, N–Pd–C≡O, $(PPh_3)_2$ — CO → Ph, N, N, Pd, $(II)(PPh_3)_2CO$, O

reductive elimination → Ph, N, N, O + $Pd(PPh_3)_3CO$

An alternative mechanism:

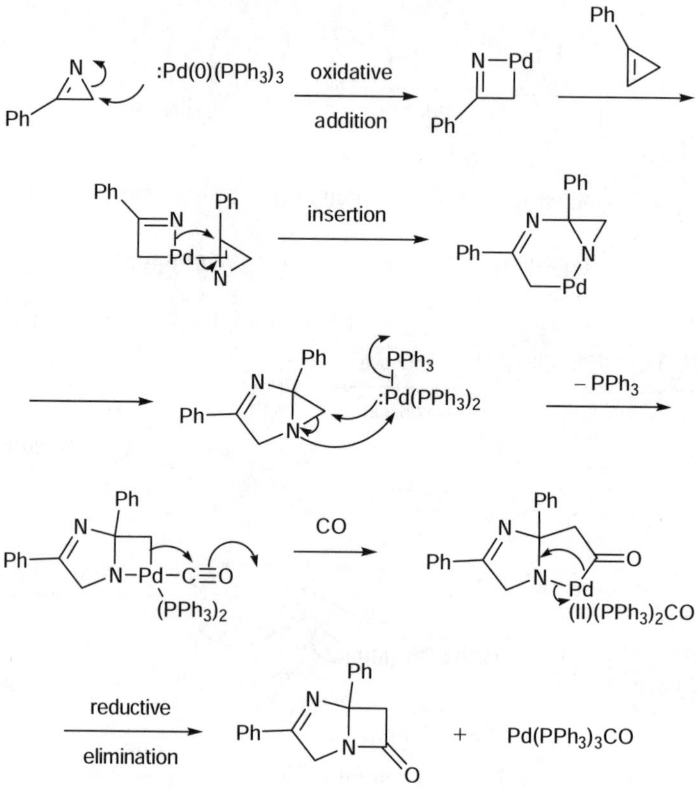

:Pd(0)(PPh$_3$)$_3$ oxidative addition

insertion

– PPh$_3$

CO

reductive elimination

+ Pd(PPh$_3$)$_3$CO

References

1. Alper, H.; Perera, C. P. *J. Am. Chem. Soc.* **1981**, *103*, 1289.
2. Alper, H.; Mahatantila, C. P. *Organometallics* **1982**, *1*, 70.
3. Alper, H. *Tetrahedron Lett.* **1987**, *28*, 3237.
4. Alper, H. *Aldrichimica Acta* **1991**, *24*, 3.

Amadori glucosamine rearrangement

Transformation of an aldose to a ketose using an amine.

HCl/H_2O
pyridine

glycosylamine

1-amino-1-deoxyketose

H^+

tautomerization

H^+

References

1. Amadori, M. *Atti Accad. Nazl. Lincei* **1925**, *2*, 337.
2. Hodges, J. E. *Adv. Carbohydrate Chem.* **1955**, *10*, 169.
3. Simon, H.; Kraus, A. *Fortschr. Chem. Forsch.* **1970**, *14*, 430.
4. Yaylayan, V. A.; Huyghues-Despointes, A. *Carbohydr. Res.* **1996**, *286*, 187.
5. Wrodnigg, T. M.; Stutz, A. E.; Withers, S. G. *Tetrahedron Lett.* **1997**, *38*, 5463.
6. Kadokawa, J.-I.; Hino, D.; Karasu, M.; Tagaya, H.; Chiba, K. *Chem. Lett.* **1998**, 383.
7. Turner, J. J.; Wilschut, N.; Overkleeft, H. S.; Klaffke, W.; Van Der Marel, G. A.; Van Boom, J. H. *Tetrahedron Lett.* **1999**, *40*, 7039.
8. Cremer, D. R.; Vollenbroeker, M.; Eichner, K. *Eur. Food Res. Technol.* **2000**, *211*, 400.

Angeli–Rimini hydroxamic acid synthesis

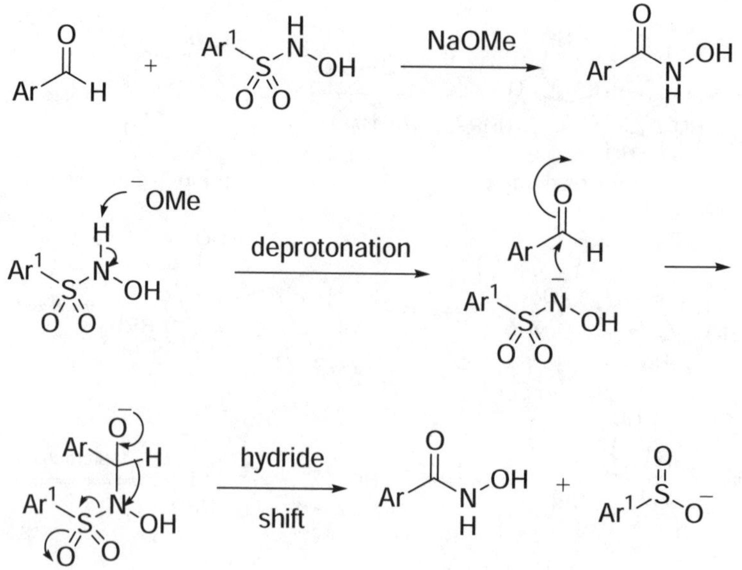

References

1. Angeli, A. *Gazz. Chim. Ital.* **1896**, *26(II)*, 17.
2. Yale H. L., *Chem. Rev.*, **1943**, *33*, 228.

ANRORC mechanism

Addition of Nucleophiles, Ring Opening and Ring Closure.

NaNH$_2$

liq. NH$_3$

+

$N^* = N^{15}$

Chichibabin amination

ring opening

common nitrile intermediate

6-*exo-dig* ring closure

tautomerization

5-*endo-trig* ring closure

– 2 H oxidation

References

1. van der Plas, H. C. *Acc. Chem. Res.* **1978**, *11*, 462.
2. Kost, A. N.; Sagitulin, R. S. *Tetrahedron* **1981**, *37*, 3423.
3. Briel, D. *Pharmazie* **1999**, *54*, 858.

Arndt–Eistert homologation

One carbon homologation of carboxylic acids using diazomethane.

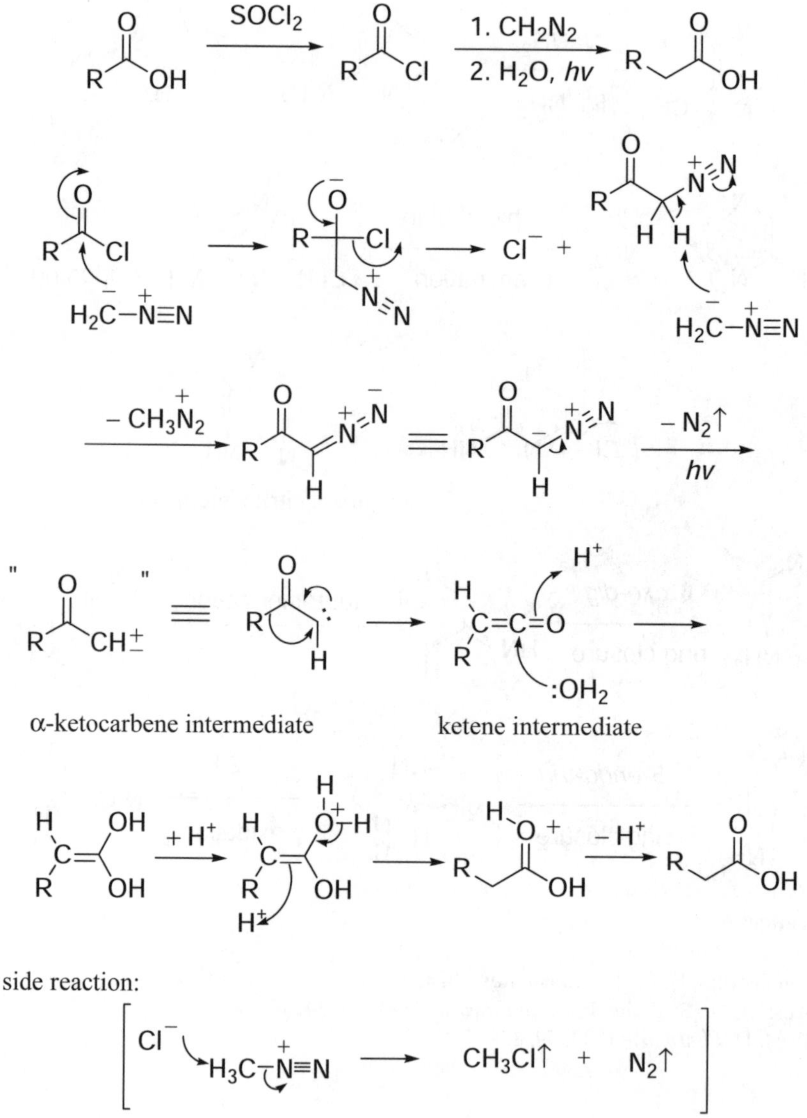
SOCl2
1. CH2N2
2. H2O, hv
– CH3N2
– N2↑
hv
α-ketocarbene intermediate
ketene intermediate
+ H+
– H+
side reaction:
CH3Cl↑ + N2↑

References

1. Arndt, F.; Eistert, B. *Ber.* **1935**, *68*, 200.
2. Kuo, Y. C.; Aoyama, T.; Shioiri, T. *Chem. Pharm. Bull.* **1982**, *30*, 899.
3. Podlech, J.; Seebach, D. *Angew. Chem., Int. Ed. Engl.* **1995**, *34*, 471.
4. Katritzky, A. R.; Zgang, S.; Fang, Y. *Org. Lett.* **2000**, *2*, 3789.

Baeyer–Drewson indigo synthesis

Applicable for the detection of *o*-nitrobenzaldehyde.

CHO
NO_2
+
O
NaOH
OH O
NO_2

O H N
N H O

O
O
OH
O
H
H
NO_2
aldol
condensation
OH O
+ N O
O

enolate
formation
OH O
+ N O
O
OH
OH
H O
+
N
O OH
$- H_2O$

OH
HO H
O
+
N
O
$- H_2O$
OH
O
N
O

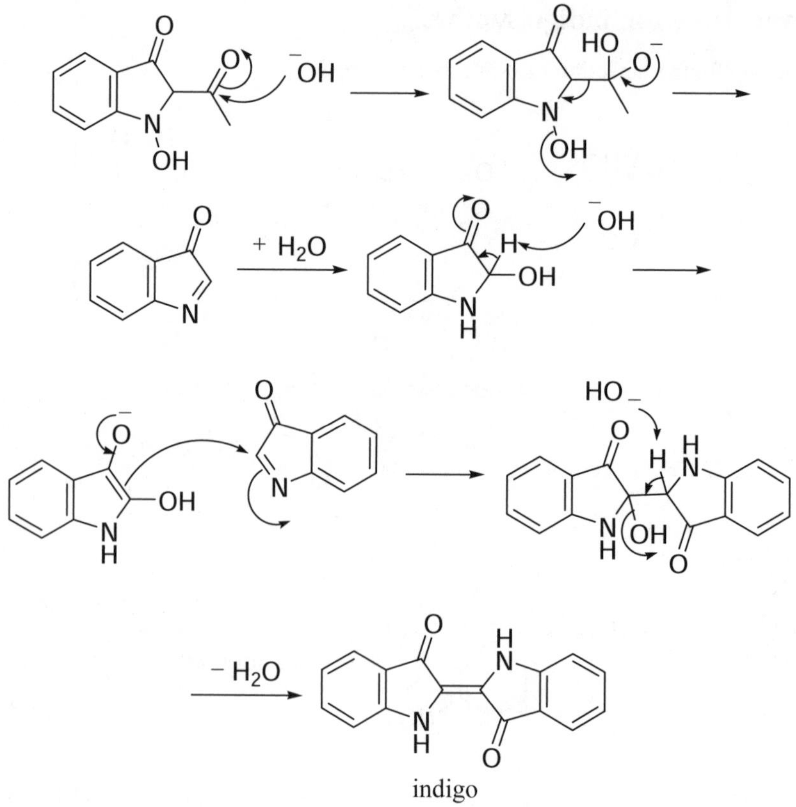

indigo

References

1. Baeyer, A.; Drewson, V. *Ber.* **1882**, *15*, 2856.
2. Hinkel, L. E.; Ayling, E. E. *J. Chem. Soc.* **1932**, 985.
3. Sainsbury, M. In *Rodd's Chemistry of Carbon Compounds IVB*, **1977**, 346.
4. McKee, J. R.; Zanger, M. *J. Chem. Educ.* **1991**, *68*, A242.

Baeyer–Villiger oxidation

General scheme:

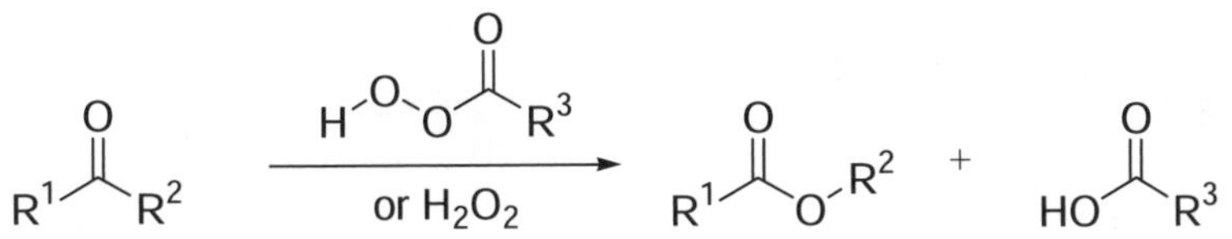

The most electron-rich alkyl group (more substituted carbon) migrates first. The general migration order: tertiary alkyl > secondary alkyl > cyclohexyl > benzyl > phenyl > primary alkyl > methyl >> H

e.g.:

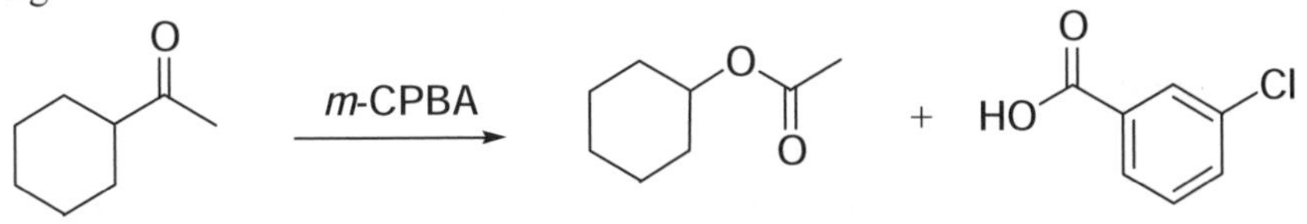

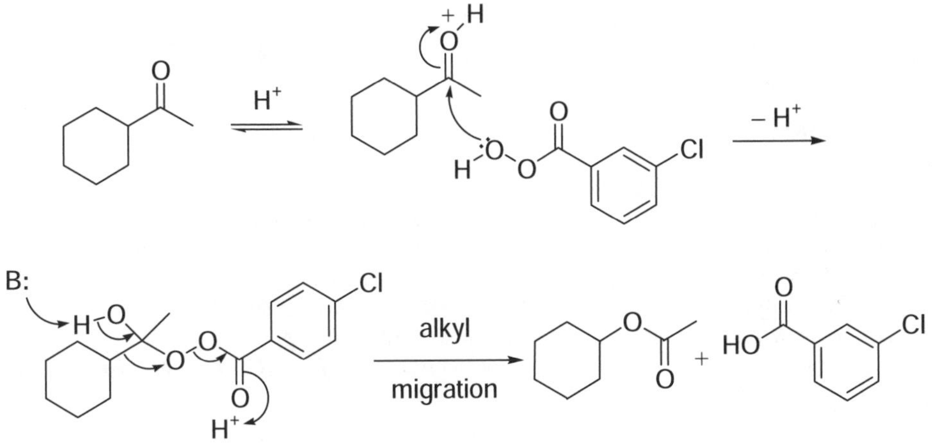

References

1. v. Baeyer, A.; Villiger, V. *Ber.* **1899**, *32*, 3625.
2. Krow, G. R. *Org. React.* **1993**, *43*, 251.
3. Renz, M.; Meunier, B. *Eur. J. Org. Chem.* **1999**, *4*, 737.
4. Bolm, C.; Beckmann, O. *Compr. Asymmetric Catal. I-III* **1999**, *2*, 803.
5. Crudden, C. M.; Chen, A. C.; Calhoun, L. A. *Angew. Chem., Int. Ed.* **2000**, *39*, 2851.
6. Hickman, Z. L.; Sturino, C. F.; Lachance, N. *Tetrahedron Lett.* **2000**, *41*, 8217.
7. Fukuda, O.; Sakaguchi, S.; Ishii, Y. *Tetrahedron Lett.* **2001**, *42*, 3479.

Baker–Venkataraman rearrangement

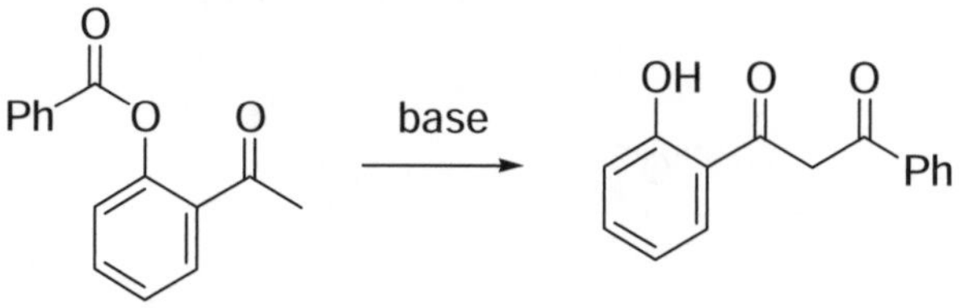

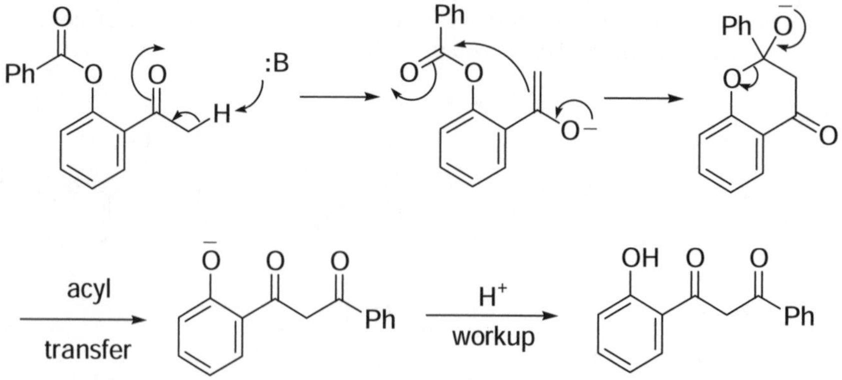

References

1. Baker, W. *J. Chem. Soc.* **1933**, 1381.
2. Kraus, G. A.; Fulton, B. S.; Wood, S. H. *J. Org. Chem.* **1984**, *49*, 3212.
3. Bowden, K.; Chehel-Amiran, M. *J. Chem. Soc., Perkin Trans. 2* **1986**, 2039.
4. Makrandi, J. K.; Kumari, V. *Synth. Commun.* **1989**, *19*, 1919.
5. Reddy, B. P.; Krupadanam, G. L. D. *J. Heterocycl. Chem.* **1996**, *33*, 1561.
6. Kalinin, A. V.; Snieckus, V. *Tetrahedron Lett.* **1998**, *39*, 4999.
7. Pinto, D. C. G. A.; Silva, A. M. S.; Cavaleiro, J. A. S. *New J. Chem.* **2000**, *24*, 85.

Bamberger rearrangement

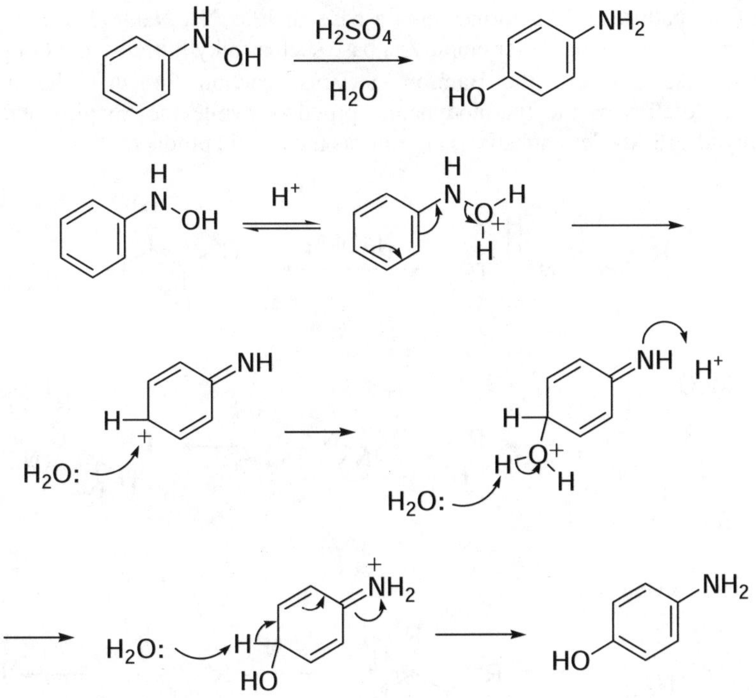

$$\text{PhNHOH} \xrightarrow[H_2O]{H_2SO_4} \text{HO-}C_6H_4\text{-}NH_2$$

References

1. Bamberger, E. *Ber.* **1894**, *27*, 1548.
2. Shine, H. J. In *Aromatic Rearrangement* Elsevier: New York, **1967**, pp 182–190.
3. Sone, T.; Tokuda, Y.; Sakai, T.; Shinkai, S.; Manabe, O. *J. Chem. Soc., Perkin Trans. 2* **1981**, 298.
4. Fishbein, J. C.; McClelland, R. A. *J. Am. Chem. Soc.* **1987**, *109*, 2824.
5. Zoran, A.; Khodzhaev, O.; Sasson, Y. *J. Chem. Soc., Chem. Commun.* **1994**, 2239.
6. Fishbein, J. C.; McClelland, R. A. *Can. J. Chem.* **1996**, *74*, 1321.
7. Naicker, K. P.; Pitchumani, K.; Varma, R. S. *Catal. Lett.* **1999**, *58*, 167.

Bamford–Stevens reaction

The Bamford–Stevens reaction and the Shapiro reaction share a similar mechanistic pathway. The former uses a base such as Na, NaOMe, LiH, NaH, $NaNH_2$, *etc.*, whereas the latter employs a base such as alkyllithiums and Grignard reagents. As a result, the Bamford–Stevens reaction furnishes the more-substituted olefins as the thermodynamic products, while the Shapiro reaction generally affords the less-substituted olefins as the kinetic products.

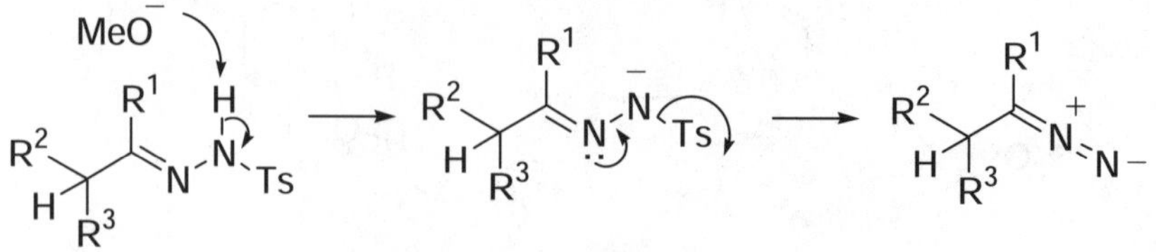

In protic solvent:

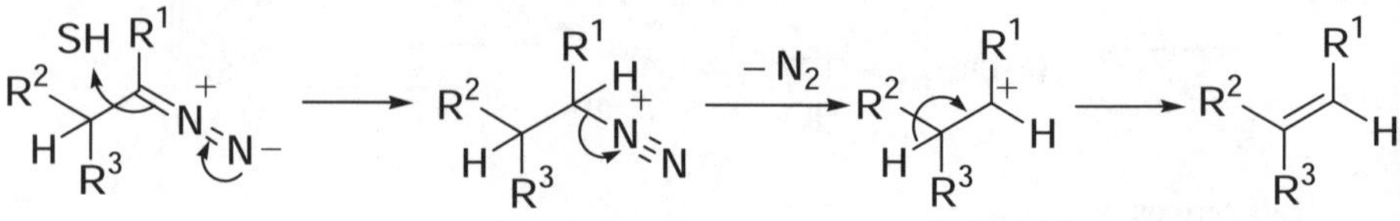

In aprotic solvent:

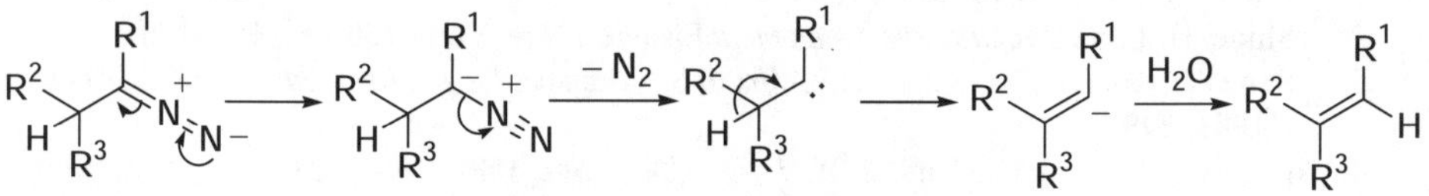

References

1. Bamford, W. R.; Stevens, T. S. M. *J. Chem. Soc.* **1952**, 4735.
2. Casanova, J.; Waegell, B. *Bull. Soc. Chim. Fr.* **1975**, 922.
3. Shapiro, R. H. *Org. React.* **1976**, *23*, 405.
4. Adlington, R. M.; Barrett, A. G. M. *Acc. Chem. Res.* **1983**, *16*, 55.
5. Sarkar, T. K.; Ghorai, B. K. *J. Chem. Soc., Chem. Commun.* **1992**, 1184.
6. Nickon, A.; Stern, A. G.; Ilao, M. C. *Tetrahedron Lett.* **1993**, *34*, 1391.
7. Olmstead, K. K.; Nickon, A. *Tetrahedron* **1998**, *54*, 12161.
8. Olmstead, K. K.; Nickon, A. *ibid.* **1999**, *55*, 7389.
9. Khripach, V. V.; Zhabinskii, V. N.; Kotyatkina, A. I.; *Mendeleev Commun.* **2001**, 144.

Bargellini reaction

Synthesis of hindered morpholinones and piperazinones from acetone and 2-amino-2-methyl-1-propanol or 1,2-diaminopropanes.

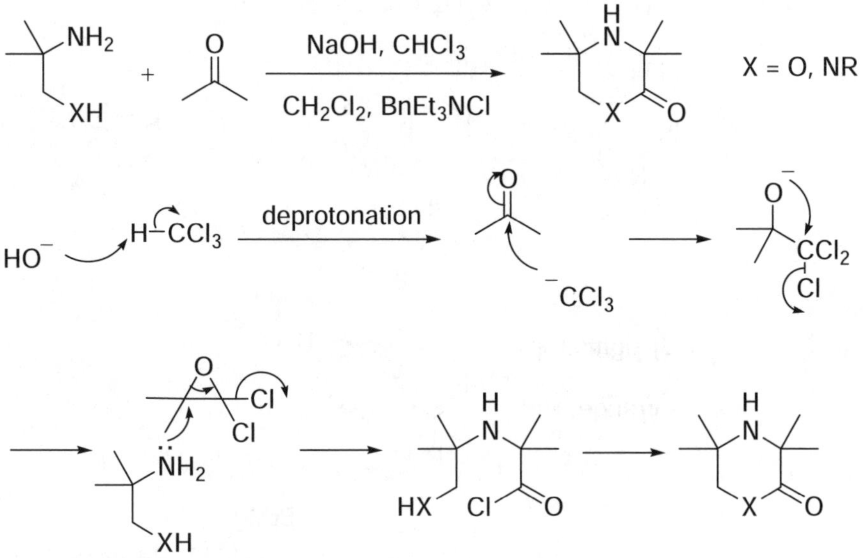

NaOH, $CHCl_3$

CH_2Cl_2, $BnEt_3NCl$

X = O, NR

deprotonation

References

1. Bargellini, G. *Gazz. Chim. Ital.* **1906**, *36*, 329.
2. Lai, J. T. *J. Org. Chem.* **1980**, *45*, 754.
3. Lai, J. T. *Synthesis* **1981**, 754.
4. Lai, J. T. *ibid.* **1984**, 122.
5. Lai, J. T. *ibid.* **1984**, 124.
6. Rychnovsky, S. D.; Beauchamp, T.; Vaidyanathan, R.; Kwan, T. *J. Org. Chem.* **1998**, *63*, 6363.

Bartoli indole synthesis

1. MgBr

2. H_2O^+

[3,3]-sigmatropic rearrangement

[3,3]-sigmatropic rearrangement

H^+ workup

References

1. Bartoli, G.; Palmieri, G.; Bosco, M.; Dalpozzo, R. *Tetrahedron Lett.* **1989**, *30*, 2129.
2. Dobson, D. R.; Gilmore, J.; Long, D. A. *Synlett* **1992**, 79.
3. Dobbs, A. P.; Voyle, M.; Whittall, N. *ibid.* **1999**, 1594.
4. Dobbs, A. *J. Org. Chem.* **2001**, *66*, 638.

Barton decarboxylation reaction

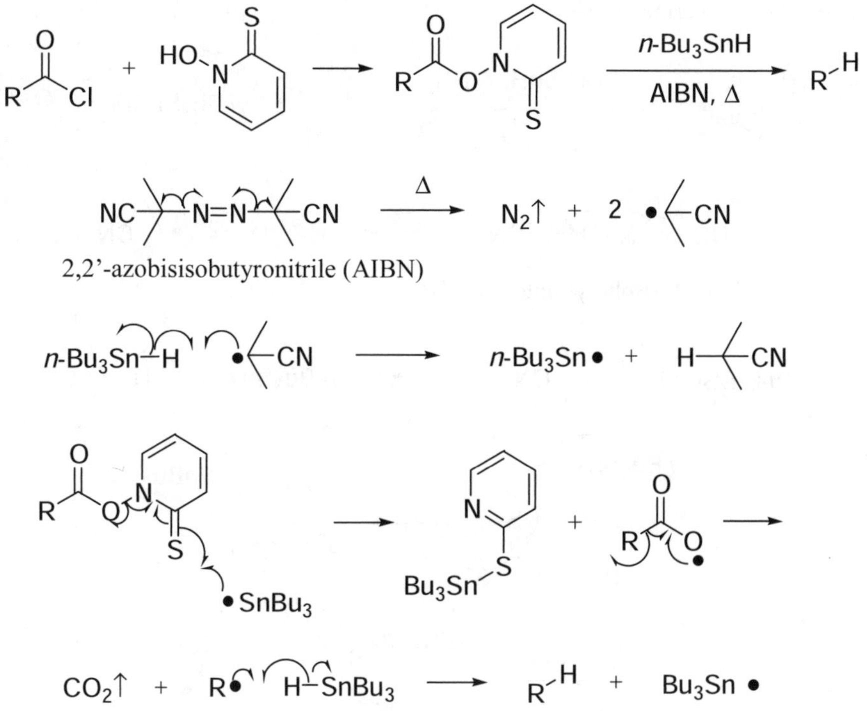

n-Bu_3SnH, AIBN, Δ

NC—N=N—CN →(Δ) N_2↑ + 2 •—CN

2,2'-azobisisobutyronitrile (AIBN)

n-Bu_3Sn—H •—CN → n-Bu_3Sn• + H—CN

• $SnBu_3$ Bu_3Sn—S

CO_2↑ + R• H—$SnBu_3$ → R—H + Bu_3Sn •

References

1. Barton, D. H. R.; Crich, D.; Motherwell, W. B. *J. Chem. Soc., Chem. Commun.* **1983**, 939.
2. Magnus, P.; Ladlow, M.; Kim, C. S.; Boniface, P. *Heterocycles* **1989**, *28*, 951.
3. Barton, D. H. R. *Aldrichimica Acta* **1990**, *23*, 3.
4. Gawronska, K.; Gawronski, J.; Walborsky, H. M. *J. Org. Chem.* **1991**, *56*, 2193.
5. Eaton, P. E.; Nordari, N.; Tsanaktsidis, J.; Upadhyaya, S. P. *Synthesis* **1995**, 501.
6. Crich, D.; Hwang, J.-T.; Yuan, H. *J. Org. Chem.* **1996**, *61*, 6189.
7. Elena, M.; Taddei, M. *Tetrahedron Lett.* **2001**, *42*, 3519.

Barton–McCombie deoxygenation reaction

Deoxygenation of alcohols by means of radical scission of their corresponding thiocarbonyl intermediates.

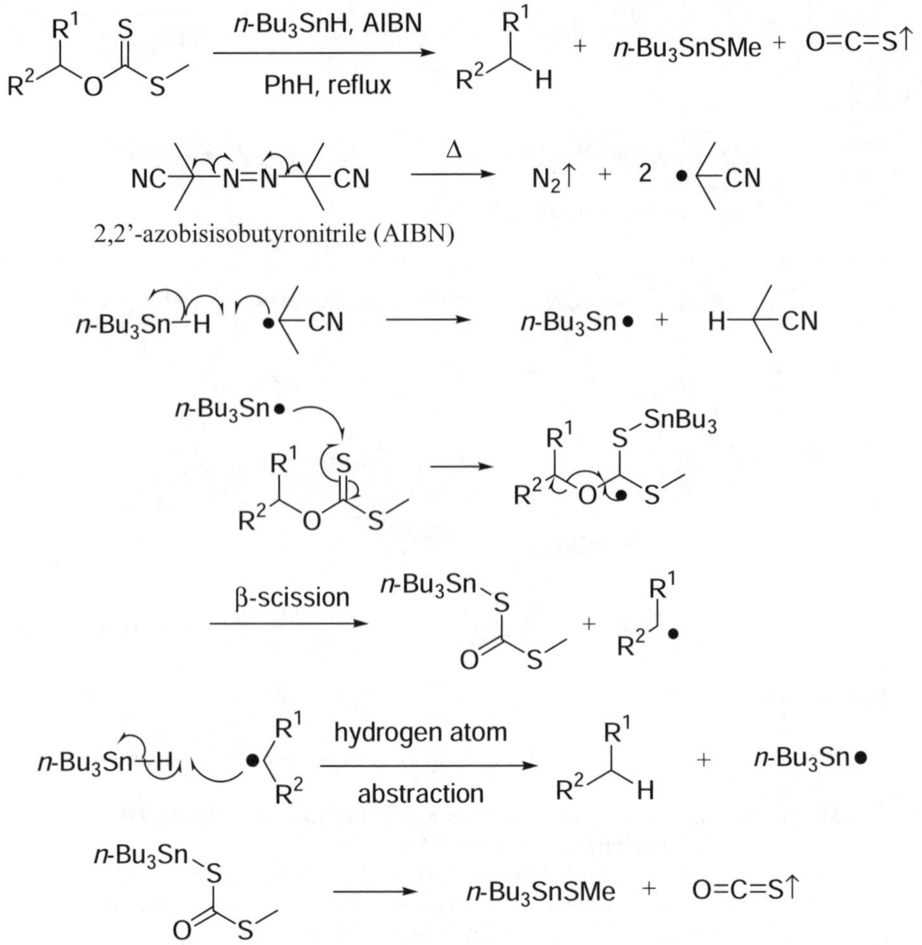

2,2'-azobisisobutyronitrile (AIBN)

References

1. Barton, D. H. R.; McCombie, S. W. *J. Chem. Soc., Perkin Trans. 1* **1975**, 1574.
2. Zard, S. Z. *Angew. Chem., Int. Ed. Engl.* **1997**, *36*, 672.
3. Lopez, R. M.; Hays, D. S.; Fu, G. C. *J. Am. Chem. Soc.* **1997**, *119*, 6949.
4. Hansen, H. I.; Kehler, J. *Synthesis* **1999**, 1925.
5. Boussaguet, P.; Delmond, B.; Dumartin, G.; Pereyre, M. *Tetrahedron Lett.* **2000**, *41*, 3377.
6. Cai, Y.; Roberts, B. P. *Tetrahedron Lett.* **2001**, *42*, 763.

Barton nitrite photolysis

NOCl

$h\nu$

– HCl

$h\nu$
homolytic
cleavage
– ON•

1,5-hydrogen
abstraction

Nitric oxide radical is a stable,
and therefore, long-lived radical

nitroso intermediate

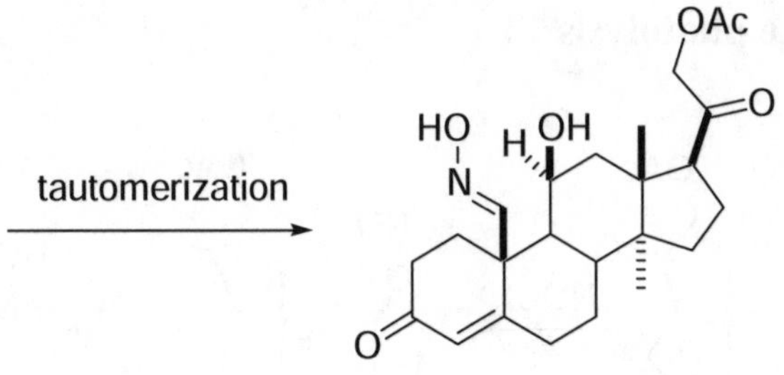

tautomerization

References

1. Barton, D. H. R.; Beaton, J. M.; Geller, L. E.; Pechet, M. M. *J. Am. Chem. Soc.* **1960**, *82*, 2640.
2. Barton, D. H. R.; Beaton, J. M.; Geller, L. E.; Pechet, M. M. *ibid.* **1960**, *82*, 2641.
3. Barton, D. H. R.; Beaton, J. M.; Geller, L. E.; Pechet, M. M. *ibid.* **1961**, *83*, 4083.
4. Barton, D. H. R.; Hesse, R. H.; Pechet, M. M.; Smith, L. C. *J. Chem. Soc., Perkin Trans. 1* **1979**, 1159.
5. Barton, D. H. R. *Aldrichimica Acta* **1990**, *23*, 3.
6. Herzog, A.; Knobler, C. B.; Hawthorne, M. F. *Angew. Chem., Int. Ed. Engl.* **1998**, *37*, 1552.

Baylis–Hillman reaction

General scheme:

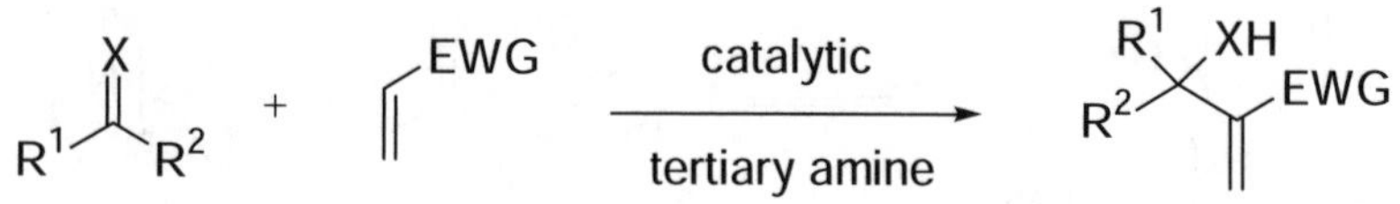

X = O, NR_2, EWR = CO_2R, COR, CHO, CN, SO_2R, SO_3R, $PO(OEt)_2$, $CONR_2$, $CH_2{=}CHCO_2Me$

e.g.:

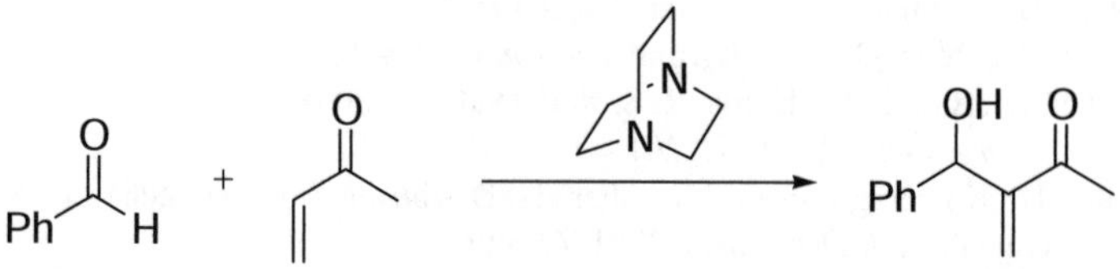

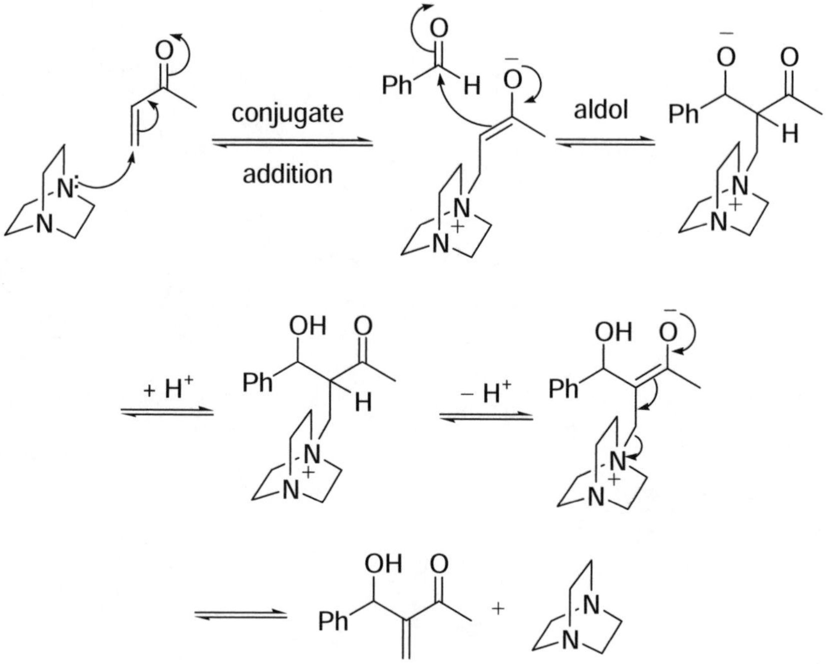

E2 (bimolecular elimination) mechanism is also operative here:

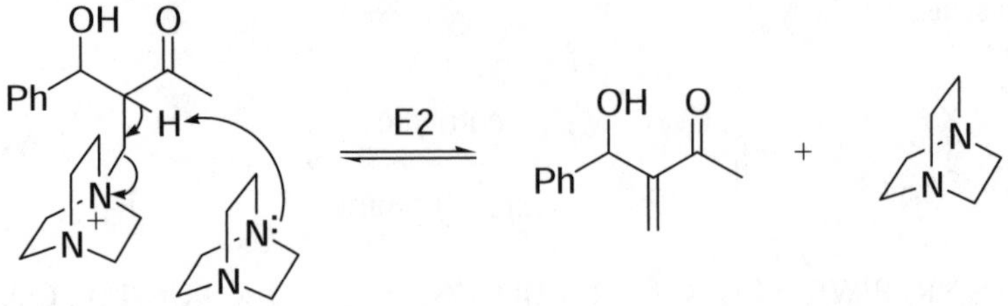

References

1. Baylis, A. B.; Hillman, M. E. D. *Ger. Pat.* 2,155,113, **1972**.
2. Drewes, S. E.; Roos, G. H. P. *Tetrahedron* **1988**, *44*, 4653.
3. Basavaiah, D.; Rao, P. D.; Hyma, R. S. *ibid.* **1996**, *52*, 8001.
4. Ciganek, E. *Org. React.* **1997**, *51,* 201.
5. Basavaih, D.; Kumaragurubaran, N.; Sharada, D. *Tetrahedron Lett.* **2001**, *42*, 85.
6. Shi, M.; Feng, Y.-S. *J. Org. Chem.* **2001**, *66*, 406.
7. Kim, J. N.; Im, Y. J.; Gong, J. H.; Imaeda, K. *Tetrahedron Lett.* **2001**, *42*, 4195.

Beckmann rearrangement

The acid-mediated isomerization of oximes to amides.

PCl_5

$-HCl$

the substituent *trans* to the leaving group migrates

$-H^+$

tautomerization

References

1. Beckmann, E. *Chem. Ber.* **1886**, *89*, 988.
2. Chatterjea, J. N.; Singh, K. R. R. P. *J. Indian Chem. Soc.* **1982**, *59*, 527.
3. Gawley, R. E. *Org. React.* **1988**, *35*, 1.
4. Catsoulacos, P.; Catsoulacos, D. *J. Heterocycl. Chem.* **1993**, *30*, 1.
5. Anilkumar, R.; Chandrasekhar, S. *Tetrahedron Lett.* **2000**, *41*, 7235.
6. Barman, D. C.; Thakur, A. J.; Sandhu, J. S. *Chem. Lett.* **2000**, 1196.
7. Khodaei, M. M.; Meybodi, F. A.; Rezai, N.; Salehi, P. *Synth. Commun.* **2001**, *31*, 2047.

Beirut reaction

Synthesis of quinoxaline-1,4-dioxide from benzofurazan oxide.

Et_3N

EtOH

Ph

CO_2Et

Et_3N:

$-CO_2$

$-Et_4N^+$

$HNEt_3$

COPh

CHO

OH

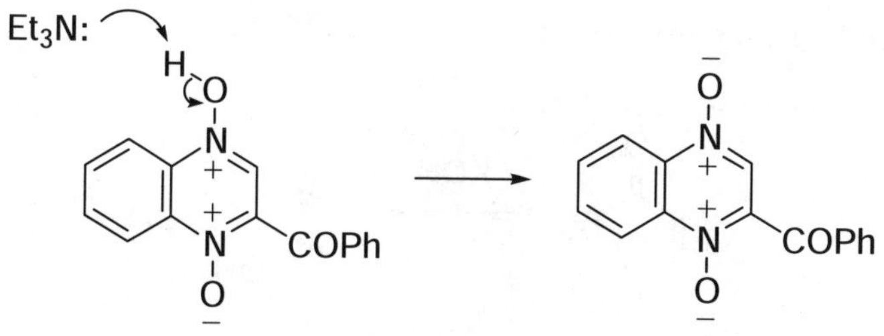

Et_3N:

COPh

COPh

References

1. Haddadin, M. J.; Issidorides, C. H. *Heterocycles* **1976**, *4*, 767.
2. Gaso, A.; Boulton, A. J. In *Advances in Heterocyclic Chem.* Vol. 29, eds, Katritzky, A. R.; Boulton, A. J., Academic Press Inc.: New York, **1981**, 251.
3. Atfah, A.; Hill, J. *J. Chem. Soc., Perkin Trans. 1* **1989**, 221.
4. Haddadin, M. J.; Issidorides, C. H. *Heterocycles* **1993**, *35*, 1503.
5. El-Abadelah, M. M.; Nazer, M. Z.; El-Abadla, N. S.; Meier, H. *Heterocycles* **1995**, *41*, 2203.

Benzilic acid rearrangement

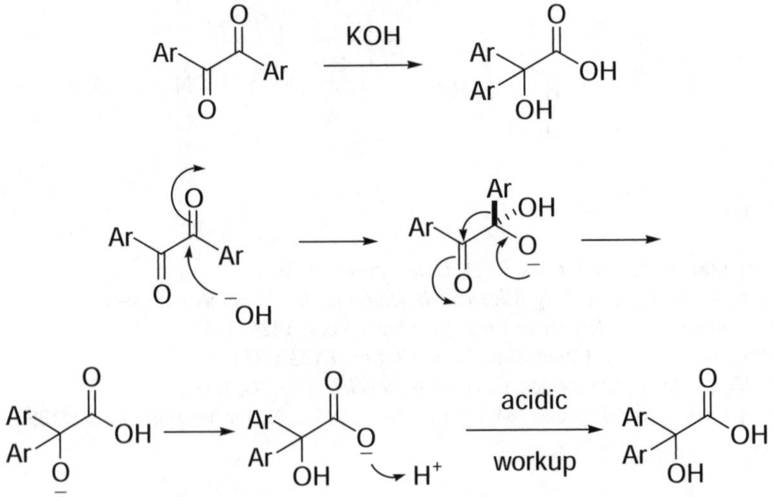

Final deprotonation of the carboxylic acid drives the reaction forward.

References

1. Liebig, J. *Liebigs Ann. Chem.* **1838**, *31*, 329.
2. Rajyaguru, I.; Rzepa, H. S. *J. Chem. Soc., Perkin Trans. 2* **1987**, 1819.
3. Toda, F.; Tanaka, K.; Kagawa, Y.; Sakaino, Y. *Chem. Lett.* **1990**, 373.
4. Robinson, J.; Flynn, E. T.; McMahan, T. L.; Simpson, S. L.; Trisler, J. C.; Conn, K. B. *J. Org. Chem.* **1991**, *56*, 6709.
5. Hatsui, T.; Wang, J.-J.; Ikeda, S.-y.; Takeshita, H. *Synlett* **1995**, 35.
6. Yu, H.-M.; Chen, S.-T.; Tseng, M.-J.; Chen, S.-T.; Wang, K.-T. *J. Chem. Res., (S)* **1999**, 62.

Benzoin condensation

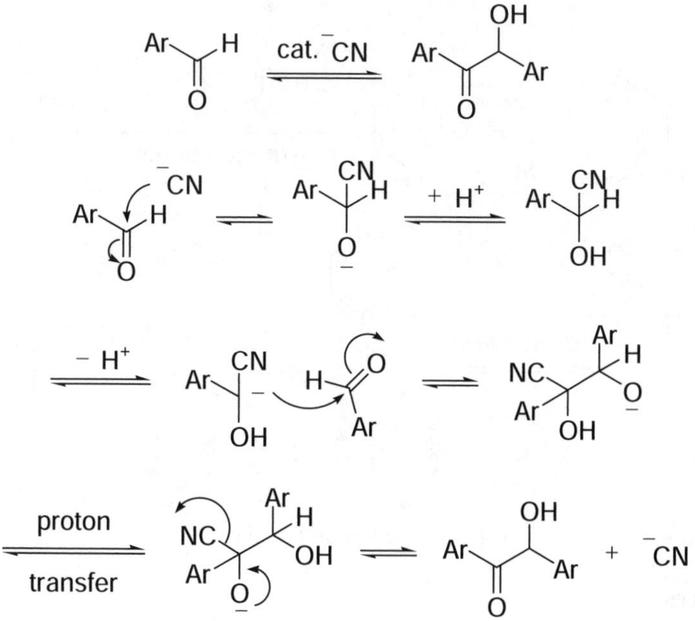

References

1. Lapworth, A. J. *J. Chem. Soc.* **1903**, *83*, 995.
2. Kluger, R. *Pure Appl. Chem.* **1997**, *69*, 1957.
3. Demir, A. S.; Dunnwald, T.; Iding, H.; Pohl, M.; Muller, M. *Tetrahedron: Asymmetry* **1999**, *10*, 4769.
4. Davis, J. H., Jr.; Forrester, K. J. *Tetrahedron Lett.* **1999**, *40*, 1621.
5. White, M. J.; Leeper, F. J. *J. Org. Chem.* **2001**, *66*, 5124.

Bergman cyclization

Δ or

hv

hydrogen donor

reversible

H

H

enediyne 1,4-benzenediyl diradical

References

1. Jones, R. R.; Bergman, R. G. *J. Am. Chem. Soc.* **1972**, *94*, 660.
2. Bergman, R. G. *Acc. Chem. Res.* **1973**, *6*, 25.
3. Evenzahav, A.; Turro, N. J. *J. Am. Chem. Soc.* **1998**, *120*, 1835.
4. Schreiner, P. R. *ibid.* **1998**, *120*, 4184.
5. McMahon, R. J.; Halter, R. J.; Fimmen, R. L.; Wilson, R. J.; Peebles, S. A.; Kuczkowski, R. L.; Stanton, J. F. *ibid.* **2000**, *122*, 939.
6. Choy, N.; Blanco, B.; wen, J.; Krishan, A.; Russell, K. C. *Org. Lett.* **2000**, *2*, 3761.
7. Rawat, D. S.; Zaleski, J. M. *Chem. Commun.* **2000**, 2493.
8. Clark, A. E.; Davidson, E. R.; Zaleski, J. M. *J. Am. Chem. Soc.* **2001**, *123*, 2650.

Biginelli pyrimidone synthesis

One-pot condensation reaction of an aromatic aldehyde, urea, and ethyl acetoacetate in acidic ethanolic solution and expansion of such a condensation thereof.

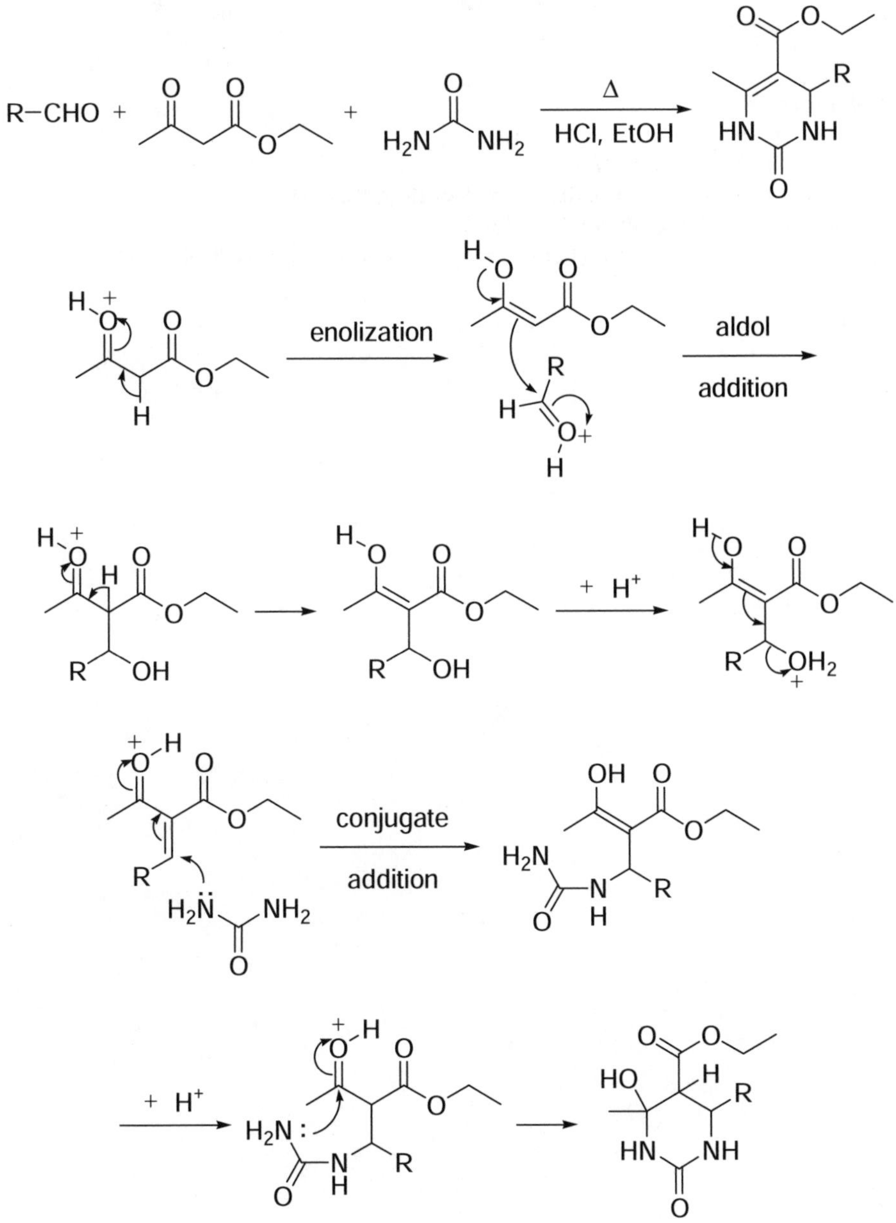

R–CHO + O O O + O H_2N NH_2 Δ HCl, EtOH O O R HN NH O

H + O O H O enolization H O O O R H O+ H aldol addition

H + O H O O R OH → H O O O R OH + H^+ H O O O R OH_2 +

+ H O O O R $H_2\ddot{N}$ NH_2 O conjugate addition OH O O H_2N N H R O

+ H^+ + H O O O H_2N : N H R O → O O HO H R HN NH O

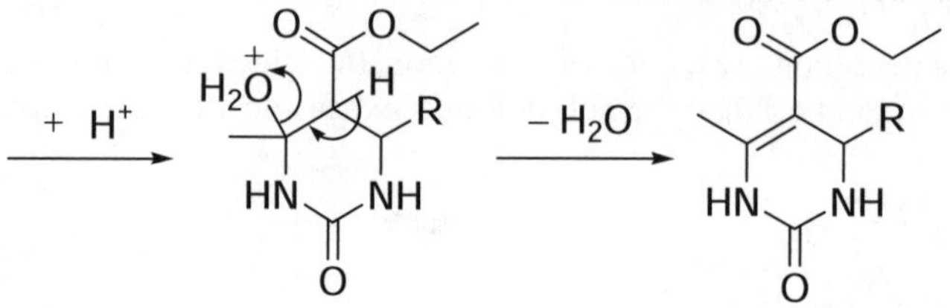

+ H^+ → – H_2O →

References

1. Biginelli, P. *Ber.* **1891**, *24*, 1317.
2. Sweet, F.; Fissekis, J. D. *J. Am. Chem. Soc.* **1973**, *95*, 8741.
3. Kappe, C. O. *Tetrahedron* **1993**, *49*, 6937.
4. Lu, J.; Bai, Y.; Wang, Z.; Yang, W.; Ma, H. *Tetrahedron Lett.* **2000**, *41*, 9075.

Birch reduction

Benzene ring bearing an electron-donating substituent:

O

Na, liq. NH_3

ROH

O

O

single electron

transfer (SET)

O

O

H

H

$H-OR$

radical anion

O

H

H

H

$+ e^-$

O

H

H

H

$H-OR$

O

Benzene ring with an electron-withdrawing substituent:

CO_2H

Na, liq. NH_3

CO_2H

CO_2H

$+ e^-$

CO_2^-

CO_2^-

H

$H-NH_2$

radical anion

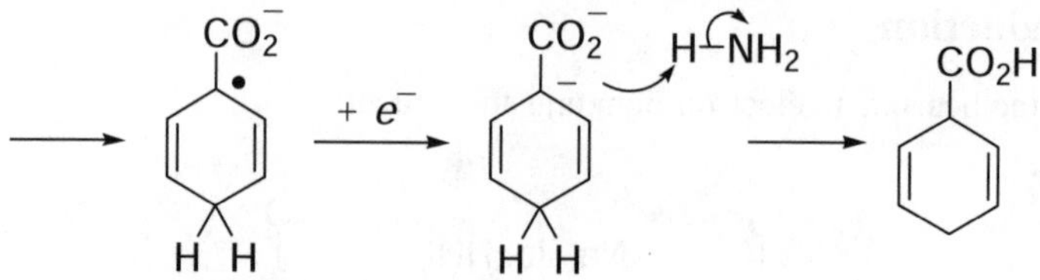

References

1. Birch, A. J. *J. Chem. Soc.* **1944**, 430.
2. Rabideau, P. W.; Marcinow, Z. *Org. React.* **1992**, *42*, 1–334.
3. Birch, A. J. *Pure Appl. Chem.* **1996**, *68*, 553.
4. Schultz, A. G. *Chem. Commun.* **1999**, 1263.
5. Ohta, Y.; Doe, M.; Morimoto, Y.; Kinoshita, T. *J. Heterocycl. Chem.* **2000**, *37*, 751.
6. Labadie, G. R.; Cravero, R. M.; Gonzalez-Sierra, M. *Synth. Commun.* **2000**, *30*, 4065.
7. Guo, Z.; Schultz, A. G. *J. Org. Chem.* **2001**, *66*, 2154.

Bischler–Möhlau indole synthesis

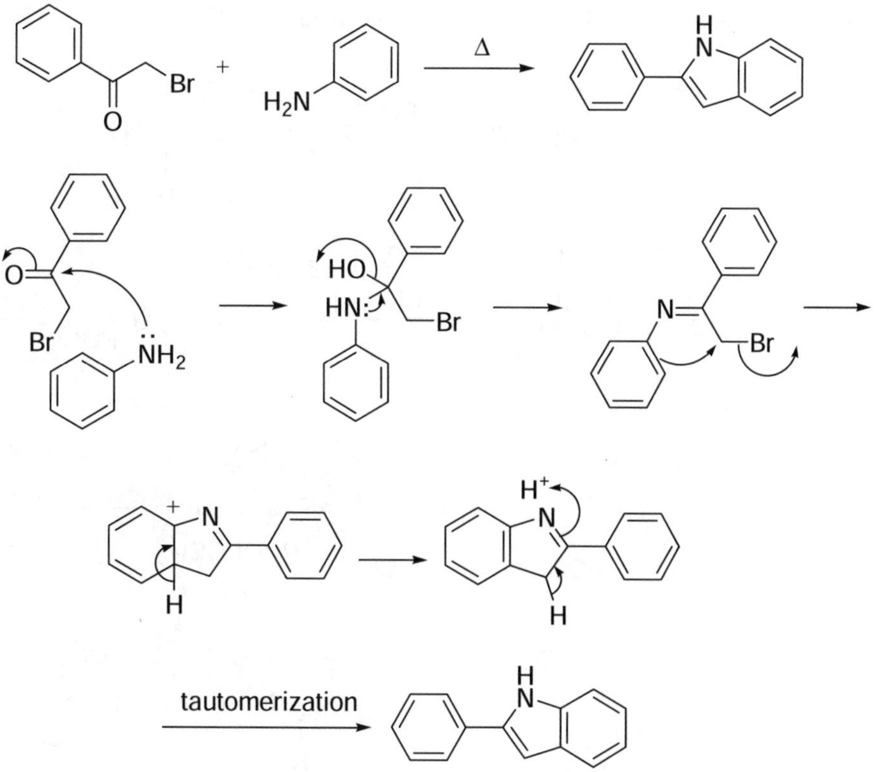

References

1. Möhlau, R. *Ber.* **1881**, *14*, 171.
2. Buu-Hoï, N. P.; Saint-Ruf, G.; Deschamps, D.; Bigot, P. *J. Chem. Soc. (C)* **1971**, 2606.
3. Bancroft, K. C. C.; Ward, T. J. *J. Chem. Soc., Perkin 1* **1974**, 1852.
4. Coic, J. P.; Saint-Ruf, G. *J. Heterocyclic Chem. Soc.* **1978**, *15*, 1367.
5. Henry, J. R.; Dodd, J. H. *Tetrahedron Lett.* **1998**, *39*, 8763.

Bischler–Napieralski reaction

Dihydroisoquinoline synthesis from β-phenethylamides.

HN O R $\xrightarrow{POCl_3}$ N R

Cl–P(=O)(Cl)–Cl, HN, O, R $\xrightarrow{-\ Cl^-}$ HN(+)=C(R)–O–$POCl_2$ ⟶

NH +, R, $OPOCl_2$ ⟶ +, N–H, Cl^-, H, R, $OPOCl_2$ ⟶

Cl^-, N–H, R, O–P(=O)(Cl)Cl ⟶ N, R + HCl + O=P(=O)Cl

References

1. Bischler, A.; Napieralski, B. *Ber.* **1893**, *26*, 1903.
2. Fodor, G.; Nagubandi, S. *Heterocycles* **1981**, *15*, 165.
3. Rozwadowska, M. D. *ibid.* **1994**, *39*, 903.
4. Sotomayor, N.; Dominguez, E.; Lete, E. *J. Org. Chem.* **1996**, *61*, 4062.
5. Doi, S.; Shirai, N.; Sato, Y. *J. Chem. Soc., Perkin Trans. 1* **1997**, 2217.
6. Sanchez-Sancho, F.; Mann, E.; Herradon, B. *Synlett* **2000**, 509.
7. Ishikawa, T.; Shimooka, K.; Narioka, T.; Noguchi, S.; Saito, T.; Ishikawa, A.; Yamazaki, E.; Harayama, T.; Seki, H.; Yamaguchi, K. *J. Org. Chem.* **2000**, *65*, 9143.
8. Miyatani, K.; Ohno, M.; Tatsumi, K.; Ohishi, Y.; Kunitomo, J.-I.; Kawasaki, I.; Yamashita, M.; Ohta, S. *Heterocycles* **2001**, *55*, 589.

Blaise reaction

β-Ketoesters from nitriles and α-haloesters using Zn.

R−CN + Br–CH(R^1)–CO_2R^2 → (1. Zn, THF, reflux; 2. H_3O^+) → R–CO–CH(R^1)–CO_2R^2

Br–CH(R^1)–CO_2R^2 → Zn(0) → R−≡N + zinc enolate (OR^2, O–ZnBr, R^1) → nucleophilic addition

OH_3^+, N–ZnBr imine, R, CO_2R^2, R^1 → workup → H, +N, H iminium, R, CO_2R^2, H_2O:, R^1 →

H_3O^+, H_2N, O–H, R, CO_2R^2, R^1 → R–CO–CH(R^1)–CO_2R^2

References

1. Blaise, E. E. *C. R. Hebd. Seances Acad. Sci.* **1901**, *132*, 478.
2. Kagan, H. B.; Suen, Y.-H. *Bull. Soc. Chim. Fr.* **1966**, 1819.
3. Hannick, S. M.; Kishi, Y. *J. Org. Chem.* **1983**, *48*, 3833.
4. Hiyama, T.; Kobayashi, K. *Tetrahedron Lett.* **1982**, *23*, 1597.
5. Krepski, L. R.; Lynch, L. E.; Heilmann, S. M.; Rasmussen, J. K. *Tetrahedron Lett.* **1985**, *26*, 981.
6. Beard, R. L.; Meyers, A. I. *J. Org. Chem.* **1991**, *56*, 2091.
7. Syed, J.; Forster, S.; Effenberger, F. *Tetrahedron: Asymmetry* **1998**, *9*, 805.
8. Narkunan, K.; Uang, B.-J. *Synthesis* **1998**, 1713.
9. Erian, A. W. *J. Prakt. Chem.* **1999**, *341*, 147.

Blanc chloromethylation reaction

formalin

$ZnCl_2$

+ HCl

Cl

+ H_2O

H^+

$- H^+$

S_N2

References

1. Blanc, G. *Bull. Soc. Chim. Fr.* **1923**, *33*, 313.
2. Franke, A. T.; Mattern, G.; Traber, W. *Helv. Chim. Acta* **1975**, *58*, 283.

Boekelheide reaction

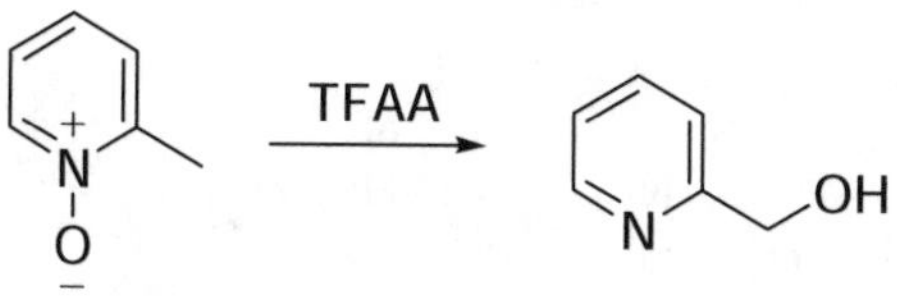

TFAA

TFAA, trifluoroacetic anhydride

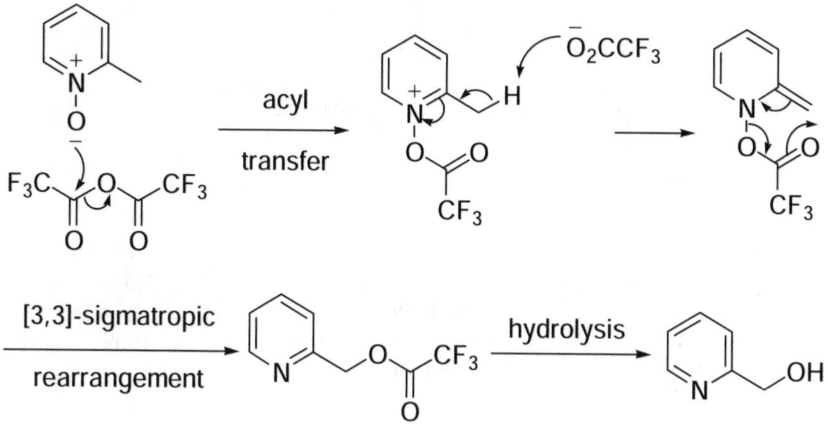

acyl transfer

$^-O_2CCF_3$

[3,3]-sigmatropic rearrangement

hydrolysis

References

1. Bell, T. W.; Firestone, A. *J. Org. Chem.* **1986**, *51*, 764.
2. Newkome, G. R.; Theriot, K. J.; Gupta, V. K.; Fronczek, F. R.; Baker, G. R. *J. Org. Chem.* **1986**, *54*, 1766.
3. Goerlitzer, K.; Schmidt, E. *Arch. Pharm.* **1991**, *324*, 359.
4. Fontenas, C.; Bejan, E.; Haddon, H. A.; Balavoine, G. G. A. *Synth. Commun.* **1995**, *25*, 629.

Boger pyridine synthesis

Hetero-Diels-Alder reaction

Retro-Diels-Alder reaction, $- N_2$

$- H_2O$

References

1. Boger, D. L.; Panek, J. S.; Meier, M. M. *J. Org. Chem.* **1982**, *47*, 895.
2. Boger, D. L. In *Comprehensive Organic Synthesis* Trost, B. M.; Fleming, I., Eds, Pergamon, **1991**, *Vol. 5*, 451–512.
3. Golka, A.; Keyte, P. J.; Paddon-Row, M. N. *Tetrahedron* **1992**, *48*, 7663.

Boord reaction

Br
R
R^1
OEt
Zn dust
R
R^1

Br
R
R^1
OEt
Zn
Oxidative addition
ZnBr
R
R^1
OEt
R
R^1 + EtO−ZnBr

References

1. Swallen, L. C.; Boord, C. E. *J. Am. Chem. Soc.* **1930**, *52*, 651.
2. Hatch, C. E., III; Baum, J. S.; Takashima, T.; Kondo, K. *J. Org. Chem.* **1980**, *45*, 3181.
3. Halton, B.; Russell, S. G. G. *ibid.* **1991**, *56*, 5553.
4. Yadav, J. S.; Ravishankar, R.; Lakshman, S. *Tetrahedron Lett.* **1994**, *35*, 3617.
5. Yadav, J. S.; Ravishankar, R.; Lakshman, S. T. *ibid.* **1994**, *35*, 3621.
6. Beusker, P. H.; Aben, R. W. M.; Seerden, J.-P. G.; Smits, J. M. M.; Scheeren, H. W. *Eur. J. Org. Chem.* **1998**, 2483.

Borsche–Drechsel cyclization

Cf. Fisher indole synthesis.

HCl

[3,3]-sigmatropic

rearrangement

H^+

References

1. Drechsel, E. *J. Prakt. Chem.* **1858**, *38*, 69.
2. Atkinson, C. M.; Biddle, B. N. *J. Chem. Soc. (C)* **1966**, 2053.
3. Rousselle, D.; Gilbert, J.; Viel, C. *C. R. Hebd. Seances Acad. Sci., Ser. C* **1977**, *284*, 377.

Boulton–Katritzky rearrangement

Rearrangement of one five-membered heterocycle into another under thermolysis.

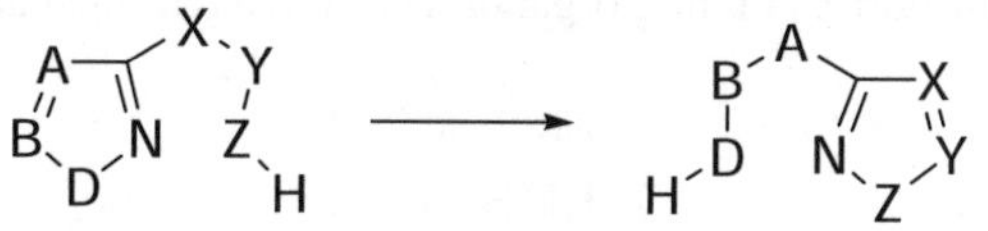

e.g. [ref. 9]:

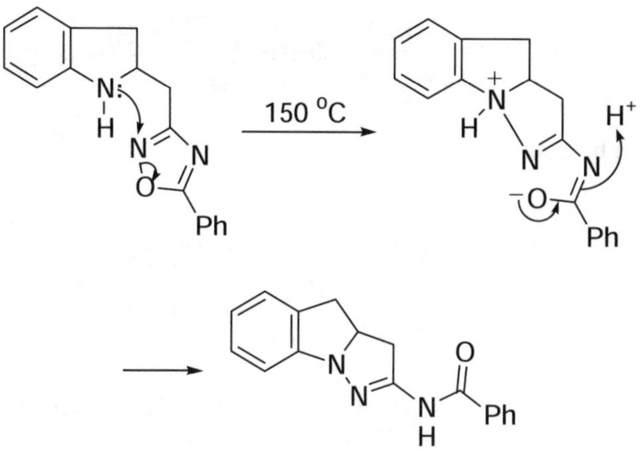

150 °C

References

1. Boulton, A. J.; Katritzky, A. R.; Hamid, A. M. *J. Chem. Soc. (C)* **1967**, 2005.
2. Balli, H.; Gunzenhauser, S. *Helv. Chim. Acta* **1978**, *61*, 2628.
3. Ruccia, M.; Vivona, N.; Spinelli, D. *Adv. Heterocyl. Chem.* **1981**, *29*, 141.
4. Butler, R. N.; Fitzgerald, K. J. *J. Chem. Soc., Perkin Trans. 1* **1988**, 1587.
5. Ostrowski, S.; Wojciechowski, K. *Can. J. Chem.* **1990**, *68*, 2239.
6. Takakis, I. M.; Hadjimihalakis, P. M.; Tsantali, G. G. *Tetrahedron* **1991**, *47*, 7157.
7. Takakis, I. M.; Hadjimihalakis, P. M. *J. Heterocycl. Chem.* **1992**, *29*, 121.
8. Vivona, N.; Buscemi, S.; Frenna, V.; Cusmano, C. *Adv. Heterocyl. Chem.* **1993**, *56*, 49.
9. Katayama, H.; Takatsu, N.; Sakurada, M.; Kawada, Y. *Heterocycles* **1993**, *35*, 453.
10. Sonnenschein, H.; Schmitz, E.; Gruendemann, E.; Schroeder, E. *Ann.* **1994**, 1177.
11. Rauhut, G. *J. Org. Chem.* **2001**, *66*, 5444.

Bouveault aldehyde synthesis

Formylation of an alkyl or aryl halide to the homologous aldehyde by transformation to the corresponding organometallic reagent then addition of DMF.

$$R{-}X \xrightarrow[\text{3. } H^+]{\text{1. M; 2. DMF}} R{-}CHO$$

$$R{-}X \xrightarrow{M} R{-}M \xrightarrow{DMF} Me_2N{-}CH(O{-}M)(R) \xrightarrow{H^+} R{-}CHO$$

A modification by Comins [3]:

$$R_2N{-}CHO \xrightarrow[\text{2. } H^+]{\text{1. } R'MgX} R'{-}CHO$$

$$R_2N{-}CHO + R'{-}MgX \longrightarrow R_2N{-}CH(O^-)(R') \longrightarrow R'{-}CHO + {}^-NR_2$$

References

1. Bouveault, L. *Bull. Soc. Chim. Fr.* **1904**, *31*, 1306.
2. Petrier, C.; Gemal, A. L.; Luche, J. L. *Tetrahedron Lett.* **1982**, *23*, 3361.
3. Comins, D. L.; Brown, J. D. *J. Org. Chem.* **1984**, *49*, 1078.
4. Einhorn, J.; Luche, J. L. *Tetrahedron Lett.* **1986**, *27*, 1791.
5. Einhorn, J.; Luche, J. L. *ibid.* **1986**, *27*, 1793.
6. Denton, S. M.; Wood, A. *Synlett* **1999**, 55.
7. Meier, H.; Aust, H. *J. Prakt. Chem.* **1999**, *341*, 466.

Bouveault–Blanc reduction

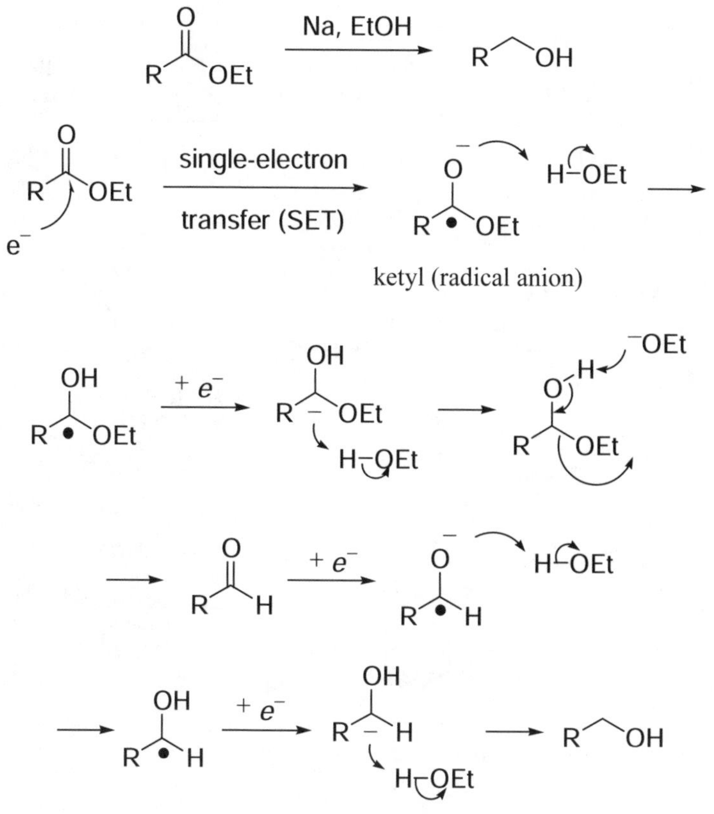

Na, EtOH

single-electron

transfer (SET)

ketyl (radical anion)

+ e⁻

+ e⁻

+ e⁻

References

1. Bouveault, L.; Blanc, G. *Compt. Rend.* **1903**, *136*, 1676.
2. Ruehlmann, K.; Seefluth, H.; Kiriakidis, T.; Michael, G.; Jancke, H.; Kriegsmann, H. *J. Organometal. Chem.* **1971**, *27*, 327.
3. Castells, J.; Grandes, D.; Moreno-Manas, M.; Virgili, A. *An. Quim.* **1976**, *72*, 74.
4. Sharda, R.; Krishnamurthy, H. G. *Indian J. Chem., Sect. B* **1980**, *19B*, 405.
5. Banerji, J.; Bose, P.; Chakrabarti, R.; Das, B. *Indian J. Chem., Sect. B* **1993**, *32B*, 709.
6. Seo, B. I.; Wall, L. K.; Lee, H.; Buttrum, J. W.; Lewis, D. E. *Synth. Commun.* **1993**, *23*, 15.
7. Zhang, Y.; Ding, C. *Huaxue Tongbao* **1997**, 36.

Boyland–Sims oxidation

Oxidation of aromatic amines to phenols using alkaline persulfate.

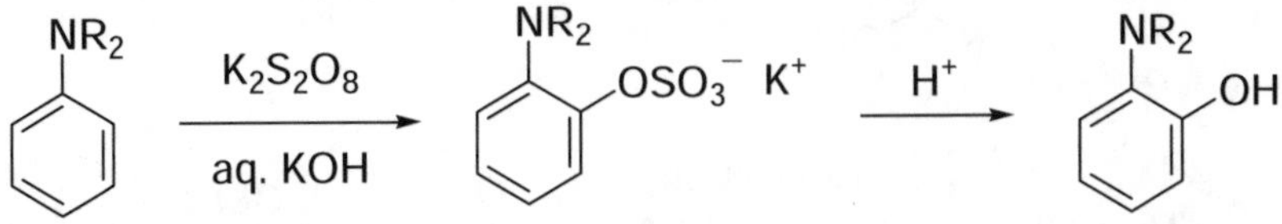

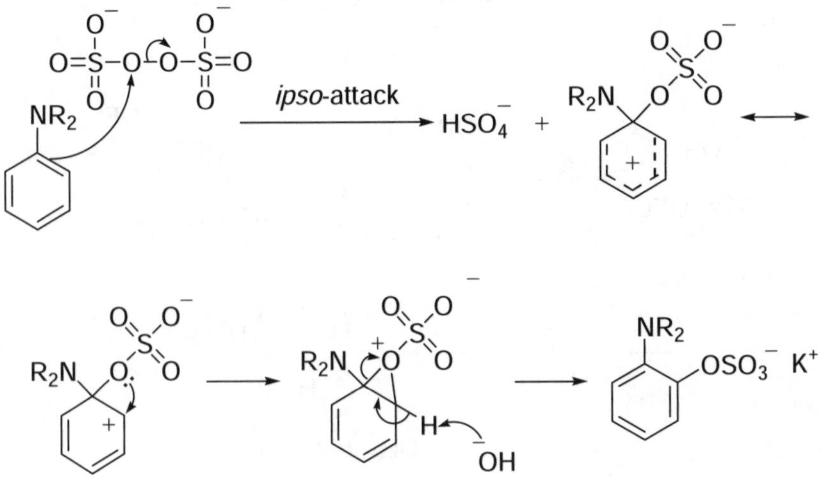

Another pathway is also operative:

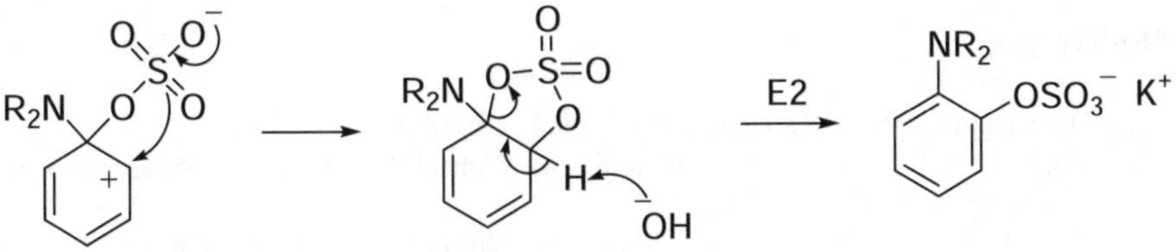

References

1. Boyland, E.; Manson, D.; Sims, P. *J. Chem. Soc.* **1953**, 3623.
2. Boyland, E.; Sims, P. *ibid.* **1954**, 980.
3. Behrman, E. J. *J. Am. Chem. Soc.* **1967**, *89*, 2424.
4. Krishnamurthi, T. K.; Venkatasubramanian, N. *Indian J. Chem., Sect. A* **1978**, *16A*, 28.
5. Behrman, E. J.; Behrman, D. M. *J. Org. Chem.* **1978**, *43*, 4551.

6. Srinivasan, C.; Perumal, S.; Arumugam, N. *J. Chem. Soc., Perkin Trans. 2* **1985**, 1855.
7. Behrman, E. J. *Org. React.* **1988**, *35*, 421.
8. Behrman, E. J. *J. Org. Chem.* **1992**, *57*, 2266.

Bradsher reaction

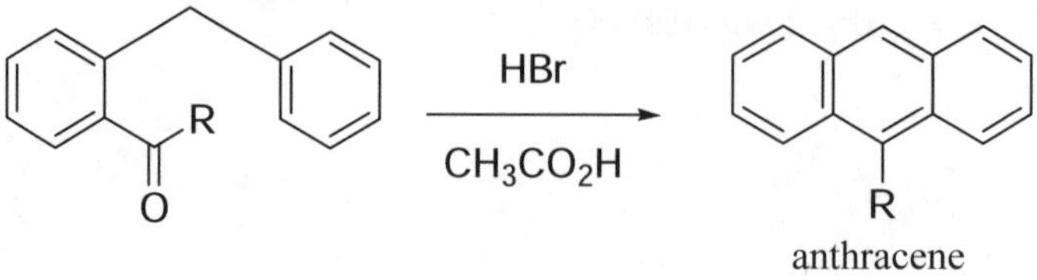

R
O
HBr
CH_3CO_2H
R
anthracene

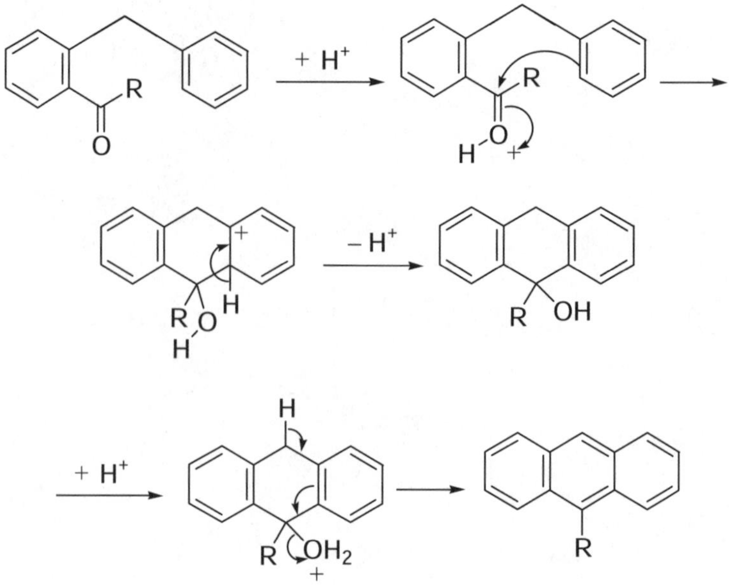

R
O
+ H^+
R
H
O
+

+
H
R
O
H
– H^+
R
OH

+ H^+
H
R
OH_2
+
R

References

1. Bradsher, C. K. *J. Am. Chem. Soc.* **1940**, *62*, 486.
2. Saraf, S. D.; Vingiello, F. A. *Synthesis* **1970**, 655.
3. Ashby, J.; Ayad, M.; Meth-Cohn, O. *J. Chem. Soc., Perkin Trans. 1* **1974**, 1744.
4. Nicolas, T. E.; Franck, R. W. *J. Org. Chem.* **1995**, *60*, 6904.
5. Magnier, E.; Langlois, Y. *Tetrahedron Lett.* **1998**, *39*, 837.

Brook rearrangement

Base-catalyzed silicon migration from carbon to oxygen.

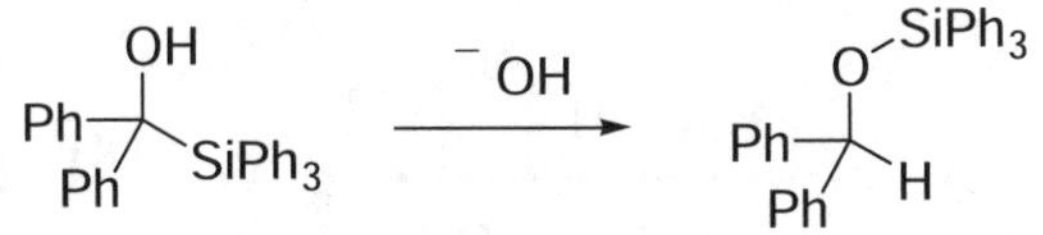

α-hydroxysilane

silyl ether

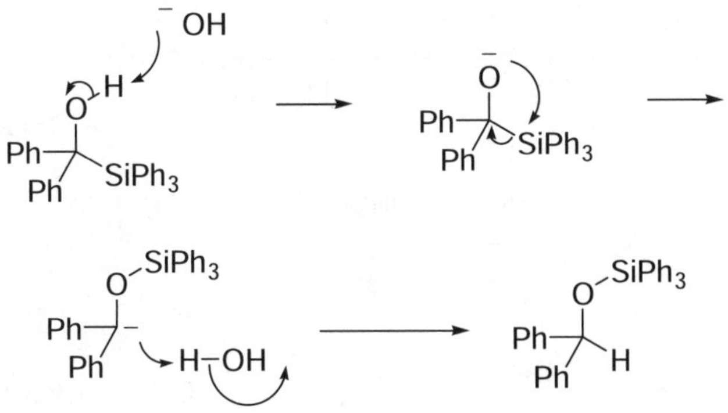

References

1. Brook, A. G. *J. Am. Chem. Soc.* **1958**, *80*, 1886.
2. Brook, A. G. *Acc. Chem. Res.* **1974**, *7*, 77.
3. Page, P. C. B.; Klair, S. S.; Rosenthal, S. *Chem. Soc. Rev.* **1990**, *19*, 147.
4. Takeda, K.; Nakatani, J.; Nakamura, H.; Yosgii, E.; Yamaguchi, K. *Synlett* **1993**, 841.
5. Fleming, I.; Ghosh, U. *J. Chem. Soc., Perkin Trans. 1* **1994**, 257.
6. Takeda, K.; Takeda, K.; Ohnishi, Y. *Tetrahedron Lett.* **2000**, *41*, 4169.
7. Sumi, K.; Hagisawa, S. *J. Organomet. Chem.* **2000**, *611*, 449.
8. Moser, W. H. *Tetrahedron* **2001**, *571*, 2065.

Brown hydroboration reaction

Addition of boranes to olefins, followed by basic oxidation of the organoboranes, resulting in alcohols.

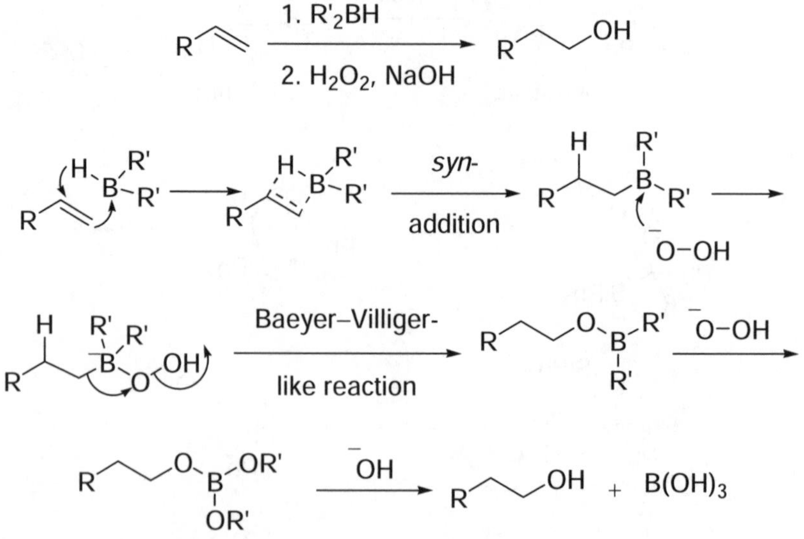

1. R'_2BH
2. H_2O_2, NaOH

syn-
addition

Baeyer–Villiger-
like reaction

^{-}OH

+ $B(OH)_3$

References

1. Brown, H. C.; Tierney, P. A. *J. Am. Chem. Soc.* **1958**, *80*, 1552.
2. Nussium, M.; Mazur, Y.; Sondheimer, F. *J. Org. Chem.* **1964**, *29*, 1120.
3. Nussium, M.; Mazur, Y.; Sondheimer, F. *ibid.* **1964**, *29*, 1131.
4. Pelter, A.; Smith, K.; Brown, H. C. *Borane Reagents* Academic Press: New York, **1972**.
5. Brown, H. C.; vara Prasad, J. V. N. *Heterocycles* **1987**, *25*, 641.

Bucherer carbazole synthesis

$NaHSO_4$

OH + NHNH₂ ⇌ NH

H⁺ OH ⇌ H H H⁺ O ⁻OSO₃H ⇌ H H OH OSO₃H

⇌ H H H⁺ O :NH₂NHPh OSO₃H ⇌ H H H⁺ OH NHNHPh OSO₃H ⇌

NHNHPh :B H OSO₃H ⇌ NH NH

H⁺

[3,3]-sigmatropic rearrangement

H NH₂⁺ NH → H NH₂ H⁺ NH → NH

References

1. Bucherer, H. T.; Seyde, F. *J. Prakt. Chem.* **1908**, *77*, 403.
2. Seeboth, H. *Angew. Chem., Int. Ed. Engl.* **1967**, *6*, 307.

Bucherer reaction

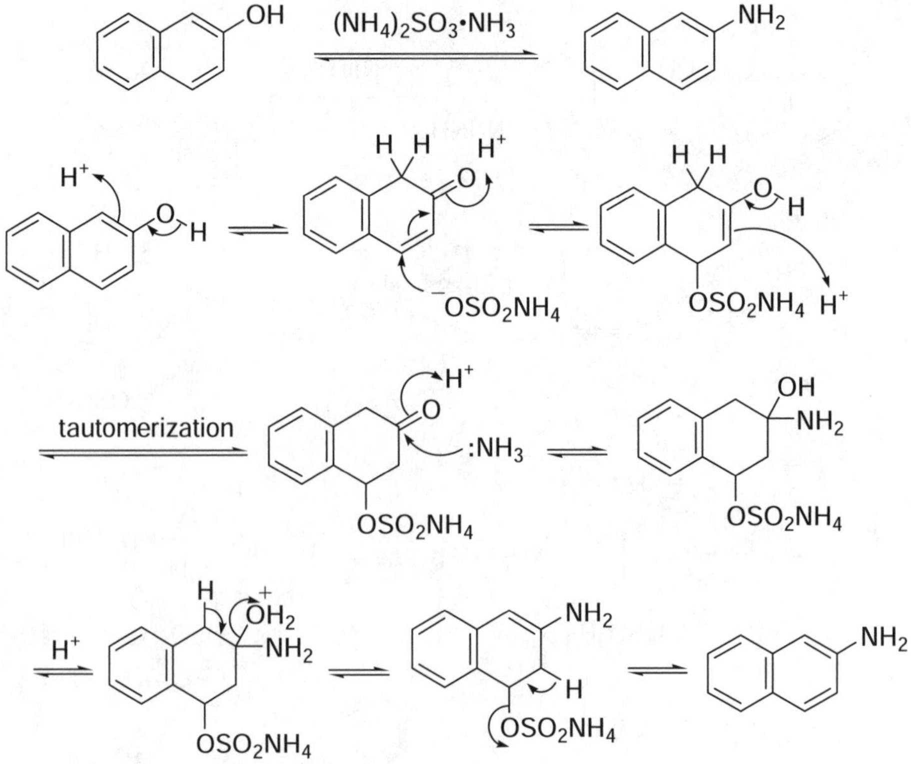
OH
(NH4)2SO3•NH3
NH2
H+
O
H
H H
H+
O
OSO2NH4
H H
O
H
OSO2NH4 H+
tautomerization
H+
O
:NH3
OSO2NH4
OH
NH2
OSO2NH4
H+
H
+
OH2
NH2
OSO2NH4
NH2
H
OSO2NH4
NH2

References

1. Bucherer, H. T. *J. Prakt. Chem.* **1904**, *69*, 49.
2. Reiche, A.; Seeboth, H. *Liebigs Ann. Chem.* **1960**, *638*, 66.
3. Gilbert, E. E. *Sulfonation and Related Reactions* Wiley: New York, **1965**, p166.
4. Seeboth, H. *Angew. Chem., Int. Ed. Engl.* **1967**, *6*, 307.
5. Canete, A.; Melendrez, M. X.; Saitz, C.; Zanocco, A. L. *Synth. Commun.* **2001**, *31*, 2143.

Bucherer–Bergs reaction

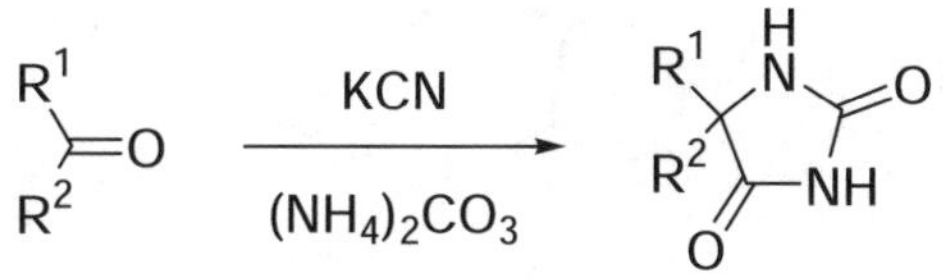

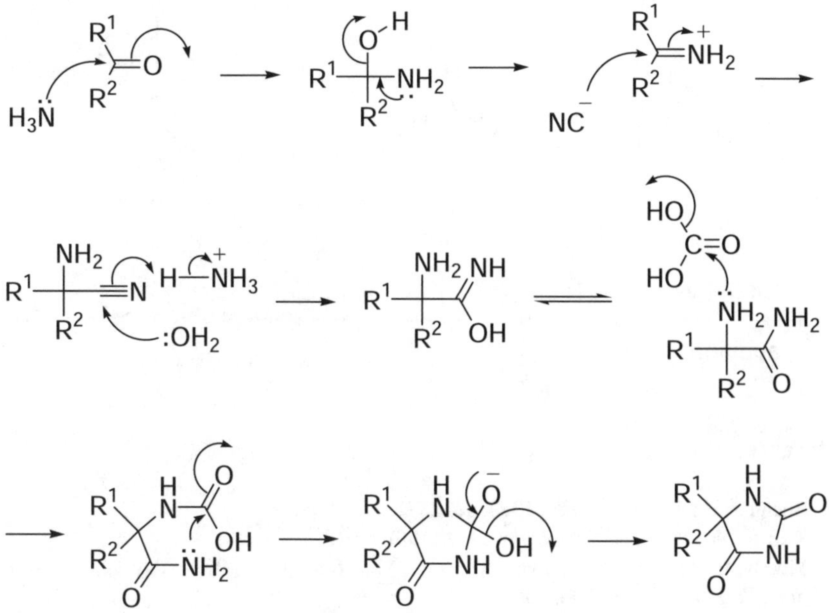

$$R^1R^2C{=}O \xrightarrow[(NH_4)_2CO_3]{KCN} \text{hydantoin}$$

References

1. Bergs, H. Ger. Pat. 566,094, **1929**.
2. Bucherer, H. T., Fischbeck, H. T. *J. Prakt. Chem.* **1934**, *140*, 69.
3. Bucherer, H. T., Steiner, W. *ibid.* **1934**, *140*, 291.
4. Chubb, F. L.; Edward, J. T.; Wong, S. C. *J. Org. Chem.* **1980**, *45*, 2315.
5. Herdeis, C.; Gebhard, R. *Heterocycles* **1986**, *24*, 1019.
6. Tanaka, K.-i.; Iwabuchi, H.; Sawanishi, H. *Tetrahedron: Asymmetry* **1995**, *6*, 2271.
7. Haroutounian, S. A.; Georgiadis, M. P.; Polissiou, M. G. *J. Heterocycl. Chem.* **1989**, *26*, 1283.
8. Hu, A.-X.; Zhao, H.-T.; Chen, S.-Z.; Zhou, Y.-P.; Kuang, C.-T. *Hecheng Huaxue* **1998**, *6*, 75.

Buchner–Curtius–Schlotterbeck reaction

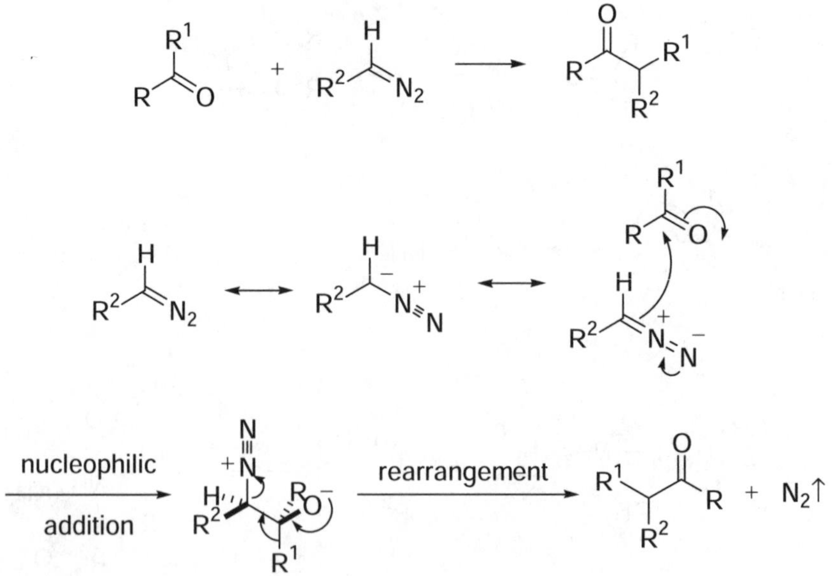

nucleophilic addition

rearrangement

$+ N_2\uparrow$

References

1. Buchner, E.; Curtius, T. *Ber.* **1989**, *18*, 2371.
2. Kirmse, W.; Horn, K. *Tetrahedron Lett.* **1967**, 1827.
3. Moody, C. J.; Miah, S.; Slawin, A. M. Z.; Mansfield, D. J.; Richards, I. C. *J. Chem. Soc., Perkin Trans. 1* **1998**, 4067.
4. Maguire, A. R.; Buckley, N. R.; O'Leary, P.; Ferguson, G. *ibid.* **1998**, 4077.

Buchner method of ring expansion

$$N_2{=}CH{-}CO_2CH_3,\ [Rh]$$

$$N_2{=}CH{-}CO_2CH_3 \xrightarrow{[Rh]} N_2\uparrow + Rh{=}CH{-}CO_2CH_3$$

Rhodium carbenoid

[2 + 2] cycloaddition

reductive elimination

electrocyclic ring opening

References

1. Buchner, E. *Ber.* **1896**, *29*, 106.
2. Von Doering, W.; Knox, L. H. *J. Am. Chem. Soc.* **1957**, *79*, 352.
3. Marchard, A. P.; Macbrockway, N. *Chem. Rev.* **1974**, *74*, 431.
4. Nakamura, A.; Konischi, A.; Tsujitani, R.; Kudo, M.; Otsuka, S. *J. Am. Chem. Soc.* **1978**, *100*, 3449.
5. Anciaux, A. J.; Noels, A. F.; Hubert, A. J.; Warin, R.; Teyssié, P. *J. Org. Chem.* **1981**, *46*, 873.
6. Doyle, M. P.; Hu, W.; Timmons, D. J. *Org. Lett.* **2001**, *3*, 933.
7. Doyle, M. P.; Phillips, I. M. *Tetrahedron Lett.* **2001**, *42*, 3155.

Buchwald–Hartwig C–N bond and C–O bond formation reactions

Direct Pd-catalyzed C–N and C–O bond formation of aryl halides with amines in the presence of stoichiometric amount of base.

Pd(OAc)$_2$, dppf

NaO*t*-Bu, tol., Δ

Pd(0)

oxidative addition

Pd(II)—Br

ligand exchange – HBr

Pd(II)—N

reductive elimination – Pd(0)

The C–O bond formation reaction follows a similar mechanistic pathway [7–9].

References

1. Paul, F.; Patt, J.; Hartwig, J. F. *J. Am. Chem. Soc.* **1994**, *116*, 5969.
2. Guram, A. S.; Buchwald, S. L. *ibid.* **1994**, *116*, 7901.
3. Wolfe, J. P.; Wagaw, S.; Marcoux, J.-F.; Buchwald, S. L. *Acc. Chem. Res.* 1**998**, *31*, 805.
4. Hartwig, J. F. *ibid.* **1998**, *31*, 852.
5. Frost, C. G.; Mendonça, P. *J. Chem. Soc., Perkin Trans. 1* **1998**, 2615.
6. Yang, B. H.; Buchwald, S. L. *J. Organomet. Chem.* **1999**, *576*, 125.
7. Palucki, M.; Wolfe, J. P.; Buchwald, S. L. *J. Am. Chem. Soc.* **1996**, *118*, 10333.
8. Mann, G.; Hartwig, J. F. *J. Org. Chem.* **1997**, *62*, 5413.
9. Mann, G.; Hartwig, J. F. *Tetrahedron Lett.* **1997**, *38*, 8005.
10. Browning, R. G.; Mahmud, H.; Badarinarayana, V.; Lovely, C. J. *Tetrahedron Lett.* **2001**, *42,* 7155.

Burgess dehydrating reagent

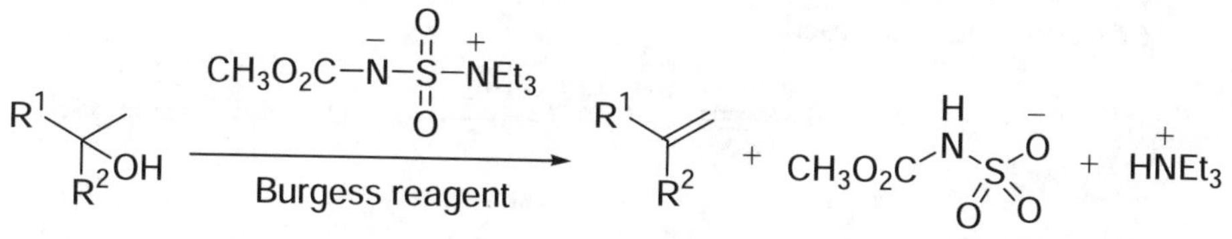

Burgess dehydrating reagent is efficient at generating olefins from secondary and tertiary alcohols where Ei (during the elimination, the two groups leave at about the same time and bond to each other concurrently) mechanism prevails:

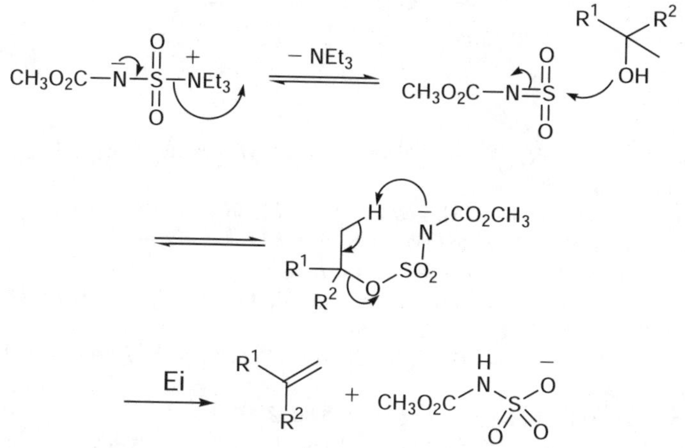

References

1. Burgess, E. M. *J. Org. Chem.* **1973**, *38*, 26.
2. Claremon, D. A.; Philips, B. T. *Tetrahedron Lett.* **1988**, *29*, 2155.
3. Lamberth, C. *J. Prakt. Chem.* **2000** *342*, 518.
4. Svenja, B. *Synlett.* **2000**, 559.
5. Miller, C. P.; Kaufman, D. H. *Synlett.* **2000**, 1169.

Cadiot–Chodkiewicz coupling

Bis-acetylene synthesis from alkynyl halides and alkynyl copper reagents.
Cf. Castro–Stephens reaction.

$$R^1\text{—}\equiv\text{—}X \; + \; Cu\text{—}\equiv\text{—}R^2 \longrightarrow R^1\text{—}\equiv\text{—}\equiv\text{—}R^2$$

$$R^1\text{—}\equiv\text{—}X \; + \; Cu\text{—}\equiv\text{—}R^2 \xrightarrow[\text{addition}]{\text{oxidative}} R^1\text{—}\equiv\text{—}Cu(X)\text{—}\equiv\text{—}R$$

Cu(III) intermediate

$$\xrightarrow[\text{elimination}]{\text{reductive}} CuX \; + \; R^1\text{—}\equiv\text{—}\equiv\text{—}R^2$$

References

1. Cadiot, P.; Chokiewicz, W. In *Chemistry of Acetylenes* Ed.: Viehe, H. G., Dekker: New York, **1969**, pp597–647.
2. Eastmond, R.; Walton, D. R. M. *Tetrahedron* **1972**, *28*, 4591.
3. Ghose, B. N.; Walton, D. R. M. *Synthesis* **1974**, 890.
4. Hopf, H.; Krause, N. *Tetrahedron Lett.* **1985**, *26*, 3323.
5. Bartik, B.; Dembinski, R.; Bartik, T.; Arif, A. M.; Gladysz, J. A. *New J. Chem.* **1997**, *21*, 739.
6. Montierth, J. M.; DeMario, D. R.; Kurth, M. J.; Schore, N. E. *Tetrahedron* **1998**, *54*, 11741.
7. Negishi, E.-i.; Hata, M.; Xu, C. *Org. Lett.* **2000**, *2*, 3687.
8. Steffen, W.; Laskoski, M.; Collins, G.; Bunz, U. H. F. *J. Organomet. Chem.* **2001**, *630*, 132.

Cannizzaro disproportionation reaction

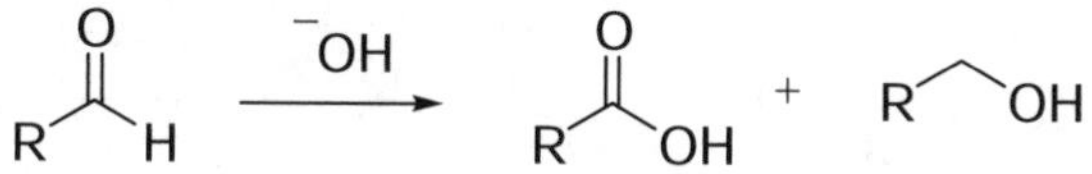

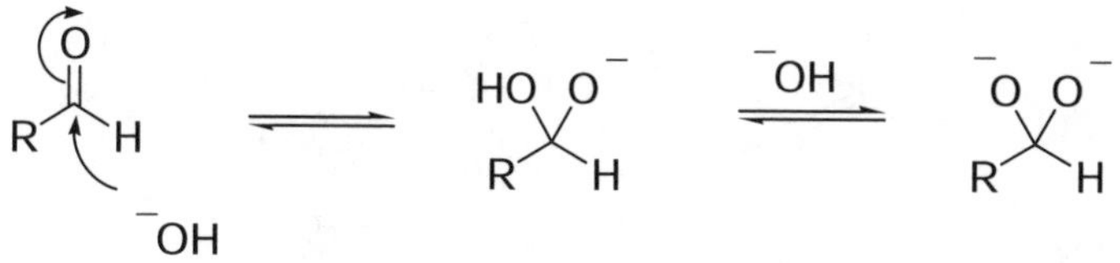

Pathway a:

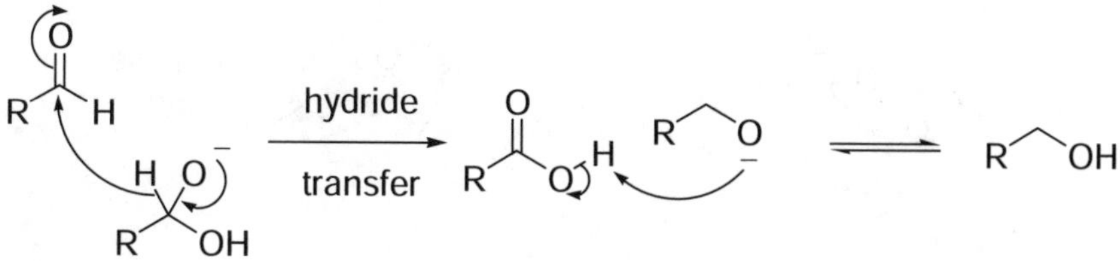

Final deprotonation of the carboxylic acid drives the reaction forward.

Pathway b:

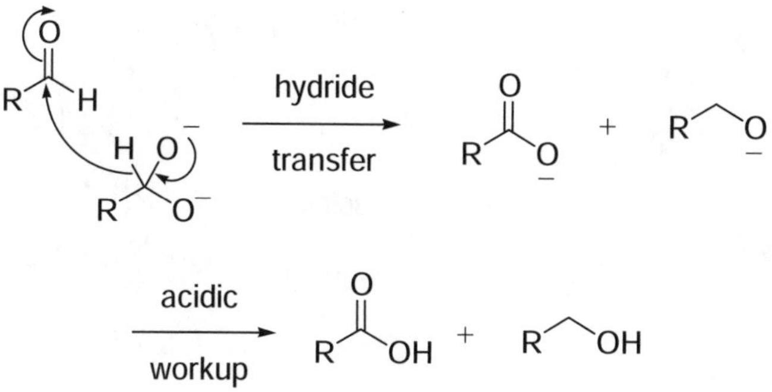

References

1. Cannizzaro, S. *Liebigs Ann. Chem.* **1853**, *88*, 129.
2. Mehta, G.; Padma, S. *J. Org. Chem.* **1991**, *56*, 1298.
3. Sheldon, J. C.; Bowie, J. H.; Dua, S.; Smith, J. D.; O'Hair, R. A. J. *ibid.* **1997**, *62*, 3931.
4. Thakuria, J. A.; Baruah, M.; Sandhu, J. S. *Chem. Lett.* **1999**, 995.
5. Russell, A. E.; Miller, S. P.; Morken, J. P. *J. Org. Chem.* **2000**, *65*, 8381.

Carroll rearrangement

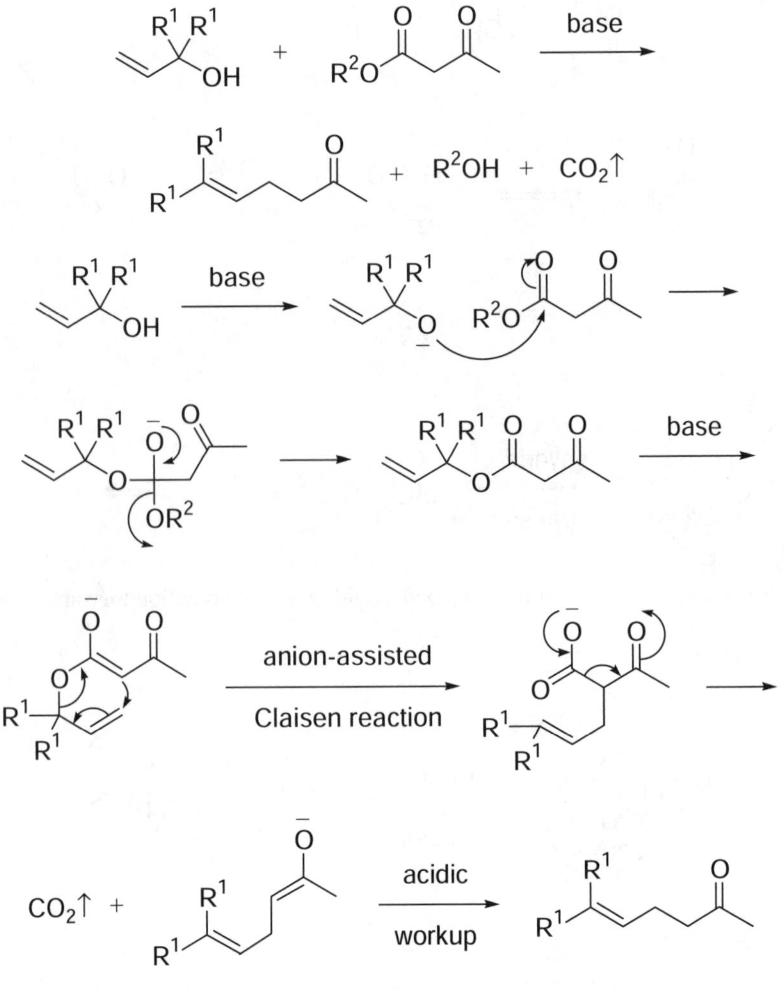

base

anion-assisted Claisen reaction

acidic workup

References

1. Carroll, M. F. *J. Chem. Soc.* **1940**, 704.
2. Wilson, S. R.; Price, M. F. *J. Org. Chem.* **1984**, *49*, 722.
3. Gilbert, J. C.; Kelly, T. A. *Tetrahedron* **1988**, *44*, 7587.
4. Enders, D.; Knopp, M.; Runsink, J.; Raabe, G. *Angew. Chem., Int. Ed. Engl.* **1995**, *34*, 2278.
5. Enders, D.; Knopp, M.; Runsink, J.; Raabe, G. *Liebigs Ann.* **1996**, 1095.

Castro–Stephens coupling

Aryl-acetylene synthesis, *Cf.* Cadiot–Chodkiewicz coupling.

$$\mathrm{Ar{-}X} + \mathrm{Cu{-}C{\equiv}C{-}R} \xrightarrow[\text{reflux}]{\text{pyridine}} \mathrm{Ar{-}C{\equiv}C{-}R}$$

$$\mathrm{Ar{-}X} + \mathrm{L_3Cu{-}C{\equiv}C{-}R} \longrightarrow \mathrm{ArX{-}CuL_2{-}C{\equiv}C{-}R}$$

$$\longrightarrow [\text{Ar, X, Cu four-centre complex with C}{\equiv}\text{C–R}] \longrightarrow \mathrm{CuX} + \mathrm{Ar{-}C{\equiv}C{-}R}$$

An alternative mechanism similar to that of the Cadiot–Chodkiewicz coupling:

$$\mathrm{Ar{-}C{\equiv}C{-}X} + \mathrm{Cu{-}C{\equiv}C{-}R} \xrightarrow[\text{addition}]{\text{oxidative}} \mathrm{Ar{-}C{\equiv}C{-}Cu(X){-}C{\equiv}C{-}R}$$

Cu(III) intermediate

$$\xrightarrow[\text{elimination}]{\text{reductive}} \mathrm{CuX} + \mathrm{Ar{-}C{\equiv}C{-}C{\equiv}C{-}R}$$

References

1. Castro, C. E.; Stephens, R. D. *J. Org. Chem.* **1963**, *28*, 2163.
2. Castro, C. E.; Stephens, R. D. *J. Org. Chem.* **1963**, *28*, 3313.
3. Staab, H. A.; Neunhoeffer, K. *Synthesis* **1974**, 424.
4. Kabbara, J.; Hoffmann, C.; Schinzer, D. *ibid.* **1995**, 299.
5. von der Ohe, F.; Bruckner, R. *New J. Chem.* **2000**, *24*, 659.
6. White, J. D.; Carter, R. G.; Sundermann, K. F.; Wartmann, M. *J. Am. Chem. Soc.* **2001**, *123*, 5407.

Chapman rearrangement

Thermal aryl rearrangement of *O*-aryliminoethers to amides.

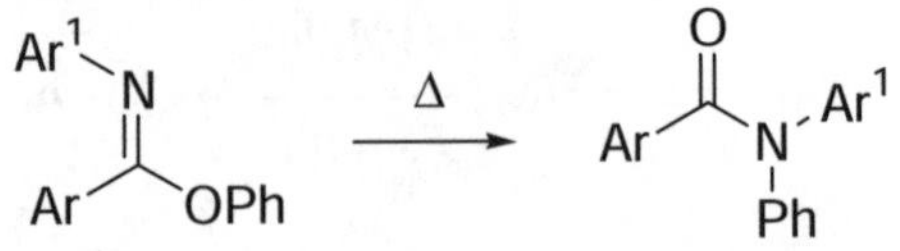

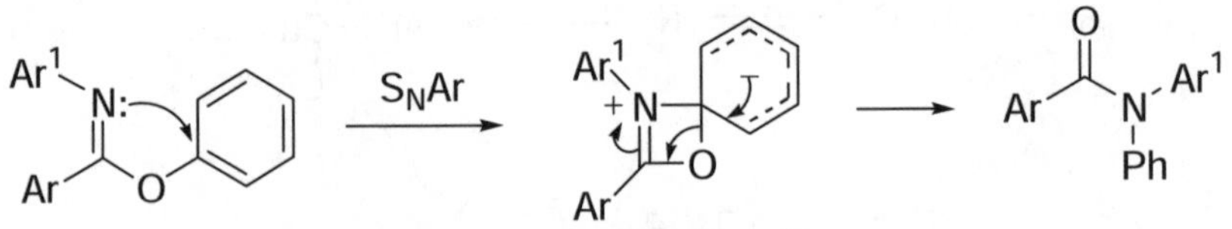

Δ

S_NAr

oxazete intermediate

References

1. Chapman, A. W. *J. Chem. Soc.* **1925**, *127*, 1992.
2. Wheeler, O. H.; Roman, F.; Rosado, O. *J. Org. Chem.* **1969**, *34*, 966.
3. Kimura, M. *J. Chem. Soc., Perkin Trans. 2* **1987**, 205.
4. Kimura, M.; Okabayashi, I.; Isogai, K. *J. Heterocyclic Chem.* **1988**, *25*. 315.
5. Dessolin, M.; Eisenstein, O.; Golfier, M.; Prange, T.; Sautet, P. *J. Chem. Soc., Chem. Commun.* **1992**, 132.

Chichibabin amination reaction

$$\text{pyridine} \xrightarrow[\Delta]{NaNH_2} \text{2-aminopyridine} + NaH$$

$$\text{pyridine} + {}^{-}NH_2 \xrightarrow{S_NAr} \text{adduct (N–Na, C(H)NH}_2\text{)} \longrightarrow \text{2-aminopyridine} + NaH$$

References

1. Chichibabin, A. E.; Zeide, O. A. *J. Russ. Phys. Chem. Soc.* **1914**, *46*, 1216.
2. McGill, C. K.; Rappa, A. *Adv. Heterocycl. Chem.* **1988**, *44*, 1.
3. Katritzky, A. R.; Qiu, G.; Long, Q.-H.; He, H.-Y.; Steel, P. J. *J. Org. Chem.* **2000**, *65*, 9201.

Chichibabin pyridine synthesis

3 R⌒CHO + NH_3 ⟶ (2-(R-CH$_2$)-3,5-di-R-pyridine)

H_3N: ... R-CH$_2$-CHO ⟶ R-CH$_2$-CH(OH)(NH$_2$) ⟶ R-CH=CH-NH$_2$ ⟶ R-CH$_2$-CH=NH

enamine imine

:NH_3 ... ⟶ Aldol addition ⟶

– H_2O ⟶ Michael addition ⟶ $H_3\overset{+}{N}$–H

:NH_3 ⟶ $H_3\overset{+}{N}$–H ⟶

⟶ autoxidation [O] ⟶

References

1. Chichibabin, A. E. *J. Russ. Phys. Chem. Soc.* **1906**, *37*, 1229.
2. Frank, R. L.; Seven, R. P. *J. Am. Chem. Soc.* **1949**, *71*, 2629.
3. Frank, R. L.; Riener, E. F. *ibid.* **1950**, *72*, 4182.
4. Weiss, M. *ibid.* **1952**, *74*, 200.
5. Herzenberg, J.; Boccato, G. *ibid.* **1958**, 248.
6. Levitt, L. S.; Levitt, B. W. *Chem. Ind.* **1963**, 1621.
7. Kessar, S. V.; Nadir, U. K.; Singh, M. *Indian J. Chem.* **1973**, *11*, 825.
8. Sagitullin, R. S.; Shkil, G. P.; Nosonova, I. I.; Ferber, A. A. *Khim. Geterotsikl. Soedin.* **1996**, 147.

Chugaev elimination

Thermal elimination of xanthates to olefins.

1. CS_2, NaOH
2. CH_3I
Δ
+ OCS + CH_3SH
xanthate

Δ
Ei

O=C=S↑ + CH_3SH↑

References

1. Chugaev, L. *Ber.* **1899**, *32*, 3332.
2. Chande, M. S.; Pranjpe, S. D. *Indian J. Chem.* **1973**, *11*, 1206.
3. Kawata, T.; Harano, K.; Taguchi, T. *Chem. Pharm. Bull.* **1973**, *21*, 604.
4. Harano, K.; Taguchi, T. *ibid.* **1975**, *23*, 467.
5. Ho, T. L.; Liu, S. H. *J. Chem. Soc., Perkin Trans. 1* **1984**, 615.
6. Meulemans, T. M.; Stork, G. A.; Macaev, F. Z.; Jansen, B. J. M.; de Groot, A. *J. Org. Chem.* **1999**, *64*, 9178.
7. Nakagawa, H.; Sugahara, T.; Ogasawara, K. *Org. Lett.* **2000**, *2*, 3181.
8. Nakagawa, H.; Sugahara, T.; Ogasawara, K. *Tetrahedron Lett.* **2001**, *42*, 4523.

Ciamician–Dennsted rearrangement

$$\text{pyrrole} \xrightarrow[\text{NaOH}]{\text{CHCl}_3} \text{3-chloropyridine}$$

$$Cl_3C{-}H + {}^{-}OH \longrightarrow H_2O + {}^{-}CCl_2{-}Cl \xrightarrow[\alpha\text{-elimination}]{-\,Cl^-} \text{"}\pm CCl_2\text{"} \equiv :CCl_2$$

carbene

$$\text{pyrrole} + \pm CCl_2 \longrightarrow \text{(Cl, Cl, N–H, HO}^-) \longrightarrow \text{3-chloropyridine}$$

References

1. Ciamician, G. L.; Dennsted, M. *Ber.* **1881**, *14*, 1153.
2. Skell, P. S.; Sandler, R. S. *J. Am. Chem. Soc.* **1958**, *80*, 2024.
3. Vogel, E. *Angew. Chem.* **1960**, *72*, 8.

Claisen, Eschenmoser–Claisen, Johnson–Claisen, and Ireland–Claisen rearrangements

The Claisen, Johnson–Claisen, Ireland–Claisen, para-Claisen rearrangements, along with the Carroll rearrangement belong to the category of *[3,3]-sigmatropic rearrangements*, which is a concerted process. The arrow-pushing here is merely illustrative. For the abnormal Claisen rearrangement, see page 1.

Claisen rearrangement

Δ, [3,3]-sigmatropic rearrangement

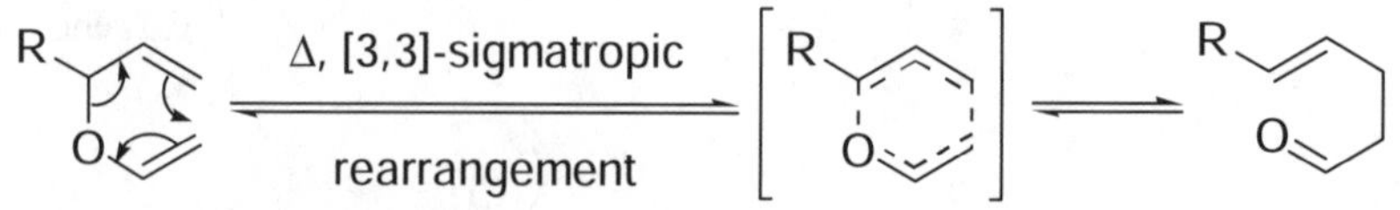

Eschenmoser–Claisen (amide acetal) rearrangements

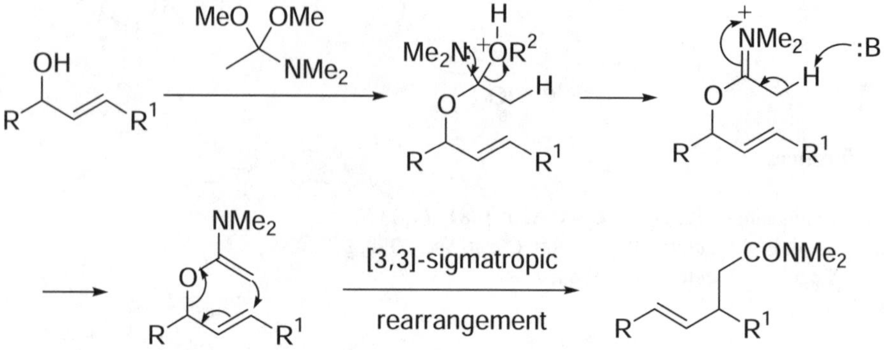

[3,3]-sigmatropic rearrangement

Johnson–Claisen (orthoester) rearrangement

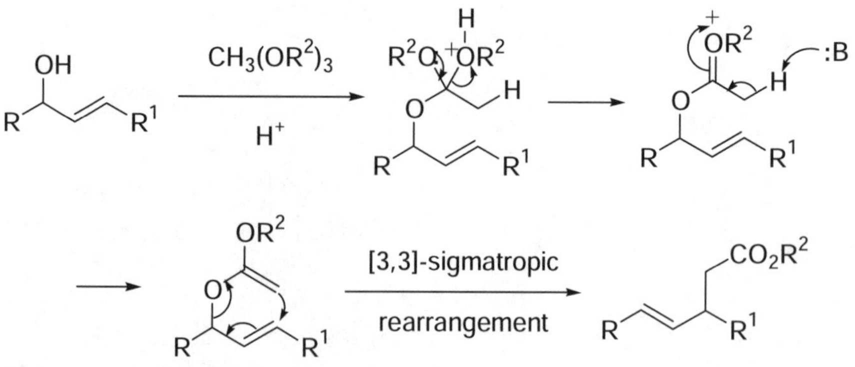

$CH_3(OR^2)_3$, H^+

[3,3]-sigmatropic rearrangement

Ireland–Claisen (silyl ester) rearrangement

LDA, then Me_3SiCl

Δ, [3,3]-sigmatropic rearrangement

H^+

References

1 Claisen, L. *Ber.* **1912**, *45*, 3157.
2 Wick, A. E.; Felix, D.; Steen, K.; Eschenmoser, A. *Helv. Chim. Acta* **1964**, *47*, 2425.
3 Johnson, W. S.; Werthemann, L.; Bartlett, W. R.; Brocksom, T. J.; Li, T.-T.; Faulkner, D. J.; Peterson, M. R. *J. Am. Chem. Soc.* **1970**, *92*, 741.
4 Ireland, R. E.; Mueller, R. H. *ibid.* **1972**, *94*, 5897.
5 Wipf, P. In *Comprehensive Organic Synthesis* Trost, B. M.; Fleming, I. Eds, Pergamon, **1991**, Vol. 5, 827–873.
6 Pereira, S.; Srebnik, M. *Aldrichimica Acta* **1993**, *26*, 17.
7 Panek, J. S.; Schaus, S.; Masse, C. E. *Chemtracts: Org. Chem.* **1995**, *8*, 238.
8 Ganem, B. *Angew. Chem., Int. Ed. Engl.* **1996**, *35*, 936.
9 Cambie, R. C.; Milbank, Jared B. J.; Rutledge, Peter S. *Org. Prep. Proced. Int.* **1997**, *29*, 365.
10 Ito, H.; Taguchi, T. *Chem. Soc. Rev.* **1999**, *28*, 43.
11 Mohamed, M.; Brook, M. A. *Tetrahedron Lett.* **2001**, *42*, 191.
12 Loh, T.-P.; Hu, Q.-Y. *Org. Lett.* **2001**, *3*, 279.

Clark–Eschweiler reductive alkylation of amines

$$R{-}NH_2 + CH_2O + HCO_2H \longrightarrow R{-}N(CH_3)_2$$

formic acid is the hydrogen source as a reducing agent

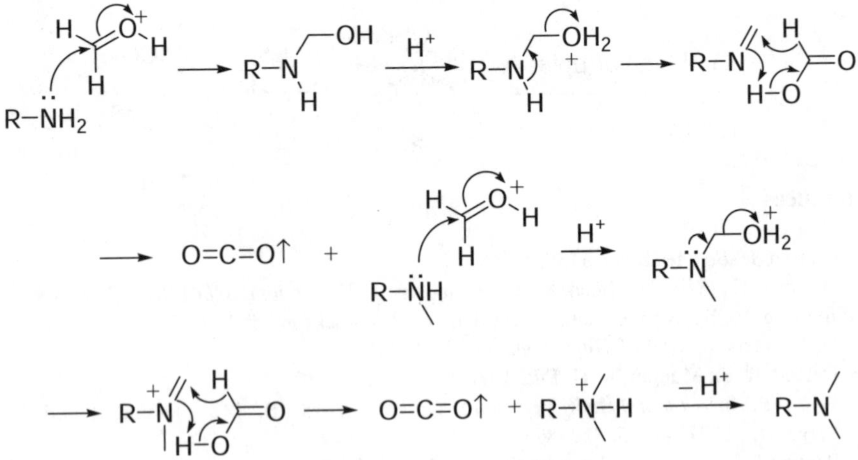

H, O+, H, H, R–NH2, R–N, OH, H, H+, R–N, OH2, +, H, R–N, H, O, H–O, O=C=O↑, +, H, O+, H, H, R–NH, H+, +, OH2, R–N, +, R–N, H, O, H–O, O=C=O↑, +, R–N–H, – H+, R–N

References

1 Moore, M. L. *Org. React.* **1949**, *5*, 301.
2 Pine, S. H.; Sanchez, B. L. *J. Org. Chem.* **1971**, *36*, 829.
3 Bobowski, G. *ibid.* **1971**, *50*, 929.
4 Alder, R. W.; Colclough, D.; Mowlam, R. W. *Tetrahedron Lett.* **1991**, *32*, 7755.
5 Fache, F.; Jacquot, L.; Lemaire, M. *ibid.* **1994**, *35*, 3313.
6 Bulman P., Philip C.; Heaney, H.; Rassias, G. A.; Reignier, S.; Sampler, E. P.; Talib, S. *Synlett* **2000**, 104.

Combes quinoline synthesis

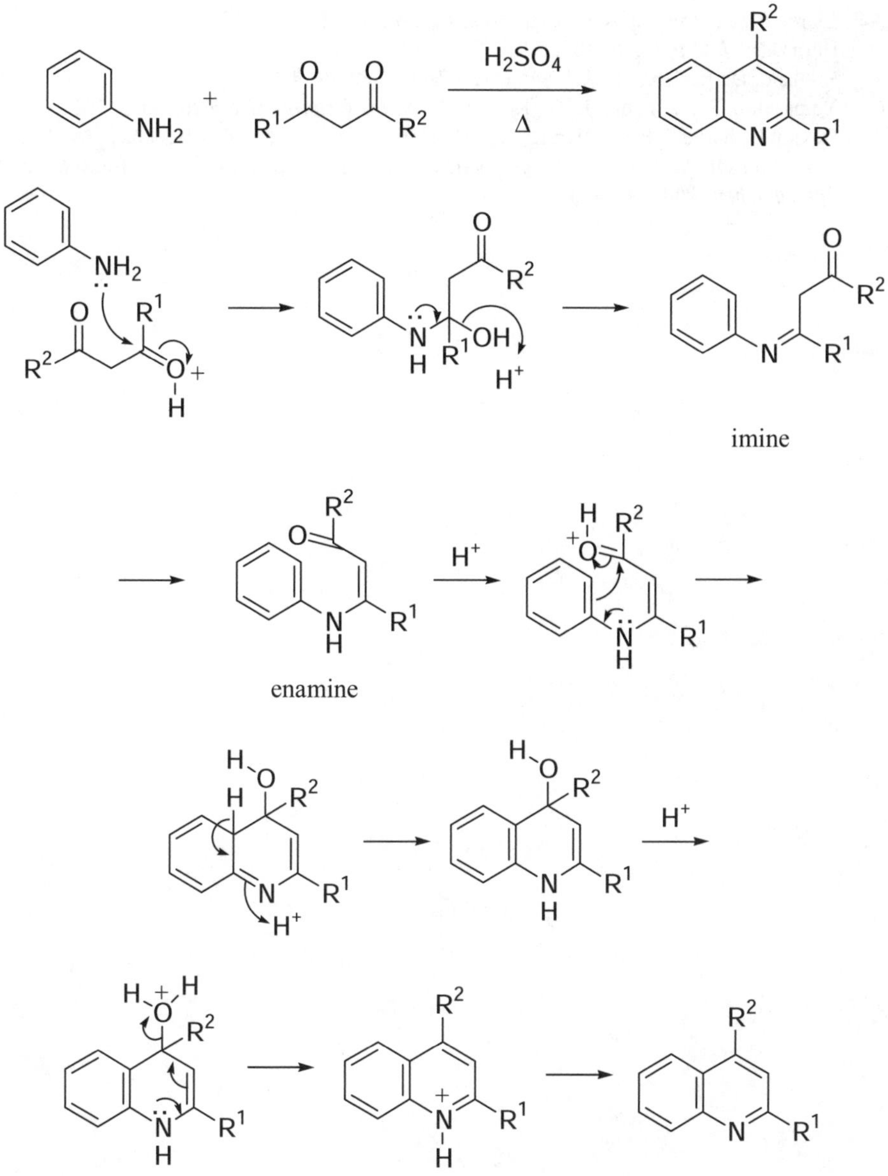

H_2SO_4

Δ

imine

enamine

H^+

References

1 Combes, A. *Bull. Soc. Chim. Fr.* **1888**, *49*, 89.
2 Coscia, A. T.; Dickerman, S. C. *J. Am. Chem. Soc.* **1959**, *81*, 3098.

3 Claret, P. A.; Osborne, A. G. *Org. Prep. Proced. Int.* **1970**, *2*, 305.
4 Born, J. L. *J. Org. Chem.* **1972**, *37*, 3952.
5 Ruhland, B.; Leclerc, G. *J. Heterocycl. Chem.* **1989**, *26*, 469.
6 Yamashkin, S. A.; Yudin, L. G.; Kost, A. N. *Khim. Geterotsikl. Soedin.* **1992**, 1011.
7 Davioud-Charvet, E.; Delarue, S.; Biot, C.; Schwoebel, B.; Boehme, C. C.; Muessigbrodt, A.; Maes, L.; Sergheraert, C.; Grellier, P.; Schirmer, R. H; Becker, K. *J. Med. Chem.* **2001**, *44*, 4268.

Conrad–Lipach reaction

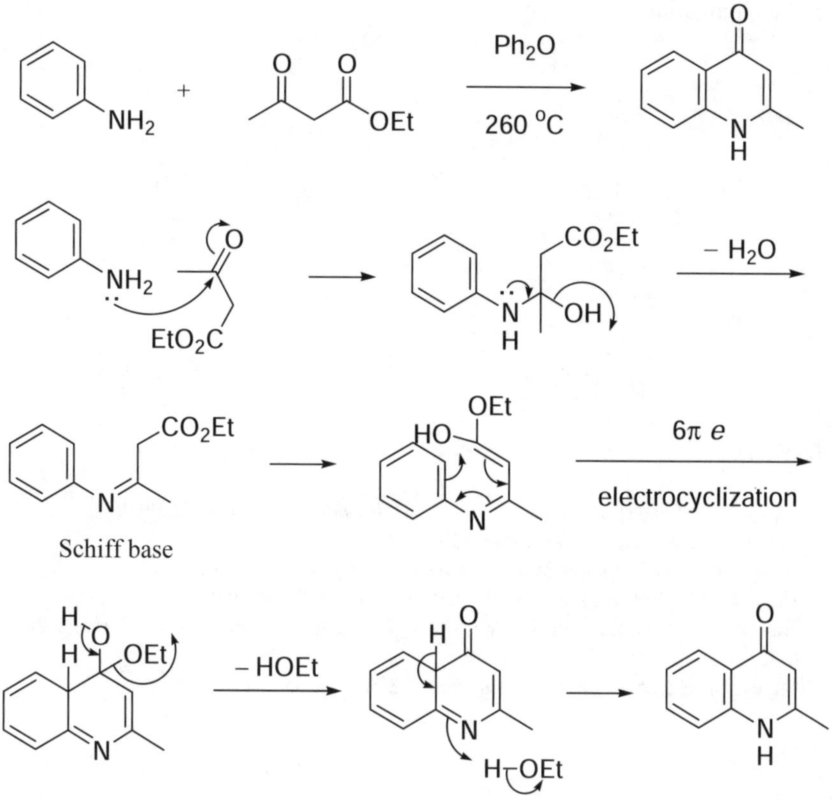

References

1 Conrad, M.; Limpach, L. *Ber.* **1891**, *20*, 944.
2 Heindel, N. D.; Bechara, I. S.; Kennewell, P. D.; Molnar, J.; Ohnmacht, C. J.; Lemke, S. M.; Lemke, T. F. *J. Med. Chem.* **1968**, *11*, 1218.

Cope elimination reaction

Thermal elimination of *N*-oxides to olefins.

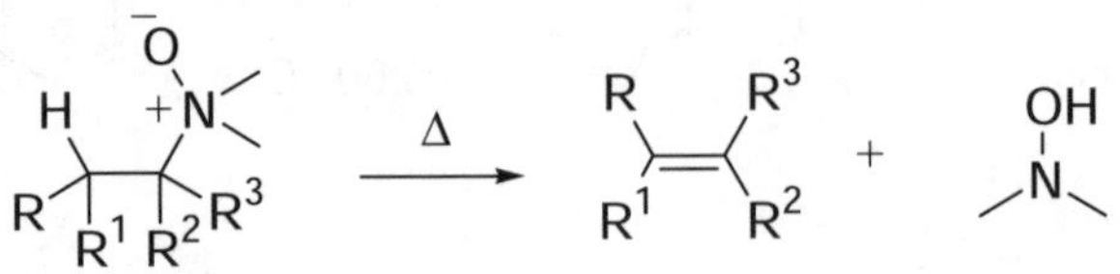

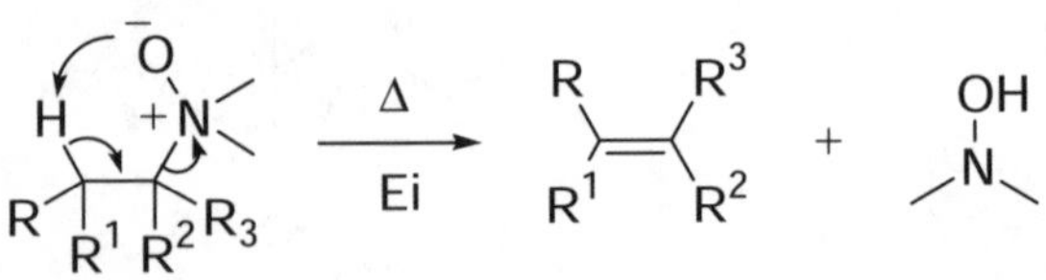

References

1 Cope, A. C.; Foster, T. T.; Towle, P. H. *J. Am. Chem. Soc.* **1949**, *71*, 3929.
2 Frey, H. M.; Walsh, R. *Chem. Rev.* **1969**, *69*, 103.
3 Gallagher, B. M.; Pearson, W. H. *Chemtracts: Org. Chem.* **1996**, *9*, 126.
4 Vidal, T.; Magnier, E.; Langlois, Y. *Tetrahedron* **1998**, *54*, 5959.
5 Gravestock, M. B.; Knight, D. W.; Malik, K. M. A.; Thornton, S. R. *Perkin 1* **2000**, 3292.
6 Bagley, M. C.; Tovey, J. *Tetrahedron Lett.* **2001**, *42*, 351.

Cope, oxy-Cope, and anionic oxy-Cope rearrangements

The Cope, oxy-Cope, and anionic oxy-Cope rearrangements belong to the category of *[3,3]-sigmatropic rearrangements*, which is a concerted process. The arrow-pushing here is only illustrative.

Cope rearrangement

R

Δ, [3,3]-sigmatropic

rearrangement

R

R

oxy-Cope rearrangement

OH

R

Δ, [3,3]-sigmatropic

rearrangement

OH

R

OH

R

O

R

anionic oxy-Cope rearrangement

OH

R

KH

THF

OK

R

Δ, [3,3]-sigmatropic

rearrangement

OK

R

OK

R

H_2O

OH

R

O

R

References

1 Cope, A. C.; Hardy, E. M. *J. Am. Chem. Soc.* **1940**, *62*, 441.
2 Evans, D. A.; Golob, A. M. *ibid.* **1975**, *97*, 4765.
3 Paquette, L. A. *Angew. Chem.* **1990**, *102*, 642.

4 Hill, R. K. In *Comprehensive Organic Synthesis* Trost, B. M.; Fleming, I., Eds, Pergamon, **1991**, *Vol. 5*, 785–826.
5 Davies, H. M. L. *Tetrahedron* **1993**, *49*, 5203.
6 Paquette, L. A. *ibid.* **1997**, *53*, 13971.
7 Miyashi, T.; Ikeda, H.; Takahashi, Y. *Acc. Chem. Res.* **1999**, *32*, 815.

Corey–Chaykovsky epoxidation

Epoxidation of carbonyls using dimethylsulfoxonium methylide or dimethylsulfonium methylide.

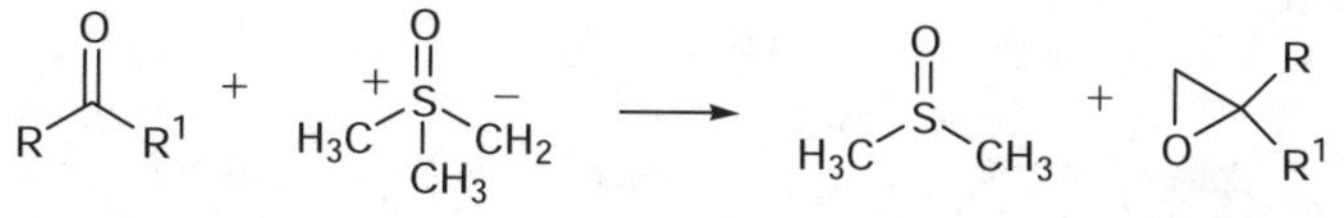

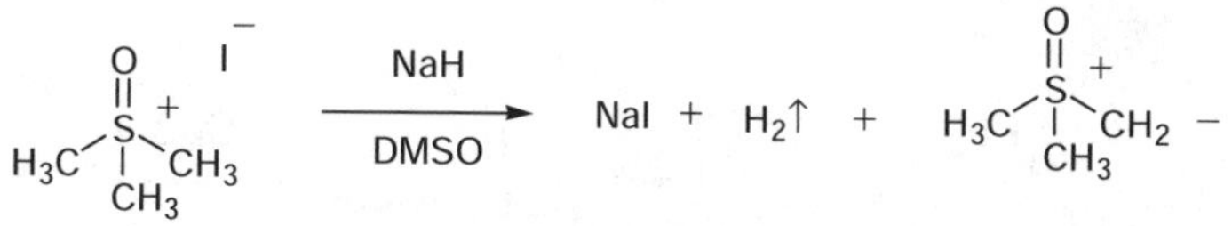

R ... R[1] + H$_3$C–S(=O)(CH$_3$)–CH$_2$ → H$_3$C–S(=O)–CH$_3$ + epoxide (R, R[1])

NaH, DMSO → NaI + H$_2$↑ + H$_3$C–S(=O)(CH$_3$)–CH$_2^-$

dimethylsulfoxonium methylide (DMSY)

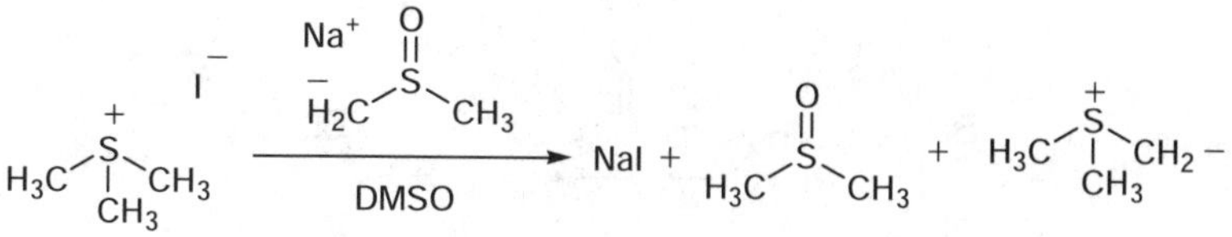

Na$^+$ $^-$H$_2$C–S(=O)–CH$_3$, DMSO → NaI + H$_3$C–S(=O)–CH$_3$ + H$_3$C–S$^+$(CH$_3$)–CH$_2^-$

dimethylsulfonium methylide

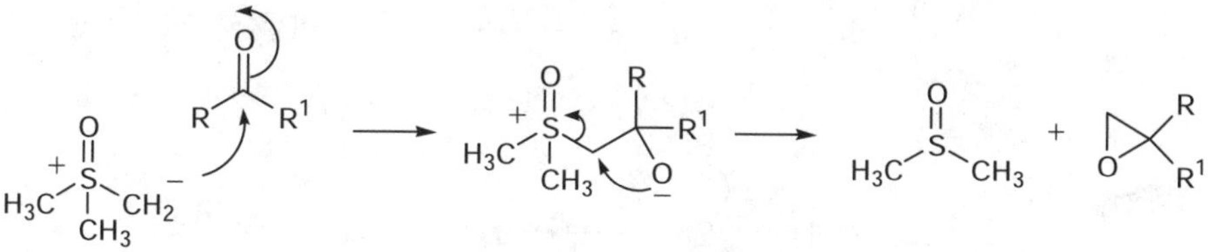

References

1 Corey, E. J.; Chaykovsky, M. *J. Am. Chem. Soc.* **1962**, *84*, 867.
2 Corey, E. J.; Chaykovsky, M. *ibid.* **1965**, *87*, 1353.
3 Trost, B. M.; Melvin, L. S., Jr. *Sulfur Ylides* Academic Press: New York, **1975**.
4 Block, E. *Reactions of Organosulfur Compounds* Academic Press: New York, **1978**.
5 Gololobov, Y. G.; Nesmeyanov, A. N. *Tetrahedron* **1987**, *43*, 2609.
6 Saito, T.; Akiba, D.; Sakairi, M.; Kanazawa, S. *Tetrahedron Lett.* **2001**, *42*, 57.

Corey–Fuchs reaction

$$R{-}CHO \xrightarrow[Zn]{CBr_4,\ PPh_3} (R)(H)C{=}CBr_2 \xrightarrow{n\text{-BuLi}} R{-}C{\equiv}C{-}H$$

$$Br_3C{-}Br + {:}PPh_3 \xrightarrow{S_N2} {}^{-}CBr_3 + Br{-}\overset{+}{P}Ph_3$$

$$Br{-}\overset{+}{P}Ph_3 + {}^{-}CBr_3 \xrightarrow{S_N2} Br_3C{-}\overset{+}{P}Ph_3 + {}^{-}Br \xrightarrow{S_N2}$$

$$Br_2 + Br_2\overset{-}{C}{-}\overset{+}{P}Ph_3 \xrightarrow{RCHO} (R)(H)C{=}CBr_2 + O{=}PPh_3$$

Wittig reaction (see page 396 for the mechanism)

$$Br_2 + Zn \longrightarrow ZnBr_2$$

$$(R)(H)C{=}CBr_2 + Bu^{-} \longrightarrow R{-}C{\equiv}C{-}Br + {}^{-}Bu \longrightarrow R{-}C{\equiv}C^{-} \xrightarrow[\text{workup}]{\text{acidic}} R{-}C{\equiv}C{-}H$$

References

1 Corey, E. J.; Fuchs, P. L. *Tetrahedron Lett.* **1972**, 3769.

2 For the synthesis of 1-bromalkynes, see, Grandjean, D.; Pale, P.; Chuche, J. *ibid.* **1994**, *35*, 3529.

3 Jiang, B.; Ma, P. *Synth. Commun.* **1995**, *25*, 3641.

4 Gilbert, A. M.; Miller, R.; Wulff, W. D. *Tetrahedron* **1999**, *55*, 1607.

5 Muller, T. J. J. *Tetrahedron Lett.* **1999**, *40*, 6563.

6 Serrat, X.; Cabarrocas, G.; Rafel, S.; Ventura, M.; Linden, A.; Villalgordo, J. M. *Tetrahedron: Asymmetry* **1999**, *10*, 3417.

Corey–Bakshi–Shibata (CBS) reduction

Enantioselective borane reduction of ketones catalyzed by chiral oxaborolidines.

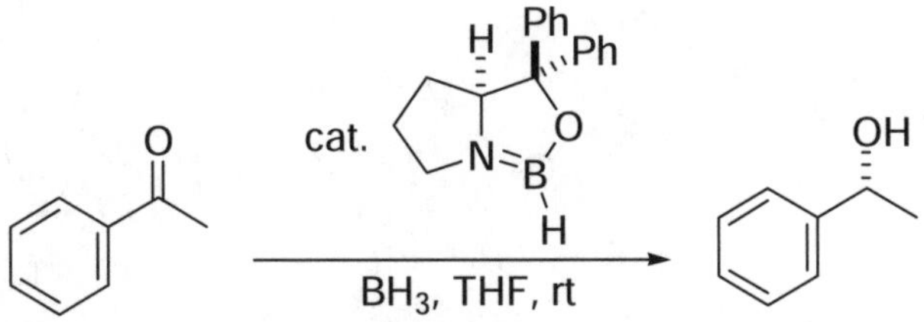

cat.

BH_3, THF, rt

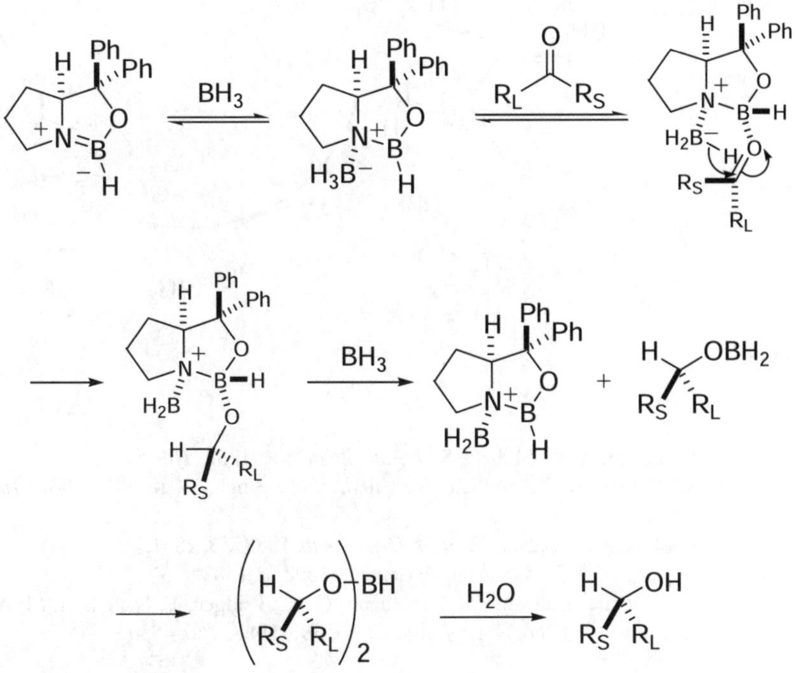

BH_3

BH_3

H_2O

The catalytic cycle:

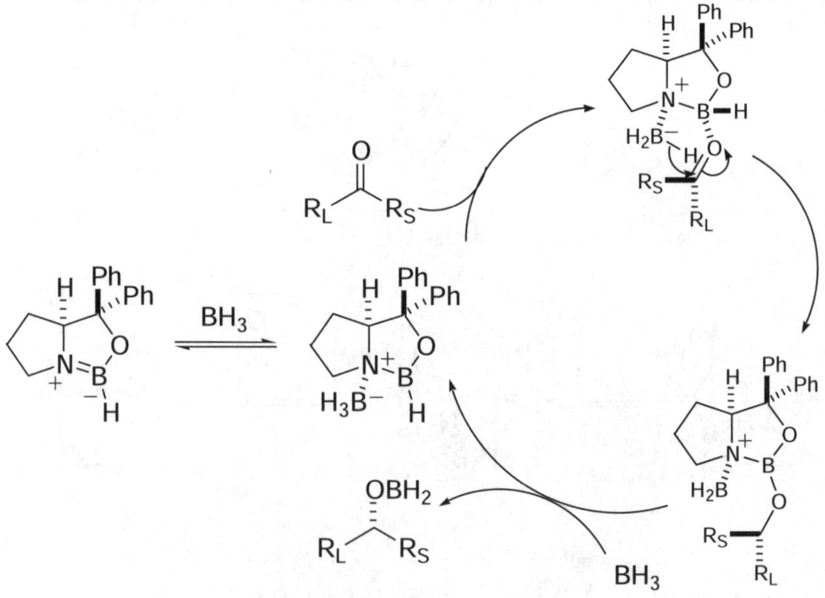

References

1 Corey, E. J.; Bakshi, R. K.; Shibata, S. *J. Am. Chem. Soc.* **1987**, *109*, 5551.

2 Corey, E. J.; Bakshi, R. K.; Shibata, S.; Chen, C.-P.; Singh, V. K. *ibid.* **1987**, *109*, 7925.

3 Corey, E. J.; Shibata, S.; Bakshi, R. K. *J. Org. Chem.* **1988**, *53*, 2861.

4 Cho, B. T.; Chun, Y. S. *Tetrahedron: Asymmetry* **1992**, *3*, 1583.

5 Clark, W. M.; Tickner-Eldridge, A. M.; Huang, G. K.; Pridgen, L. N.; Olsen, M. A.; Mills, R. J.; Lantos, I.; Baine, N. H. *J. Am. Chem. Soc.* **1998**, *120*, 4550.

Corey–Kim oxidation

Oxidation of alcohols to the corresponding aldehyde or ketone using NCS/DMF, followed by treatment with a base.

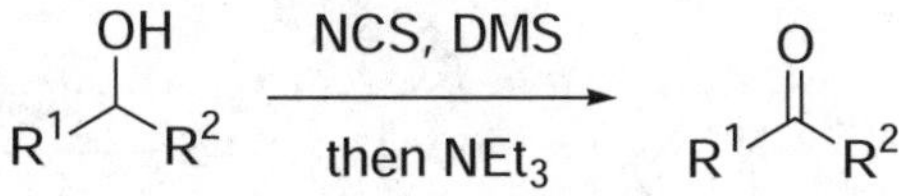

NCS = ***N*-C**hlorosuccinamide; DMS = **Di**methyl**s**ulfide.

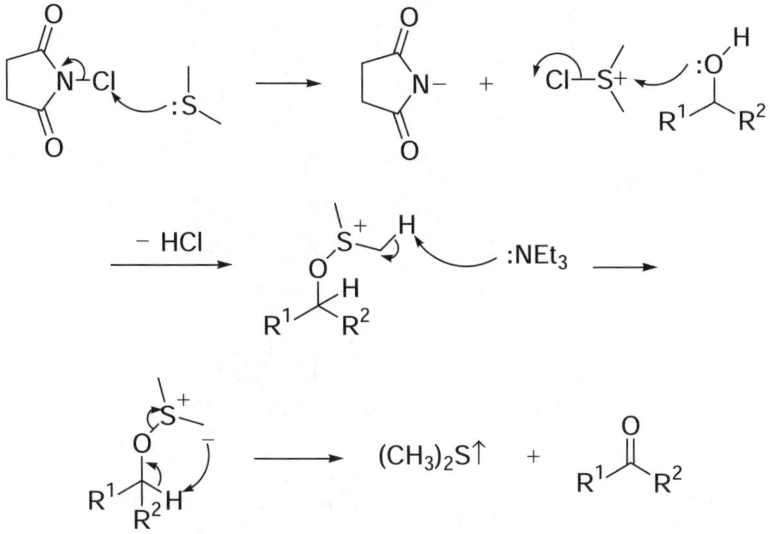

Alternatively:

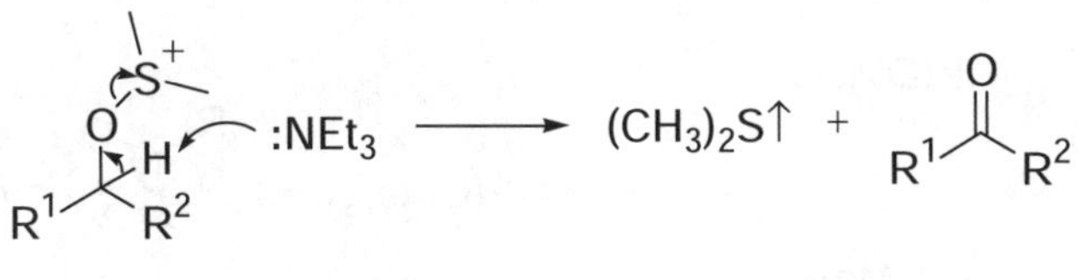

References

1 Corey, E. J.; Kim, C. U. *J. Am. Chem. Soc.* **1972**, *94*, 7586.
2 Katayama, S.; Fukuda, K.; Watanabe, T.; Yamauchi, M. *Synthesis* **1988**, 178.
3 Shapiro, G.; Lavi, Y. *Heterocycles* **1990**, *31*, 2099.
4 Pulkkinen, J. T.; Vepsalainen, J. J. *J. Org. Chem.* **1996**, *61*, 8604.

Corey–Winter olefin synthesis

Transformation of diols to the corresponding olefins by sequential treatment with 1,1'-thiocarbonyldiimidazole and trimethylphosphite.

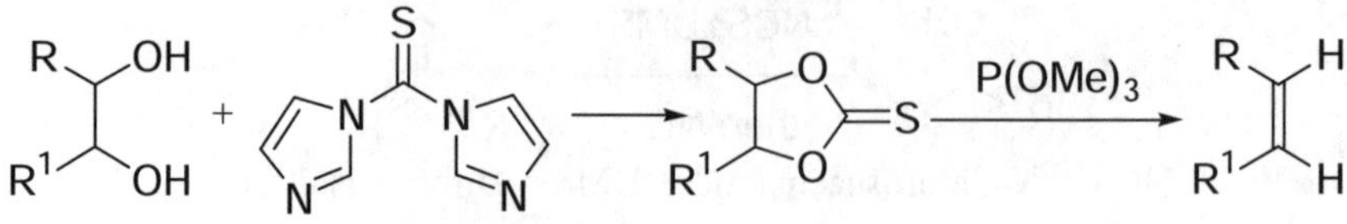

P(OMe)$_3$

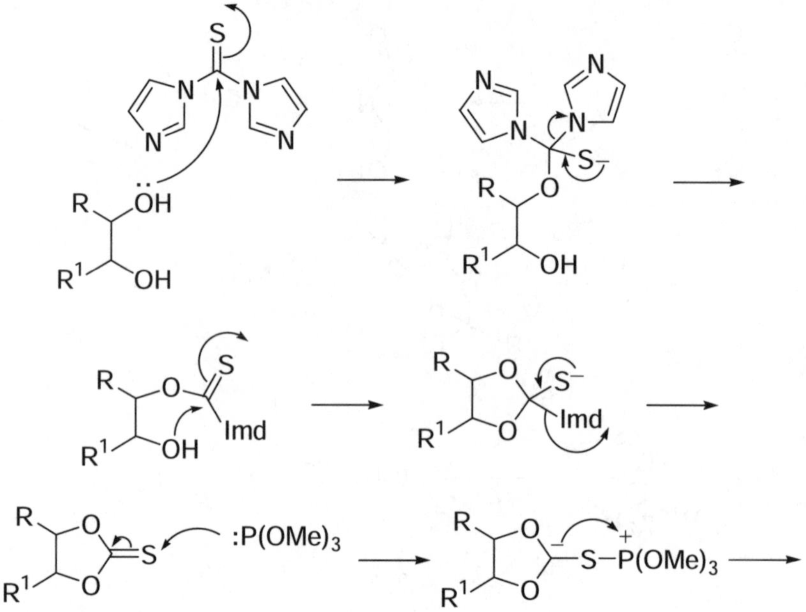

Imd

Imd

:P(OMe)$_3$

S–P(OMe)$_3$

1,3-dioxolane-2-thione (cyclic thionocarbonate)

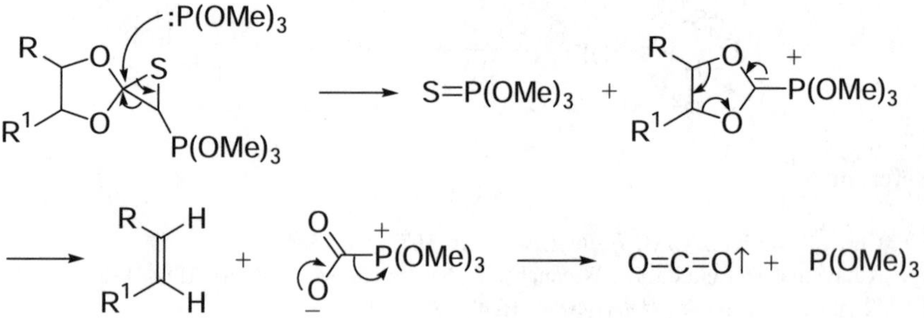

:P(OMe)$_3$

P(OMe)$_3$

S=P(OMe)$_3$ +

P(OMe)$_3$

P(OMe)$_3$

O=C=O↑ + P(OMe)$_3$

A mechanism involving a carbene intermediate is also viable as it is supported by pyrolysis studies:

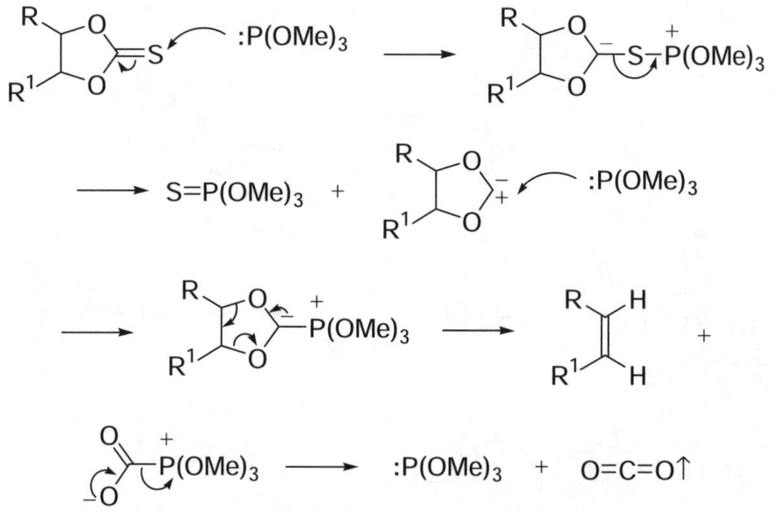

References

1 Corey, E. J.; Winter, E. *J. Am. Chem. Soc.* **1963**, *85*, 2677.
2 Horton, D.; Tindall, C. G., Jr. *J. Org. Chem.* **1970**, *35*, 3558.
3 Hartmann, W.; Fischler, H. M.; Heine, H. G. *Tetrahedron Lett.* **1972**, 853.
4 Block, E. *Org. Recat.* **1984**, *30*, 457.
5 Dudycz, L. W. *Nucleosides Nucleotides* **1989**, *8*, 35.
6 Carr, R. L. K.; Donovan, T. A., Jr.; Sharma, M. N.; Vizine, C. D.; Wovkulich, M. J. *Org. Prep. Proced. Int.* **1990**, *22*, 245.
7 Crich, D.; Pavlovic, A. B.; Wink, D. J. *Synth. Commun.* **1999**, *29*, 359.

Cornforth rearrangement

Thermal rearrangement of keto-oxazoles.

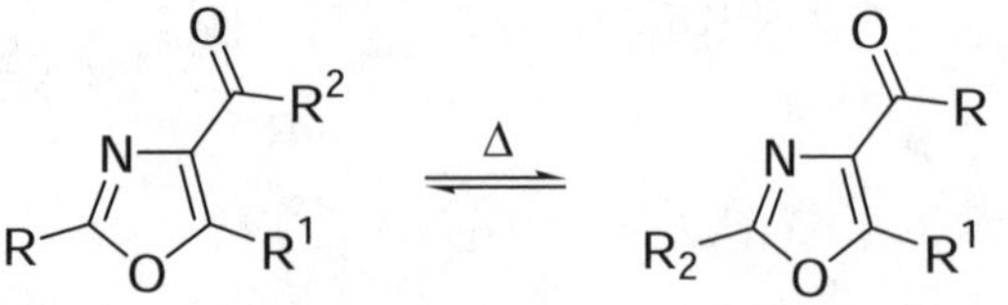

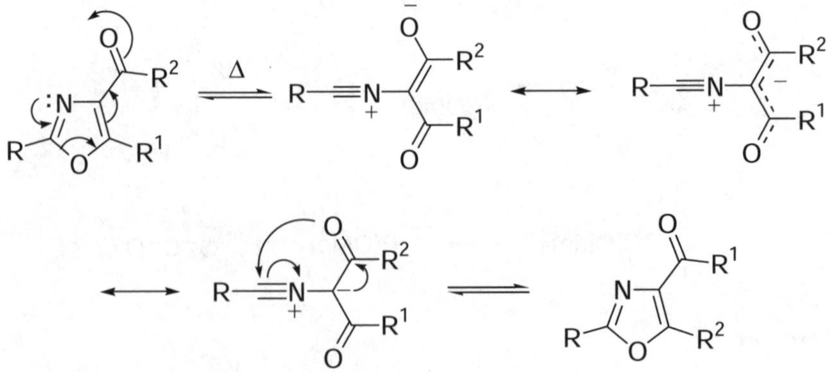

dicarbonyl nitrile ylide intermediate

References

1 Cornforth, J. W. In *The Chemistry of Penicillin* Princeton University Press: New Jersey, **1949**, 700.
2 Dewar, M. J. S.; Spanninger, P. A.; Turchi, I. J. *J. Chem. Soc., Chem. Commun.* **1973**, 925.
3 Dewar, M. J. S. *J. Am. Chem. Soc.* **1974**, *96*, 6148.
4 Dewar, M. J. S.; Turchi, I. J. *J. Org. Chem.* **1975**, *40*, 1521.
5 Williams, D. R.; McClymont, E. L. *Tetrahedron Lett.* **1993**, *34*, 7705.

Criegee glycol cleavage

Vicinal diol is oxidized to the two corresponding carbonyl compounds using $Pb(OAc)_4$.

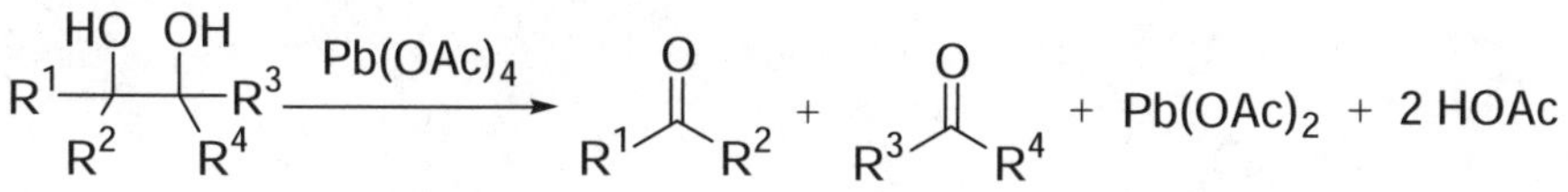

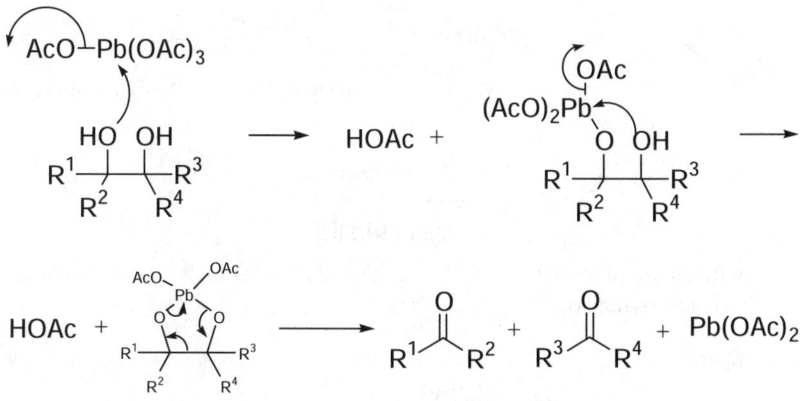

References

1 Criegee, R. *Ber.* **1931**, *64*, 260.
2 Michailovici, M. L. *Synthesis* **1970**, 209.
3 Hatakeyama, S.; Numata, H.; Osanai, K.; Takano, S. *J. Org. Chem.* **1989**, *54*, 3515.

Criegee mechanism of ozonolysis

O_3

[3 + 2]
cycloaddition

primary ozonide (1,2,3-trioxolane)

1,3-dipolar
cycloaddition

zwitterionic peroxide
(Criegee zwitterion)

secondary ozonide (1,2,4-trioxolane)

References

1 Criegee, R.; Werner, G. *Liebigs Ann. Chem.* **1949**, *9*, 564.
2 Criegee, R. *Rec. Chem. Proc.* **1957**, *18*, 111.
3 Criegee, R. *Angew. Chem.* **1975**, *87*, 765.
4 Kuczkowski, R. L. *Chem. Soc. Rev.* **1992**, *21*, 79.
5 Ponec, R.; Yuzhakov, G.; Haas, Y.; Samuni, U. *J. Org. Chem.* **1997**, *62*, 2757.
6 Anglada, J. M.; Crehuet, R.; Maria Bofill, J. *Chem.--Eur. J.* **1999**, *5*, 1809.
7 Dussault, P. H.; Raible, J. M. *Org. Lett.* **2000**, *2*, 3377.

Curtius rearrangement

$$RCOCl \xrightarrow{NaN_3} RCON_3 \xrightarrow{\Delta}$$

$$N_2\uparrow + R{-}N{=}C{=}O \xrightarrow{H_2O} R{-}NH_2 + CO_2\uparrow$$

$$RCOCl + {}^{-}N_3 \longrightarrow RC(O^{-})(N_3)Cl \longrightarrow RCON_3 \equiv RC(O){-}N{=}\overset{+}{N}{=}\overset{-}{N}$$

$$RC(O){-}\overset{-}{N}{-}\overset{+}{N}{\equiv}N \xrightarrow{\Delta} N_2\uparrow + R{-}N{=}C{=}O \quad (H^+,\ :OH_2)$$

isocyanate intermediate

$$\longrightarrow R{-}NH{-}C(O){-}O{-}H \ (:B) \longrightarrow R{-}NH_2 + CO_2\uparrow$$

References

1 Curtius, T. *Ber.* **1890**, *23*, 3023.

2 Chen, J. J.; Hinkley, J. M.; Wise, D. S.; Townsend, L. B. *Synth. Commun.* **1996**, *26*, 617.

3 Am Ende, D. J.; DeVries, K. M.; Clifford, P. J.; Brenek, S. J. *Org. Process Res. Dev.* **1998**, *2*, 382.

4 Braibante, M. E. F.; Braibante, H. S.; Costenaro, E. R. *Synthesis* **1999**, 943.

5 Migawa, M. T.; Swayze, E. E. *Org. Lett.* **2000**, *2*, 3309.

6 Haddad, M. E.; Soukri, M.; Lazar, S.; Bennamara, A.; Guillaumet, G.; Akssira, M. *J. Heterocycl. Chem.* **2000**, *37*, 1247.

Dakin reaction

Cf. Baeyer–Villiger oxidation

$$HO\text{-}C_6H_4\text{-}CHO \xrightarrow[45\text{–}50\ ^{o}C]{H_2O_2,\ NaOH} HO\text{-}C_6H_4\text{-}OH + HCO_2H$$

References:

1. Dakin, H. D. *J. Am. Chem. Soc.* **1909**, *42*, 477.
2. Jung, M. E.; Lazarova, T. I. *J. Org. Chem.* **1997**, *62*, 1553.
3. Varma, R. S.; Naicker, K. P. *Org. Lett.* **1999**, *1*, 189.

Dakin–West reaction

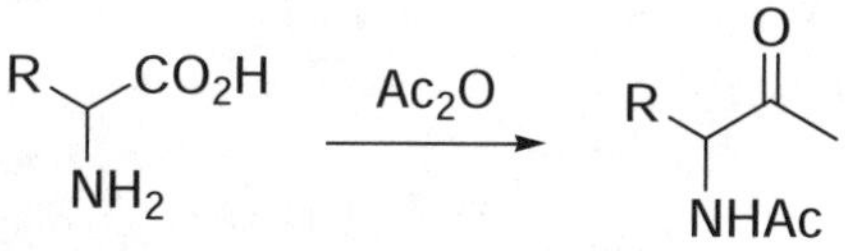

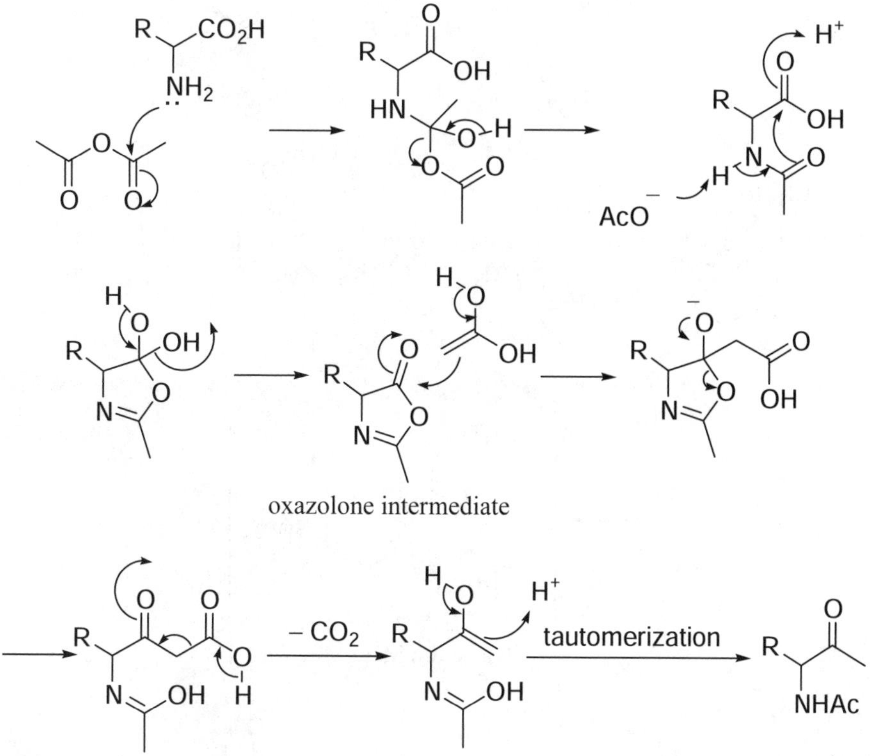

oxazolone intermediate

References:

1. Dakin, H. D.; West, R. *J. Biol. Chem.* **1928**, 78.
2. Buchanan, G. L. *Chem. Soc. Rev.* **1988**, *17*, 91.
3. Jung, M. E.; Lazarova, T. I. *J. Org. Chem.* **1997**, *62*, 1553.
4. Kawase, M.; Hirabayashi, M.; Koiwai, H.; Yamamoto, K.; Miyamae, H. *Chem. Commun.* **1998**, 641.
5. Kawase, M.; Okada, Y.; Miyamae, H. *Heterocycles* **1998**, *48*, 285.
6. Kawase, M.; Hirabayashi, M.; Kumakura, H.; Saito, S.; Yamamoto, K. *Chem. Pharm. Bull.* **2000**, *48*, 114.

Danheiser annulation

Trimethylsilylcyclopentene annulation from an α,β-unsaturated ketone and trimethylsilylallene in the presence of a Lewis acid.

$TiCl_4$, CH_2Cl_2

–75 °C, 82%

SiMe3

– Cl⁻

1,2-shift of silyl group

Transition State

References

1. Danheiser, R. L; Carini, D. J.; Basak, A. *J. Am. Chem. Soc.* **1981**, *103*, 1604.
2. Danheiser, R. L; Carini, D. J.; Fink, D. M.; Basak, A. *Tetrahedron* **1983**, *39*, 935.
3. Danheiser, R. L; Fink, D. M.; Tsai, Y.-M. *Org. Synth.* **1988**, *66*, 8.
4. Iwasawa, N.; Matsuo, T.; Iwamoto, M.; Ikeno, T. *J. Am. Chem. Soc.* **1998**, *120*, 3903.
5. Smith, A. B. III; Adams, C. M.; Kozmin, S. A.; Paone, D. V. *ibid.* **2001**, *123*, 5925.

Darzens glycidic ester condensation

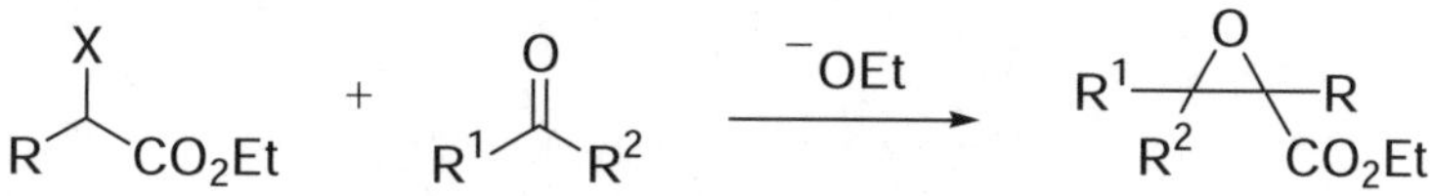

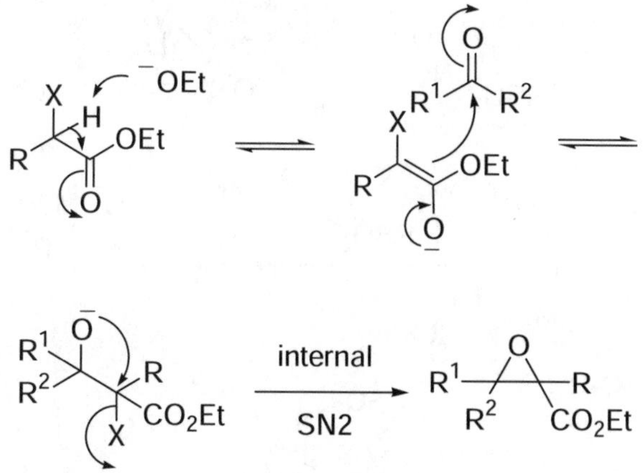

References

1. Darzens, G. *Compt. Rend.* **1904**, *139,* 1214.
2. Bauman, J. G.; Hawley, R. C.; Rapoport, H. *J. Org. Chem.* **1984**, *49*, 3791.
3. Takahashi, T.; Muraoki, M.; Capo, M.; Koga, K. *Chem. Pharm. Bull.* **1995**, *43*, 1821.
4. Ohkata, K.; Kimura, J.; Shinohara, Y.; Takagi, R.; Hiraga, Y. *Chem. Commun.* **1996**, 2411.
5. Takagi, R.i; Kimura, J.; Shinohara, Y.; Ohba, Y.; Takezono, K.; Hiraga, Y.; Kojima, S.; Ohkata, K. *J. Chem. Soc., Perkin Trans. 1* **1998**, 689.
6. Hirashita, T.; Kinoshita, K.; Yamamura, H.; Kawai, M.; Araki, S. *ibid.* **2000**, 825.
7. Shinohara, Y.; Ohba, Y.; Takagi, R.; Kojima, S.; Ohkata, K. *Heterocycles* **2001**, *55*, 9.

Davis chiral oxaziridine reagents

Chiral *N*-sulfonyloxaziridines employed for asymmetric hydroxylation *etc.*

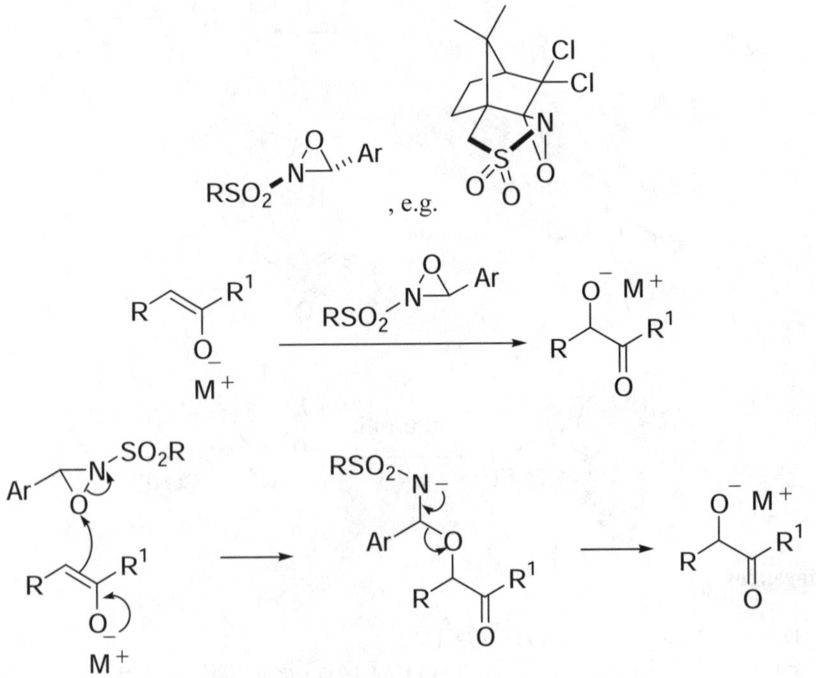

, e.g.

References

1. Davis, F. A.; Vishwakarma, L. C.; Billmers, J. M.; Finn, J. *J. Org. Chem.* **1984**, *49*, 3241.
2. Davis, F. A.; Billmers, J. M.; Gosciniak, D. J.; Towson, J. C.; Bach, R. D. *ibid.* **1986**, *51*, 4240.
3. Davis, F. A.; Chen, B.-C. *Chem. Rev.* **1992**, *92*, 919.
4. Davis, F. A.; ThimmaReddy, R.; Weismiller, M. C. *J. Am. Chem. Soc.* **1989**, *111*, 5964.
5. Davis, F. A.; Kumar, A.; Chen, B. C. *J. Org. Chem.* **1991**, *56*, 1143.
6. Davis, F. A.; Reddy, R. T.; Han, W.; Carroll, P J. *J. Am. Chem. Soc.* **1992**, *114*, 1428.
7. Tagami, K.; Nakazawa, N.; Sano, S.; Nagao, Y. *Heterocycles* **2000**, *53*, 771.

de Mayo reaction

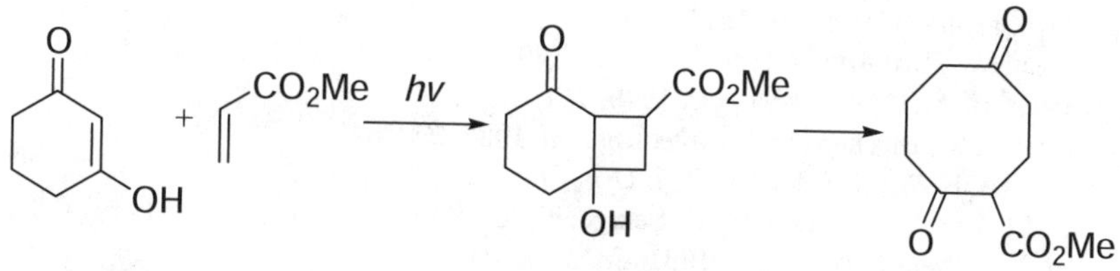

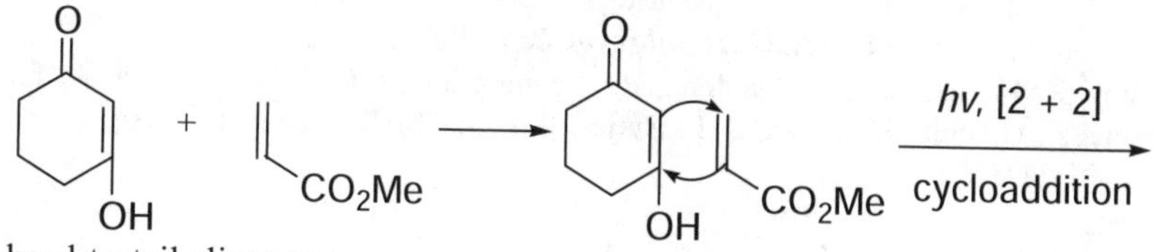

O
+ CO_2Me hv →
OH
O CO_2Me
OH
→
O
O CO_2Me

O
+
OH CO_2Me →
head-to-tail alignment
O
OH CO_2Me
hv, [2 + 2]
cycloaddition

O
retro-aldol
O H CO_2Me
cyclobutane cleavage
O–H
H^+ →
O CO_2Me
O
O CO_2Me

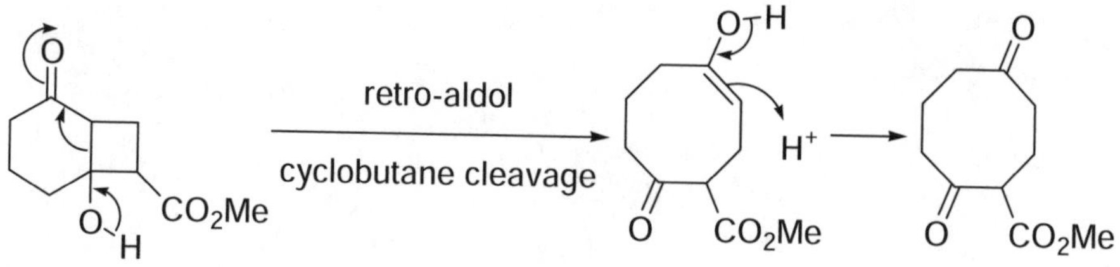

A minor regioisomer:

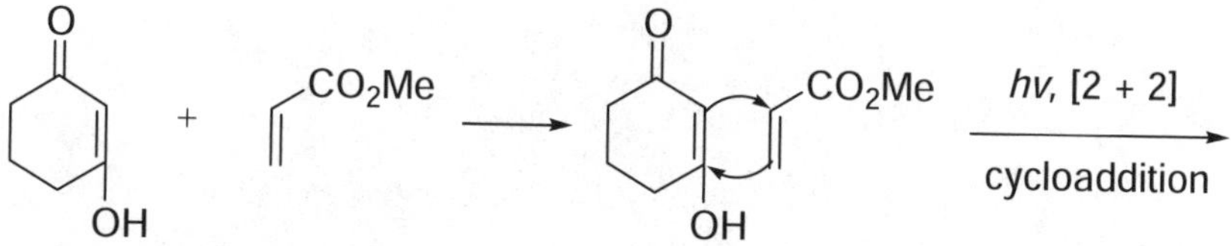

O
+ CO_2Me →
OH
head-to-head alignment
O CO_2Me
OH
hv, [2 + 2]
cycloaddition

O CO_2Me
retro-aldol
O H
cyclobutane cleavage
O–H
H^+
CO_2Me
O
→
O
CO_2Me
O

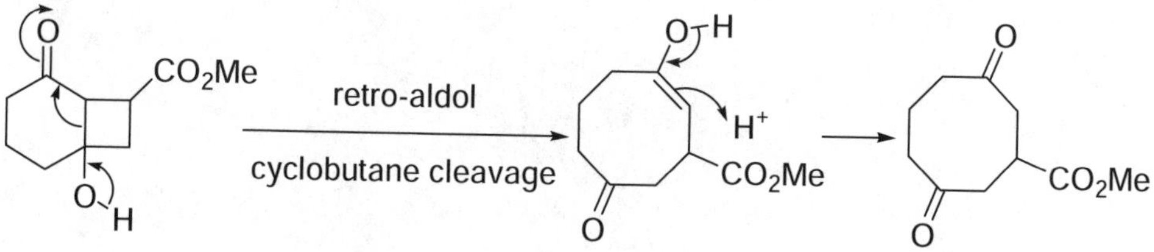

References

1. de Mayo, P.; Takeshita, H.; Sattar, A. B. M. A. *Proc. Chem. Soc., London* **1962**, 119.
2. de Mayo, P. *Acc. Chem. Res.* **1971**, *4*, 49.
3. Oppolzer, W. *Pure Appl. Chem.* **1981**, *53*, 1189.
4. Pearlman, B. A. *J. Am. Chem. Soc.* **1979**, *101*, 6398.
5. Kaczmarek, R.; Blechert, S. *Tetrahedron Lett.* **1986**, *27*, 2845.
6. Disanayaka, B. W.; Weedon, A. C. *J. Org. Chem.* **1987**, *52*, 2905.
7. Sato, M.; Abe, Y.; Takayama, K.; Sekiguchi, K.; Kaneko, C.; Inoue, N.; Furuya, T.; Inukai, N. *J. Heterocycl. Chem.* **1991**, *28*, 241.
8. Sato, M.; Sunami, S.; Kogawa, T.; Kaneko, C. *Chem. Lett.* **1994**, 2191.
9. Quevillon, T. M.; Weedon, A. C. *Tetrahedron Lett.* **1996**, *37*, 3939.
10. Blaauw, R. H.; Briere, J.-F.; de Jong, R.; Benningshof, J. C. J.; van Ginkel, A. E.; Fraanje, J.; Goubitz, K.; Schenk, H.; Rutjes, F. P. J. T.; Hiemstra, H. *J. Org. Chem.* **2001**, *66*, 233.

Demjanov rearrangement

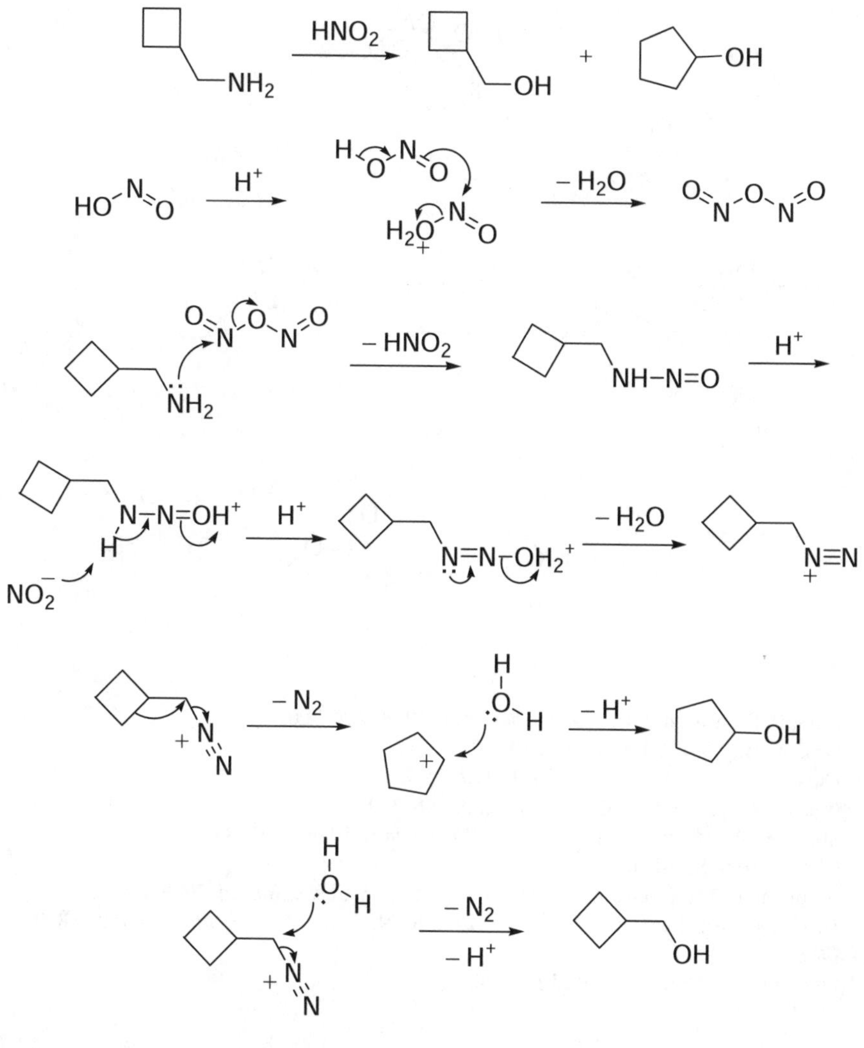

References

1. Demjanov, N. J.; Lushnikov, M. *J. Russ. Phys. Chem. Soc.* **1903**, *35*, 26.
2. Uyehara, T.; Kabasawa, Y.; Furuta, Toshiaki; K., T. *Bull. Chem. Soc. Jpn.* **1986**, *59*, 539.
3. Fattori, D.; Henry, S.; Vogel, P. *Tetrahedron* **1993**, *49*, 1649.
4. Boeckman, R. K. *Org. Synth.* **1999**, *77*, 141.

Dess–Martin periodinane oxidation

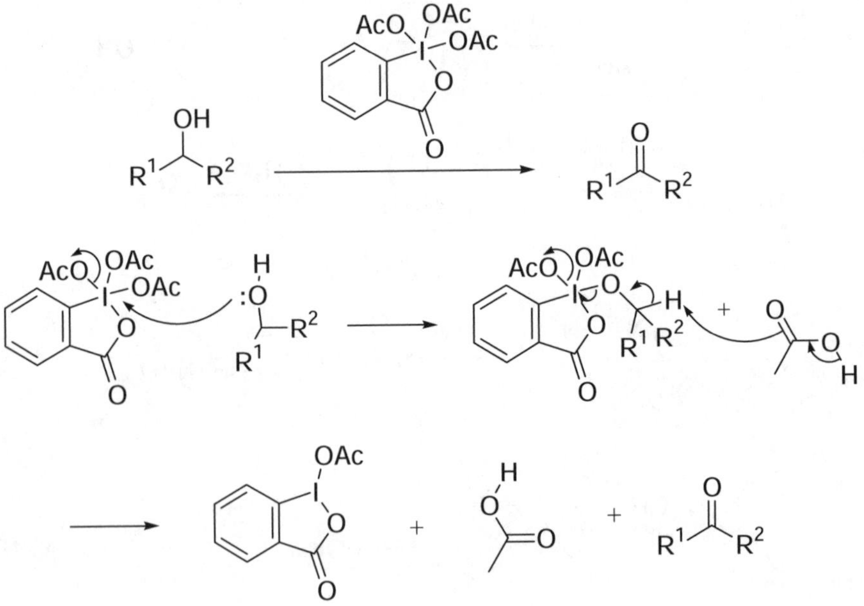

References

1. Dess, P. B.; Martin, J. C. *J. Am. Chem. Soc.* **1978**, *100*, 300.
2. Dess, P. B.; Martin, J. C. *ibid.* **1979**, *101*, 5294.
3. Dess, P. B.; Martin, J. C. *ibid.* **1991**, *113*, 7277.
4. Ireland, R. E.; Liu, L. *J. Org. Chem.* **1993**, *58*, 2899.
5. Speicher, A.; Bomm, V.; Eicher, T. *J. Prakt. Chem.* **1996**, *338*, 588.
6. Chaudhari, S. S.; Akamanchi, K. G. *Synthesis* **1999**, 760.
7. Nicolaou, K. C.; Zhong, Y.-L.; Baran, P. S. *Angew. Chem., Int. Ed.* **2000**, *39*, 622.
8. Jenkins, N. E.; Ware, R. W., Jr.; Atkinson, R. N.; King, S. B. *Synth. Commun.* **2000**, *30*, 947.
9. Promarak, V.; Burn, P. L. *Perkin 1* **2001**, *1*, 14.

Dieckmann condensation

The Dieckmann condensation is the intramolecular version of the Claisen condensation.

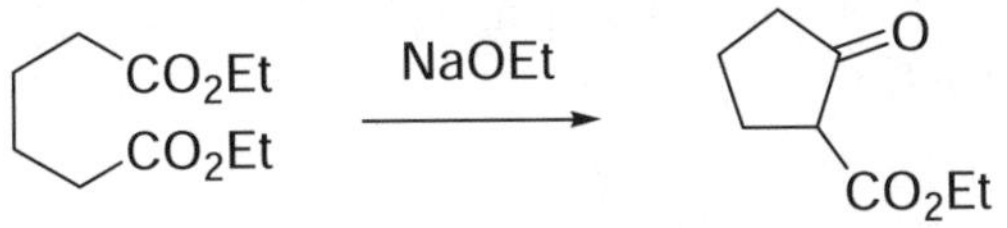

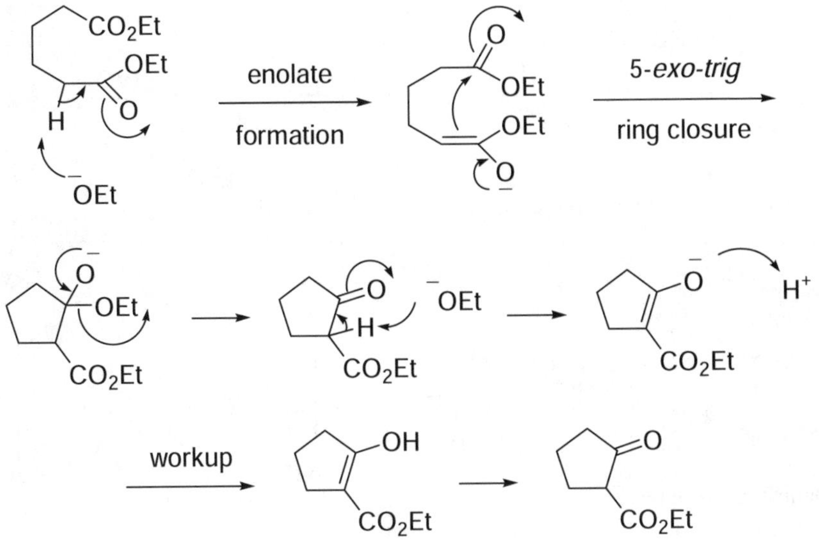

References

1. Dieckmann, W. *Ber.* **1894**, *27*, 102.
2. Davis, B. R.; Garrett, P. J. *Comp. Org. Synth.* **1991**, *2*, pp 806–829.
3. Toda, F.; Suzuki, T.; Higa, S. *J. Chem. Soc., Perkin Trans. 1* **1998**, 3521.
4. Shindo, M.; Sato, Y.; Shishido, K. *J. Am. Chem. Soc.* **1999**, *121*, 6507.
5. Balo, C.; Fernandez, F.; Garcia-Mera, X.; Lopez, C. *Org. Prep. Proced. Int.* **2001**, *32*, 563.

Diels–Alder reaction, inverse electronic demand Diels–Alder reaction, hetero-Diels–Alder reaction

The Diels–Alder reaction, reverse electronic demand Diels–Alder reaction, as well as the hetero-Diels–Alder reaction, belong to the category of *[4+2]-cycloaddition reactions*, which is a concerted process. The arrow-pushing here is merely illustrative.

Normal Diels–Alder reaction

EDG, EWG, Δ, EDG, EWG

diene dienophile adduct

EDG = electron-donating group; EWG = electron-withdrawing group

e.g.

OMe, CO_2Et, Δ, Me_3SiO, OMe, EtO_2C, H, H, H, Me_3SiO

Danishefsky diene

Alder's *endo* rule

OMe, CO_2Et, Me_3SiO, OMe, CO_2Et, Me_3SiO

Inverse electronic demand Diels–Alder reaction

EWG, EDG, Δ, EWG, EDG

diene dienophile adduct

e.g.

Hetero-Diels–Alder reaction

a. Heterodiene addition to dienophile

b. Heterodienophile addition to diene

References

1. Diels, O.; Alder, K. *Liebigs Ann. Chem.* **1928**, *460*, 98.
2. Danishefsky, S.; Kitahara, T. *J. Am. Chem. Soc.* **1974**, *96*, 7807.
3. Oppolzer, W. In *Comprehensive Organic Synthesis*Trost, B. M.; Fleming, I., Eds, Pergamon, **1991**, *Vol. 5*, 315–399.
4. Boger, D. L. In *Comprehensive Organic Synthesis* Trost, B. M.; Fleming, I., Eds, Pergamon, **1991**, *Vol. 5*, 451–512.
5. Weinreb, S. M. In *Comprehensive Organic Synthesis* Trost, B. M.; Fleming, I., Eds, Pergamon, **1991**, *Vol. 5*, 401–499.
6. Mehta, G.; Uma, R. *Acc. Chem. Res.* **2000**, *33*, 278.
7. Behforouz, M.; Ahmadian, M. *Tetrahedron* **2000**, *56*, 5259.
8. Bernath, G.; Stajer, G.; Fulop, F.; Sohar, P. *J. Heterocycl. Chem.* **2000**, *37*, 439.
9. Jorgensen, K. A. *Angew. Chem., Int. Ed.* **2000**, *39*, 3558.
10. Evans, D. A.; Johnson, J. S.; Olhava, E. J. *J. Am. Chem. Soc.* **2000**, *122*, 1635.
11. Huang, Y.; Rawal, V. H. *Org. Lett.* **2000**, *2*, 3321.
12. Doyle, M. P.; Phillips, I. M.; Hu, W. *J. Am. Chem. Soc.* **2001**, *123*, 5366.

Dienone–phenol rearrangement

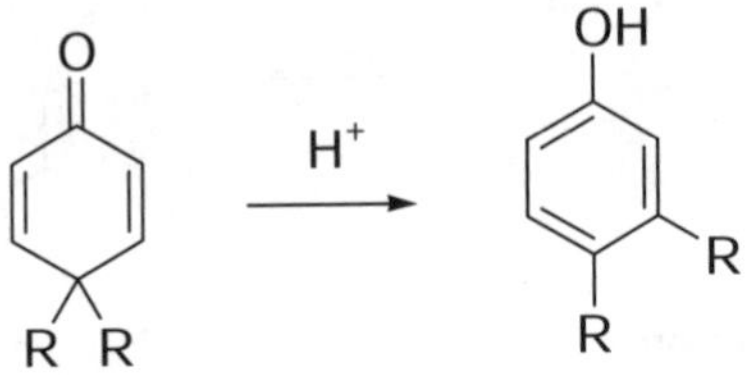

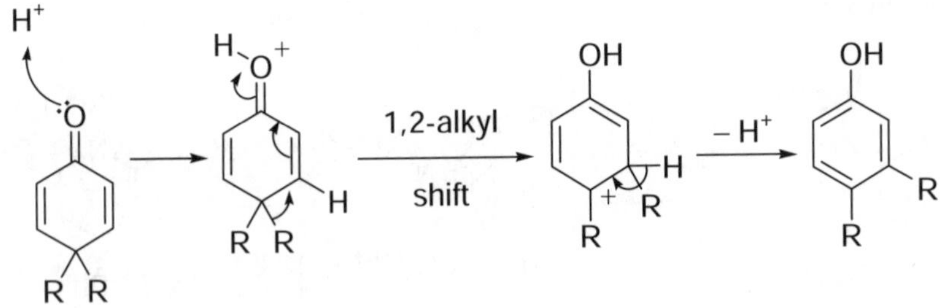

$$\text{4,4-R}_2\text{-cyclohexa-2,5-dienone} \xrightarrow{H^+} \text{3,4-R}_2\text{-phenol}$$

H⁺ ; 1,2-alkyl shift ; – H⁺

References

1. Shine, H. J. In *Aromatic Rearrangement* Elsevier: New York, **1967**, pp 55–68.
2. Schultz, A. G.; Hardinger, S. A. *J. Org. Chem.* **1991**, *56*, 1105.
3. Schultz, A. G.; Green, N. J. *J. Am. Chem. Soc.* **1991**, *114*, 1824.
4. Hart, D. J.; Kim, A.; Krishnamurthy, R.; Merriman, G. H.; Waltos, A. M. *Tetrahedron* **1992**, *48*, 8179.
5. Frimer, A. A.; Marks, V.; Sprecher, M.; Gilinsky-Sharon, P. *J. Org. Chem.* **1994**, *59*, 1831.
6. Oshima, T.; Nakajima, Y.-i.; Nagai, T. *Heterocycles* **1996**, *43*, 619.
7. Draper, R. W.; Puar, M. S.; Vater, E. J.; Mcphail, A. T. *Steroids* **1998**, *63*, 135.
8. Banerjee, A. K.; Castillo-Melendez, J. A.; Vera, W.; Azocar, J. A.; Laya, M. S. *J. Chem. Res., (S)* **2000**, 324.
9. Zimmerman, H E.; Cirkva, V. *J. Org. Chem.* **2001**, *66*, 1839.

Di-π-methane rearrangement

1,4-Dienes to vinylcyclopropanes under photolysis.

$h\nu$

1,4-diene

vinylcyclopropane

$h\nu$

diradical

diradical

References

1. Zimmerman, H. E.; Grunewald, G. L. *J. Am. Chem. Soc.* **1966**, *88*, 183.
2. Janz, K. M.; Scheffer, J. R. *Tetrahedron Lett.* **1999**, *40*, 8725.
3. Zimmerman, H. E.; Cirkva, V. *Org. Lett.* **2000**, *2*, 2365.
4. Tu, Y. Q.; Fan, C. A.; Ren, S. K.; Chan, A. S. C. *Perkin 1* **2000**, 3791.
5. Jimenez, M. C.; Miranda, M. A.; Tormos, R. *Chem. Commun.* **2000**, 2341.
6. Ihmels, H.; Mohrschladt, C. J.; Grimme, J. W.; Quast, H. *Synthesis* **2001**, 1175.

Doebner reaction

Three-component reaction yielding isoquinolines.

CO_2H NH$_2$ OHC + + O CO_2H N

+ O–H H$^+$ NH$_2$ H OH – H_2O N H

H O + CO_2H N H N H$^+$

H$^+$ O CO_2H N H B: HO CO_2H H N+ H H$^+$ + H_2O CO_2H H N H

CO_2H CO_2H – H_2O air oxidation N H – 2 H N

References

1. Doebner, O. *Liebigs Ann. Chem.* **1887**, *242*, 256.
2. Allen, C. F. H.; Spangler, F. W.; Webster, E. R. *J. Org. Chem.* **1951**, *16*, 17.

3. Nitidandhaprabhas, O. *Nature* **1966**, *212*, 5061.
4. Zhdanov, Y. A.; Alekseev, Y. E.; Dorofeenko, G. N. *Carbohyd. Res.* **1968**, *8*, 121.
5. Mitra, A. K.; De, A.; Karchaudhuri, N. *Synth. Commun.* **1999**, *29*, 573.

Doebner–von Miller reaction

Doebner–von Miller reaction is a variant of the Skraup reaction. Therefore, the mechanism for the Skraup reaction is also operative for the Doebner–von Miller reaction. An alternative mechanism shown below is based on the fact that the preformed imine (Schiff base) also gave 2-methylquinoline:

NH_2 + OHC HCl or $ZnCl_2$ N

NH_2 H $\overset{+}{O}$–H N H $\overset{+}{O}H_2$ N H N

[2 + 2] cycloaddition N N Ph H N N H

B: H N N H N N H H^+

N H N H HN N H

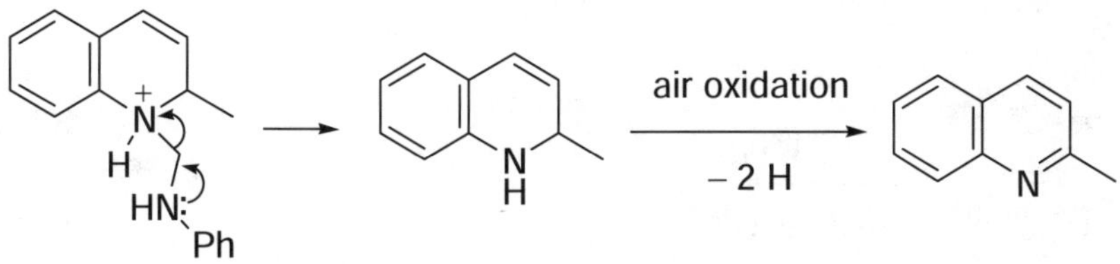

References

1. Doebner, O.; von Miller, W. *Ber.* **1883**, *16*, 2464.
2. Eisch, J. J.; Dluzniewski, T. *J. Org. Chem.* **1989**, *54*, 1269.
3. Zhang, Z. P.; Tillekeratne, L. M. V.; Hudson, Richard A. *Tetrahedron Lett.* **1998**, *39*, 5133.
4. Matsugi, M.; Tabusa, F.; Minamikawa, J.-i. *ibid.* **2000**, *41*, 8523.
5. Fürstner, A.; Thiel, O. R.; Blanda, G. *Org. Lett.* **2000**, *2*, 3731.
6. Kavitha, J.; Vanisree, M.; Subbaraju, G. V. *Indian J. Chem.* **2001**, *40B*, 522.

Doering–LaFlamme allene synthesis

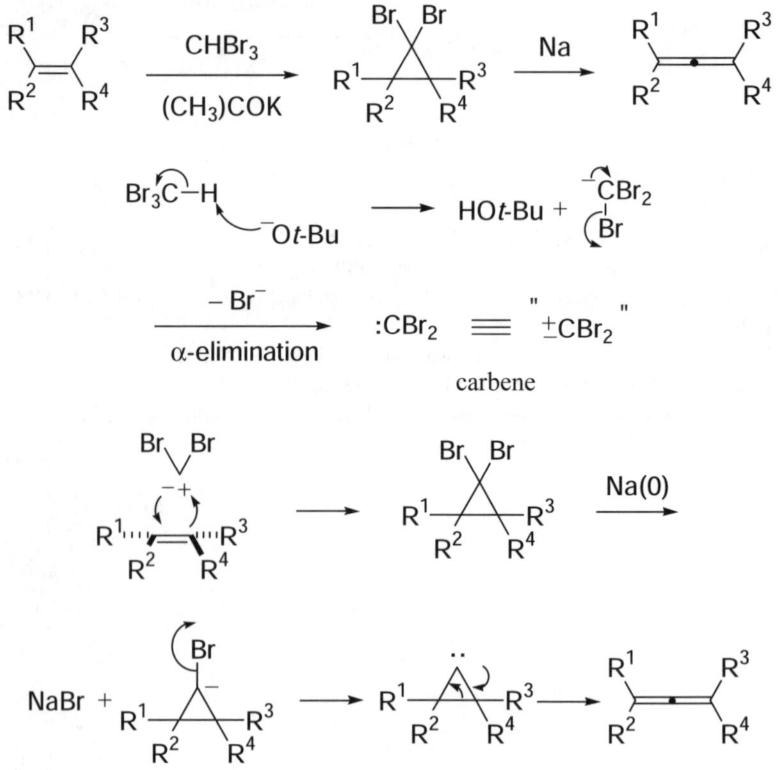

References

1. Doering, W. von E.; LaFlamme, P. M. *Tetrahedron* **1958**, *2*, 75.
2. Skattebol, L. *Tetrahedron Lett.* **1961**, 167.
3. Christl, M.; Braun, M.; Wolz, E.; Wagner, W. *Ber.* **1994**, *127*, 1137.
4. Magid, R. M.; Jones, M., Jr. *Tetrahedron* **1997**, *53*, xiii-xvi (Preface).

Dornow–Wiehler isoxazole synthesis

O

H + O_2N CO_2Et

O_2N

piperidine

EtOH, reflux

EtO_2C O N

CO_2Et

O_2N

H

O_2N CO_2Et + N

enolate

formation

H + H

N

O

O_2N OEt

O

H

O_2N

OH

CO_2Et

H

:B

NO_2

O_2N

dehydration

$- H_2O$

O

O_2N OEt

CO_2Et

H B+

NO_2

O_2N

Michael

addition

EtO_2C NO_2

NO_2

H

CO_2Et

O_2N

:B

$- BH^+NO_2^-$

EtO_2C

O–

N+

O

H

CO_2Et

O_2N

:B

tautomerization

:B

EtO_2C H

O

N+

O–

CO_2Et

O_2N

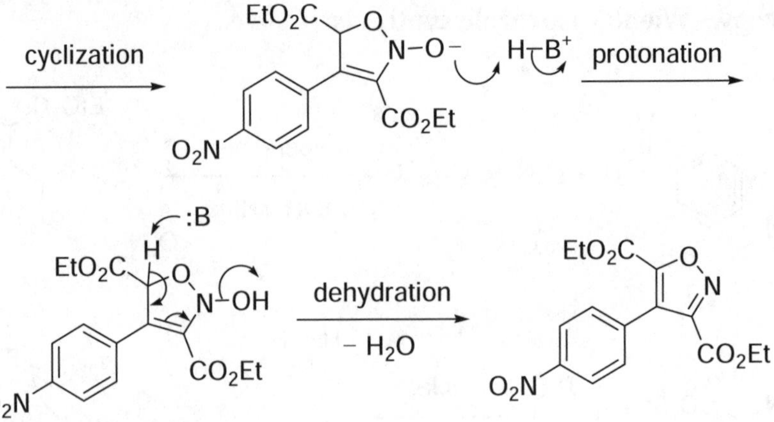
cyclization
EtO2C
O
N–O–
H
B+
protonation
CO2Et
O2N
:B
H
EtO2C
O
N
OH
CO2Et
O2N
dehydration
– H2O
EtO2C
O
N
CO2Et
O2N

References

1. Dornow, A.; Wiehler, G. *Liebigs Ann. Chem.* **1952**, *578*, 113.
2. Dornow, A.; Wiehler, G. *ibid.* **1952**, *578*, 122.
3. Umezawa, S.; Zen, S. *Bull. Chem. Soc. Jpn.* **1963**, *36*, 1150.

Dötz reaction

O

$(CO)_5Cr$ + R—≡—R Δ → HO — O

$Cr(CO)_3$

R R

O O

$(CO)_5Cr$ Δ − CO $(CO)_4Cr$ coordination R—≡—R

O O

$(CO)_4Cr$ cycloaddition $(CO)_4Cr$ CO insertion

R—≡—R R R

metallacyclobutene

$(CO)_4Cr$ O fragmentation O C O electrocyclization

O R R R R

R $Cr(CO)_4$

metallacyclopentenone

chromium-vinyl ketene

H H $Cr(CO)_3$

O enolization HO O

R O − CO

$(CO)_4Cr$ R R R

References

1. Dötz, K. H. *Angew. Chem., Int. Ed. Engl.* **1975**, *14*, 644.
2. Torrent, M. *Chem. Commun.* **1998**, 999.
3. Torrent, M.; Sola, M.; Frenking, G. *Chem. Rev.* **2000**, *100*, 439.
4. Barluenga, J.; Lopez, L. A.; Martinez, S.; Tomas, M. *Tetrahedron* **2000**, *56*, 4967.
5. Jackson, T. J.; Herndon, J. W. *Tetrahedron* **2001**, *57*, 3859.

Dutt–Wormall reaction

$$\mathrm{Ar{-}NH_2} \xrightarrow{\mathrm{HNO_2}} \mathrm{Ar{-}N_2^+} \xrightarrow[\text{2. } ^{-}\mathrm{OH}]{\text{1. TsNH}_2} \mathrm{Ar{-}N_3}$$

$$\mathrm{HO{-}N{=}O} \xrightarrow{\mathrm{H^+}} \mathrm{H{-}O{-}N{=}O} \;\; \mathrm{H_2O^+{-}N{=}O} \xrightarrow{-\,\mathrm{H_2O}} \mathrm{O{=}N{-}O{-}N{=}O}$$

$$\mathrm{O{=}N{-}O{-}N{=}O} \;\; \mathrm{Ar{-}\ddot{N}H_2} \xrightarrow{-\,\mathrm{NO_2^-}} \mathrm{Ar{-}NH{-}N{=}O} \xrightarrow{\mathrm{H^+}} \mathrm{Ar{-}NH{-}N{=}O^+{-}H} \;\; \mathrm{:B}$$

$$\xrightarrow{\mathrm{H^+}} \mathrm{Ar{-}\ddot{N}{=}N{-}OH_2^+} \xrightarrow{-\,\mathrm{H_2O}} \mathrm{Ar{-}\overset{+}{N}{\equiv}N} \;\; \mathrm{H_2\ddot{N}{-}SO_2{-}C_6H_4{-}CH_3}$$

$$\longrightarrow \mathrm{Ar{-}N{=}N{-}NH{-}SO_2{-}Tol} \;\; ^{-}\mathrm{OH} \longrightarrow \mathrm{Ar{-}\overset{-}{N}{-}\ddot{N}{=}N{-}SO_2{-}Tol} \longrightarrow$$

$$^{-}\mathrm{O{-}SO{-}Tol}\;(\mathrm{O{=}S(O^-){=}O,\ Tol}) + \mathrm{Ar{-}\overset{-}{N}{-}\overset{+}{N}{\equiv}N} \longleftrightarrow \mathrm{Ar{-}N{=}\overset{+}{N}{=}\overset{-}{N}} \longleftrightarrow \mathrm{Ar{-}N_3}$$

References

1. Dutt, J. C.; Whitehead, H. R.; Wormall, A. *J. Chem. Soc.* **1921**, *119*, 2088.
2. Bretschneider, H.; Rager, H. *Monatsh.* **1950**, *81*, 970.
3. Laing, I. G. In *Rodd's Chemistry of Carbon Compounds IIIC* **1973**, 107.

Eschenmoser coupling reaction

Enamine from thiamide and alkyl halide.

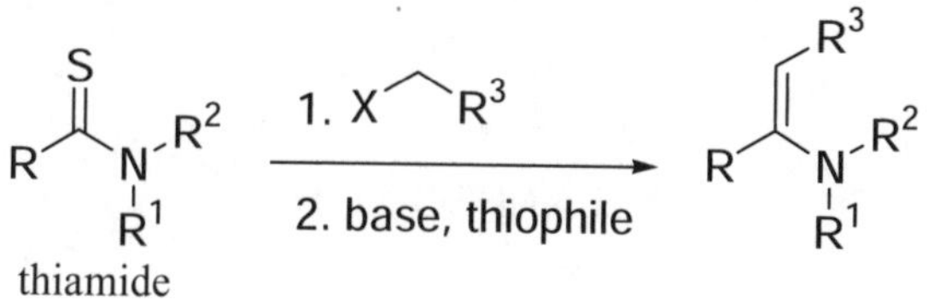

thiamide

1. $X\frown R^3$
2. base, thiophile

e.g.

1. $BrCH_2CO_2CH_3$
2. PPh_3, Et_3N, 90%

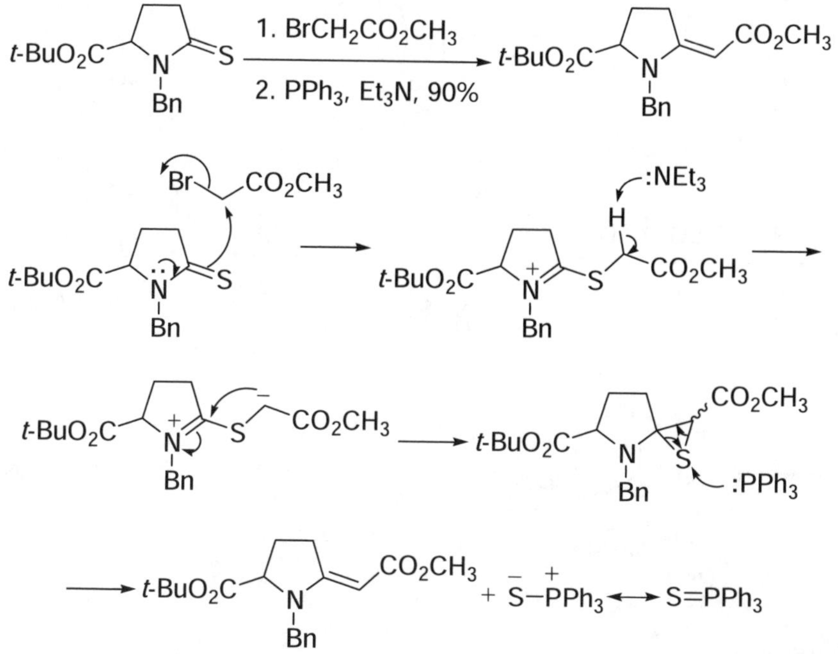

References

1. Roth, M.; Dubs, P.; Götschi, E.; Eschenmoser, A. *Helv. Chim. Acta* **1971**, *54*, 710.
2. Peterson, J. S.; Fels, G.; Rapoport, H. *J. Am. Chem. Soc.* **1984**, *106*, 4539.
3. Shiosaki, K. In *Comprehensive Organic Synthesis* Trost, B. M.; Fleming, I., Eds, Pergamon, **1991**, *Vol. 2*, 865–892.
4. Levillain, J.; Vazeux, M. *Synthesis* **1995**, 56.
5. Mulzer, J.; List, B.; Bats, Jan W. *J. Am. Chem. Soc.* **1997**, *119*, 5512.
6. Hodgkinson, T. J.; Kelland, L. R.; Shipman, M.; Vile, J. *Tetrahedron* **1998**, *54*, 6029.

Eschenmoser–Tanabe fragmentation

1. H_2O_2, ^-OH
2. H_2NNHSO_2Ar, H^+
3. ^-OH

$^-O{-}OH$

H_2NNHSO_2Ar, H^+

^-OH

+ $^-SO_2Ar$ + $N_2\uparrow$

References

1. Eschenmoser, A.; Felix, D.; Ohloff, G. *Helv. Chim. Acta* **1967**, *50*, 708.
2. Tanabe, M.; Crowe, D. F.; Dehn, R. L. *Tetrahedron Lett.* **1967**, 3943.
3. Felix, D.; Müller, R. K.; Horn, U.; Joos, R.; Schreiber, J.; Eschenmoser, A. *Helv. Chim. Acta* **1972**, *55*, 1276.
4. Kasal, A.; Kohout, L.; Filip, J. *Collect. Czech. Chem. Commun.* **1985**, *50*, 1402.
5. Dai, W.; Katzenellenbogen, J. A. *J. Org. Chem.* **1993**, *58*, 1900.
6. Abad, A.; Arno, M.; Agullo, C.; Cunat, A. C.; Meseguer, B.; Zaragoza, R. J. *J. Nat. Prod.* **1993**, *56*, 2133.
7. Mueck-Lichtenfeld, C. *J. Org. Chem.* **2000**, *65*, 1366.

Étard reaction

Oxidation of an arylmethyl group to the corresponding aryl aldehyde using chromyl chloride.

CrO_2Cl_2

ene reaction

[2,3]-sigmatropic rearrangement

Cl⁻ H Cl Cl Cr O OH

$- Cl^-$

O H + Cl-Cr—OH

References

1. Étard, A. L. *Compt. Rend.* **1880**, *90*, 524.
2. Rentea, C. N.; Necsoiu, I.; Rentea, M.; Ghenciulescu, A.; Nenitzescu, C. D. *Tetrahedron* **1966**, *22*, 3501.
3. Schildknecht, H.; Hatzmann, G. *Angew. Chem., Int. Ed. Engl.* **1968**, *7*, 293.
4. Duffin, H. C.; Tucker, R. B. *Tetrahedron* **1968**, *24*, 6999.
5. Schiketanz, I. I.; Badea, F.; Hanes, A; Necsoiu, I. *Rev. Roum. Chim.* **1984**, *29*, 353.
6. Luzzio, F. A.; Moore, W. J. *J. Org. Chem.* **1993**, *58*, 512.

Evans aldol reaction

Asymmetric aldol condensation using an acyl oxazolidinone, the Evans chiral auxiliary.

1. Bu_2BOTf, R_3N

2. PhCHO, – 78 °C

Bu_2B–OTf

R_3N:

Z-(O)-boron enolate formation

PhCHO

aldol condensation

workup

Refrences

1. Evans, D. A.; Bartroli, J.; Shih, T. L. *J. Am. Chem. Soc.* **1981**, *103*, 2127.
2. Evans, D. A.; McGee, L. R. *ibid.* **1981**, *103*, 2876.
3. Gage, J. R.; Evans, D. A. **1990**, *68*, 83
4. Allin, S. M.; Shuttleworth, S J. *Tetrahedron Lett.* **1996**, *37*, 8023.
5. Ager, D. J.; Prakash, I.; Schaad, D. R. *Aldrichimica Acta* **1997**, *30*, 3.
6. Faita, G.; Paio, A.; Quadrelli, P.; Rancati, F.; Seneci, P. *Tetrahedron Lett.* **2000**, *41*, 1265.

7. Braddock, D. C.; Brown, J. M. *Tetrahedron: Asymmetry* **2000**, *11*, 3591.
8. Lu, Y.; Schiller, P. W. *Synthesis* **2001**, 1639.
9. Kamino, T.; Murata, Y.; Kawai, N.; Hosokawa, S.; Kobayashi, S. *Tetrahedron Lett.* **2001**, *42*, 5249.
10. Williams, D. R.; Patnaik, S.; Clark, M. P. *J. Org. Chem* **2001**, *66*, 8463.

Favorskii rearrangement and Quasi-Favorskii rearrangement

Favorskii rearrangement

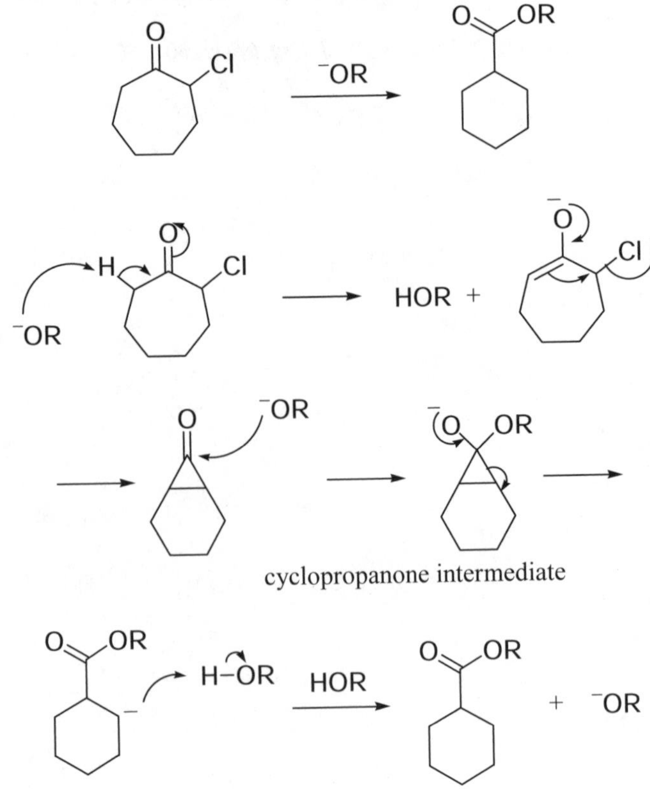

cyclopropanone intermediate

Quasi-Favorskii rearrangement

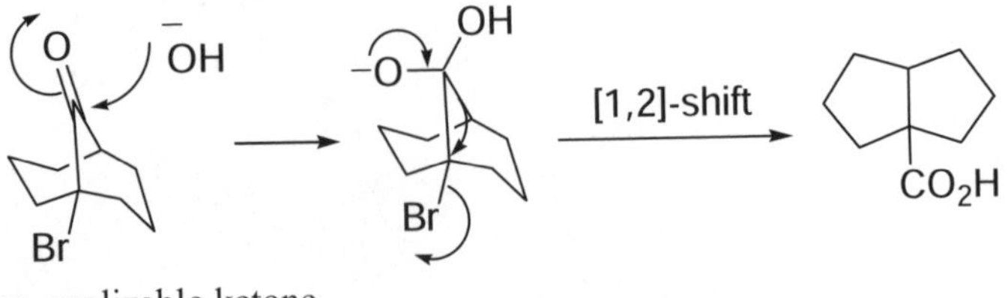

non-enolizable ketone

References

1. Favorskii, A. *J. Prakt. Chem.* **1895**, *51*, 533.
2. Chenier, P. J. *J. Chem. Ed.* **1978**, *55*, 286.

3. Barreta, A.; Waegill, B. In *Reactive Intermediates*, Abramovitch, R. A., ed. Plenum Press: New York, **1982**, pp 527–585.
4. Gambacorta, A.; Turchetta, S.; Bovivelli, P.; Botta, M. *Tetrahedron* **1991**, *47*, 9097.
5. El-Wareth, A.; Sarhan, A. O.; Hoffmann, H. M. R. *J. Prakt. Chem./Chem.- Ztg.* **1997**, *339*, 390.
6. Dhavale, D. D.; Mali, V. P.; Sudrik, S. G.; Sonawane, H. R. *Tetrahedron* **1997**, *53*, 16789.
7. Braverman, S.; Cherkinsky, M.; Kumar, E. V. K. S.; Gottlieb, H. E. *ibid.* **2000**, *56*, 4521.
8. Mamedov, V. A.; Tsuboi, S.; Mustakimova, L. V.; Hamamoto, H.; Gubaidullin, A. T.; Litvinov, I. A.; Levin, Y. A. *Chem. Heterocycl. Compd.* **2001**, *36*, 911.

Feist–Bénary furan synthesis

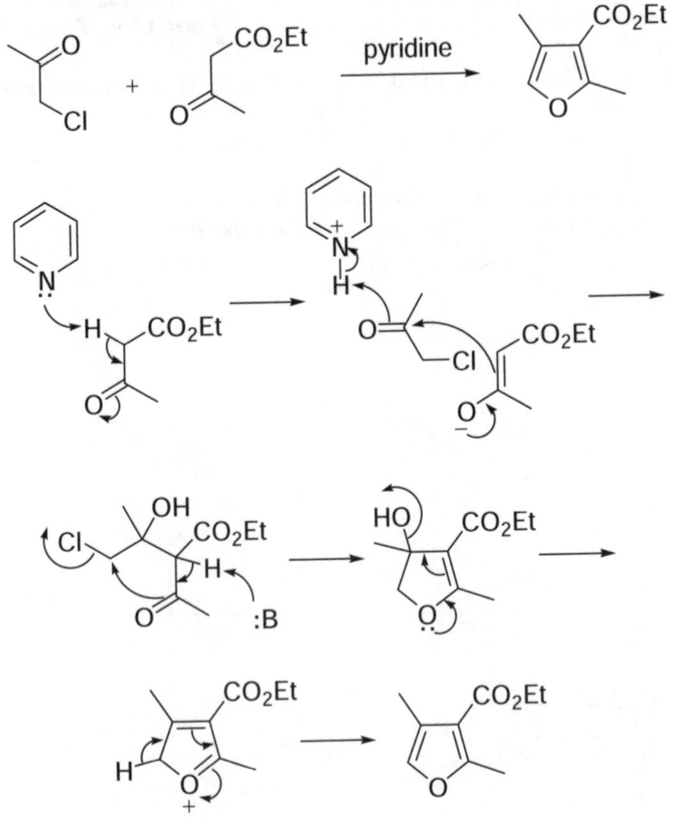

pyridine

CO_2Et

References

1. Feist, F. *Ber.* **1902**, *35*, 1537.
2. Bénary, E. **1911**, *44*, 489.
3. Bisagni, E.; Marquet, J. P.; Andre-Louisfert, J.; Cheutin, A.; Feinte, F. *Bull. Soc. Chim. Fr.* **1967**, 2796.
4. Cambie, R. C.; Moratti, S. C.; Rutledge, P. S.; Woodgate, P. D. *Synth. Commun.* **1990**, *20*, 1923.
5. Calter, M.; Zhu, C. *Abstr. Pap.-Am. Chem. Soc.* **2001**, 221st ORGN-574.

Ferrier rearrangement

Lewis-acid (such as $BF_3{\bullet}OEt_2$, $SnCl_4$, *etc.*) promoted rearrangement of unsaturated carbohydrates.

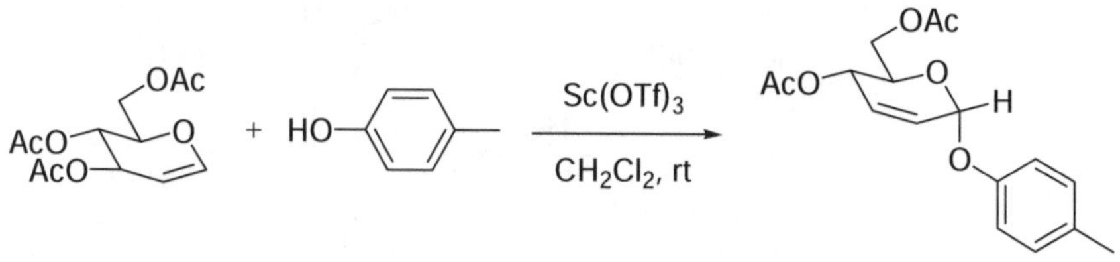

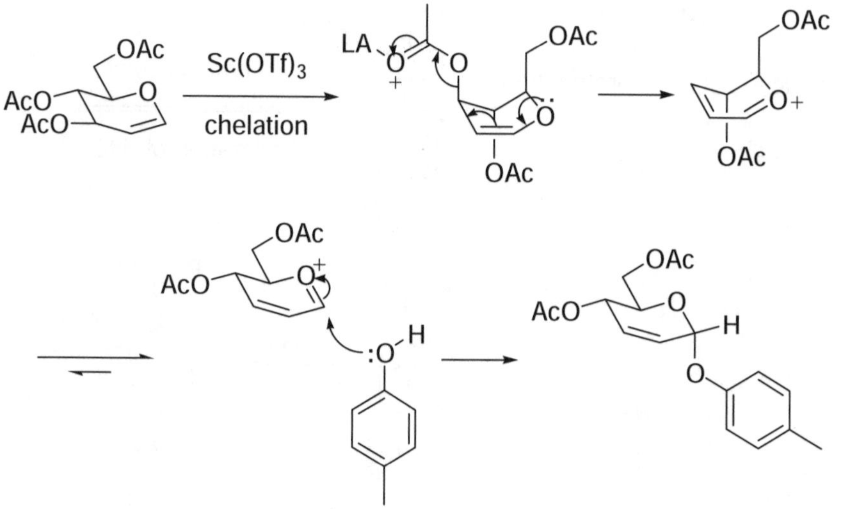

The axial addition is favored due to the anomeric effect.

References

1. Ferrier, R. J. *J. Chem. Soc. (C)* **1968**, 974.
2. Ferrier, R. J. *J. Chem. Soc., Perkin. Trans. 1* **1979**, 1455.
3. Fraser-Reid, B. *Acc. Chem. Res.* **1996**, *29*, 57.
4. Paquette, Leo A. *Recent Res. Dev. Chem. Sci.* **1997**, *1*, 1.
5. Linker, T.; Sommermann, T.; Gimisis, T.; Chatgilialoglu, C. *Tetrahedron Lett.* **1998**, *39*, 9637.
6. Smith, A. B., III; Verhoest, P. R.; Minbiole, K. P.; Lim, J. J. *Org. Lett.* **1999**, *1*, 909.
7. Babu, B. S.; Balasubramanian, K. K. *Synth. Commun.* **1999**, *29*, 4299.
8. Taillefumier, C.; Chapleur, Y. *Can. J. Chem.* **2000**, *78*, 708.
9. Yadav, J. S.; Reddy, B. V. S.; Murthy, C. V. S. R.; Kumar, G. M. *Synlett* **2000**, 1450.

Fischer–Hepp rearrangement

Transformation of *N*-nitroso-anilines to the corresponding *para*-nitroso anilines. *Cf.* Orton rearrangement.

HCl

NO

H^+

protonation

*intra*molecular

migration of ^+NO

N=O

H NO

$-H^+$

NO

References

1. Fischer, O.; Hepp, E. *Ber.* **1886**, *19*, 2991.
2. Williams, D. L. H. *Tetrahedron* **1975**, *31*, 1343.
3. Biggs, I. D.; Williams, D. L. H. *J. Chem. Soc., Perkin Trans. 2* **1976**, 691.
4. Biggs, I. D.; Williams, D. L. H. *ibid.* **1977**, 44.
5. Williams, D. L. H. *ibid.* **1982**, 801.
6. Morris, P. I. *Chem. Ind.* **1999**, 968.

Fischer indole synthesis

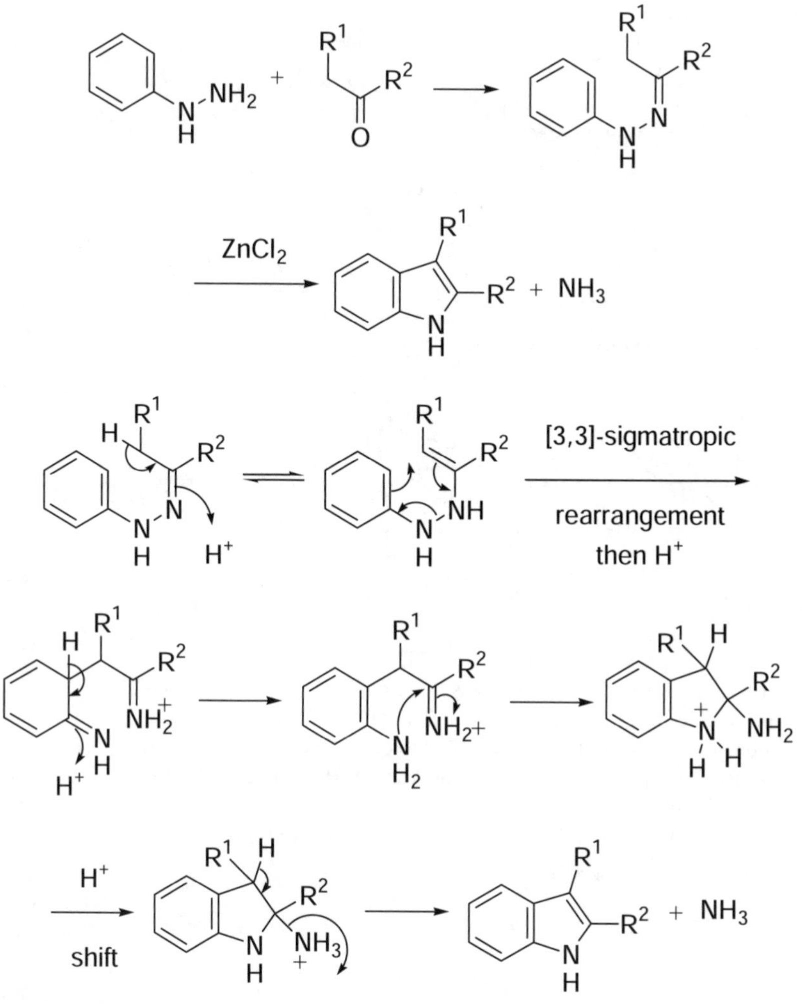

ZnCl$_2$

+ NH$_3$

[3,3]-sigmatropic

rearrangement

then H$^+$

H$^+$

shift

+ NH$_3$

References

1. Fischer, E.; Jourdan, F. *Ber.* **1883**, *16*, 2241.
2. Fischer, E.; Hess. O. *ibid.* **1884**, *17*, 559.
3. Robinson, B. *Chem. Rev.* **1969**, *69*, 227.
4. Hughes, D. L. *Org. Prep. Proc. Int.* **1993**, *25*, 607.
5. Burchak, O. N.; Chibiryaev, A. M.; Tkachev, A. V. *Heterocycl. Commun.* **2000**, *6*, 73.
6. Da Settimo, A.; Marini, A. M.; Primofiore, G.; Da Settimo, F.; Salerno, S.; La Motta, C.; Pardi, G.; Ferrarini, P. L.; Mori, C. *J. Heterocycl. Chem.* **2000**, *37*, 379.
7. Bhattacharya, G.; Su, T.-L.; Chia, C.-M.; Chen, K.-T. *J. Org. Chem.* **2001**, *66*, 426.

Fischer–Speier esterification

Often known as "Fischer esterification", protic acid-catalyzed esterification of an acid and an alcohol.

$$RC(=O)OH + R^1\text{—}OH \xrightarrow{\text{cat. HCl}} RC(=O)OR^1$$

$$RC(=O)OH \underset{\text{protonation}}{\overset{H^+}{\rightleftharpoons}} RC(=O^+H)OH \underset{\text{addition}}{\overset{\text{nucleophilic}}{\rightleftharpoons}} RC(OH)_2(O^+HR^1)$$

$$\overset{-H^+}{\rightleftharpoons} RC(OH)_2(OR^1) \overset{+H^+}{\rightleftharpoons} RC(OH)(OH_2^+)(OR^1) \overset{-H_2O,\ -H^+}{\rightleftharpoons} RC(=O)OR^1$$

References

1. Fischer, E.; Speier, A. *Ber.* **1895**, *28*, 3252.
2. Hardy, J. P.; Kerrin, S. L.; Manatt, S. L. *J. Org. Chem.* **1973**, *38*, 4196.
3. Fujii, T.; Yoshifuji, S. *Chem. Pharm. Bull.* **1978**, *26*, 2253.
4. Pcolinski, M. J.; O'Mathuna, D. P.; Doskotch, R. W. *J. Nat. Prod.* **1978**, *58*, 209.
5. Kai, T.; Sun, X.-L.; Tanaka, M.; Takayanagi, H.; Furuhata, K. *Chem. Pharm. Bull.* **1996**, *44*, 208.
6. Birney, D. M.; Starnes, S. D. *J. Chem. Educ.* **1996**, *76*, 1560.

Fleming oxidation

Cf. Tamao–Kumada oxidation

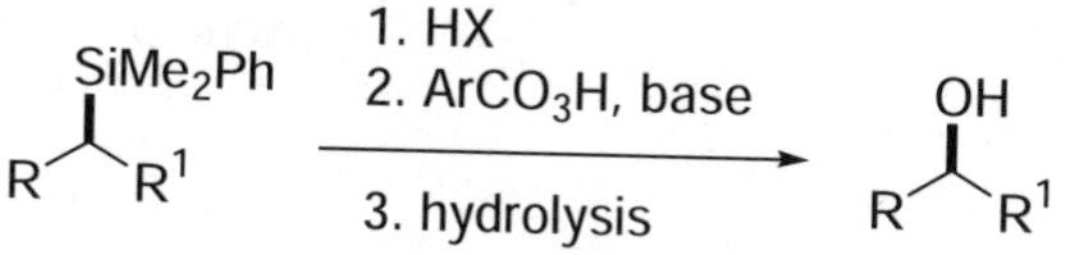

SiMe$_2$Ph, R, R^1 — 1. HX; 2. ArCO$_3$H, base; 3. hydrolysis → OH, R, R^1

retention of configuration

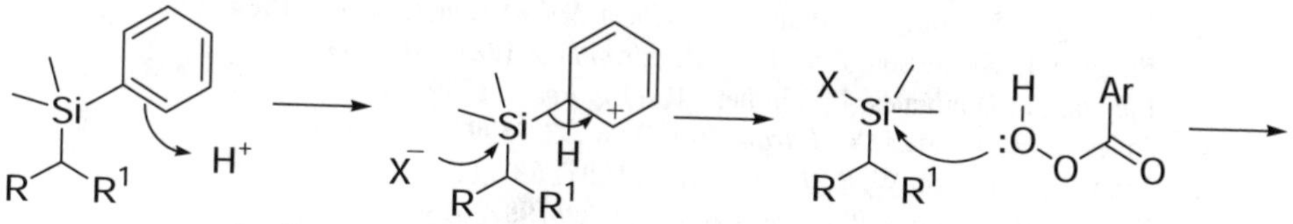

the β-carbocation is stabilized by the silicon group

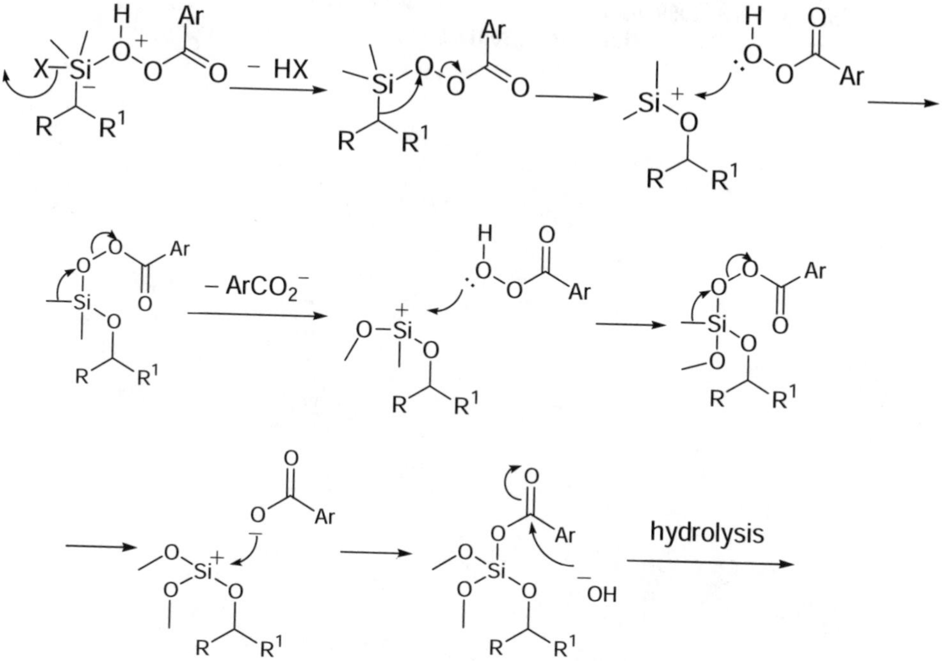

– HX

– $ArCO_2^-$

hydrolysis

HO O⁻ Ar O Si O O R R1 → ⁻O H+ R R1 acidic workup → OH R R1

References

1. Fleming, I.; Henning, R.; Plaut, H. *J. Chem. Soc., Chem. Commun.* **1984**, 29.
2. Fleming, I.; Sanderson, P. E. J. *Tetrahedron Lett.* **1987**, *28*, 4229.
3. Fleming, I.; Dunogues, J.; Smithers, R. *Org. React.* **1989**, *37*, 57.
4. Jones, G. R.; Landais, Y. *Tetrahedron* **1996**, *52*, 7599.
5. Hunt, J. A.; Roush, W. R. *J. Org. Chem.* **1997**, *62*, 1112.
6. Knölker, H.-J.; Jones, P. G.; Wanzl, G. *Synlett* **1997**, 613.
7. Studer, A.; Steen, H. *Chem.--Eur. J.* **1999**, *5*, 759.
8. Barrett, A. G. M.; Head, J.; Smith, M. L.; Stock, N. S.; White, A. J. P.; Williams, D. J. *J. Org. Chem.* **1999**, *64*, 6005.
9. Lee, T. W.; Corey, E. J. *Org. Lett.* **2001**, *3*, 3337.

Forster reaction

Diazoketone formation from α-oximinoketones.

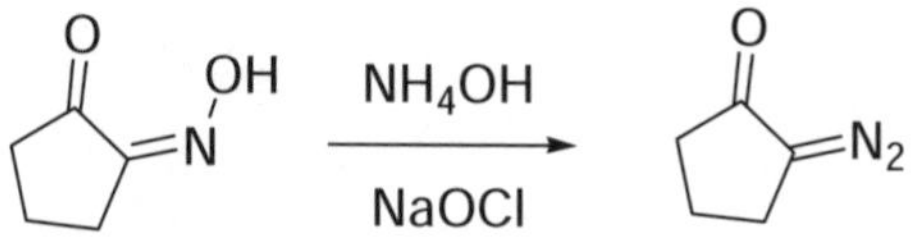

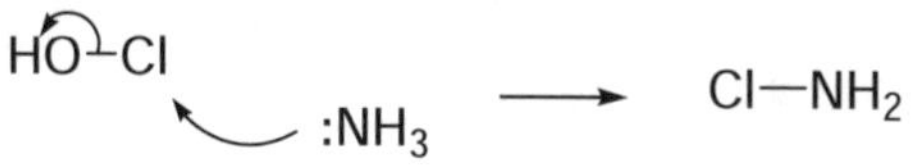

NH_4OH

NaOCl

$HO-Cl$ $:NH_3$ ⟶ $Cl-NH_2$

chloramine

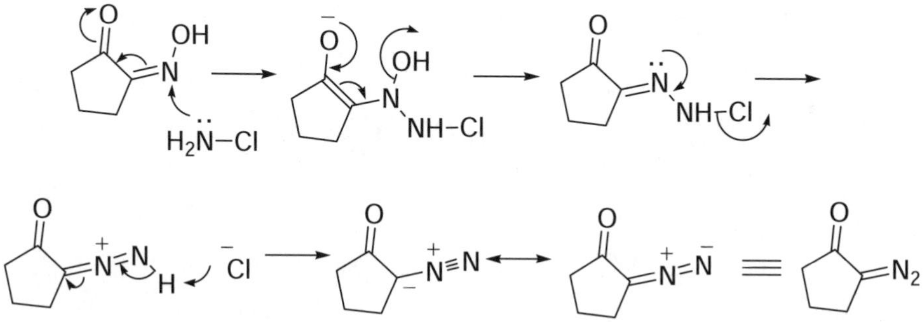

Alternatively:

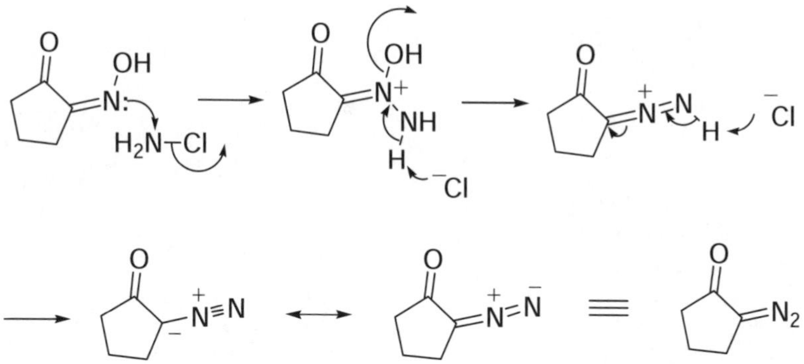

References

1. Forster, M. O. *J. Chem. Soc.* **1915**, *107*, 260.
2. Meinwald, J.; Gassman, P. G.; Miller, E. G. *J. Am. Chem. Soc.* **1959**, *81*, 4751.
3. Rundel, W. *Angew. Chem.* . **1962**, *74*, 469.
4. Huneck, S. *Chem. Ber.* **1965**, *98*, 3204.

5. Overberger, C. G.; Anselme, J. P. *Tetrahedron Lett.* **1963**, 1405.
6. Van Leusen, A. M.; Strating, J.; Van Leusen, D. *ibid.* **1973**, 5207.
7. L'abbe, G.; Dekerk, J. P.; Deketele, M. *J. Chem. Soc., Chem. Commun.* **1983**, 588.
8. L'abbe, G.; Luyten, I.; Toppet, S. *J. Heterocycl. Chem.* **1992**, *29*, 713.

Frater–Seebach alkylation

Asymmetric alkylation of β-hydroxylesters.

$LiN(TMS)_2$, THF

then CH_3I

References

1. Frater, G.; Muller, U.; Gunter, W. *Tetrahedron* **1984**, *48*, 1269.
2. Seebach, D.; Imwinkelried, R.; Weber, T. *Modern Synth. Method* **1986**, *4*, 125.
3. Heathcock, C. H.; Kath, J. C.; Ruggeri, R. B. *J. Org. Chem.* **1995**, *60*, 1120.

Friedel–Crafts reaction

Friedel–Crafts *acylation* reaction:

$$C_6H_6 \xrightarrow[AlCl_3]{RCOCl} C_6H_5COR + HCl$$

$$RCOCl + AlCl_3 \xrightarrow{\text{complexation}} RC(=O)Cl^{+}\text{–}\overset{-}{A}lCl_3 \longrightarrow$$

$$AlCl_4^- + R\text{–}C^{+}{=}O \longleftrightarrow R\text{–}C{\equiv}O^{+} \xrightarrow[\text{substitution}]{\text{electrophilic}}$$

acylium ion

$$\text{Cl}_4\overset{-}{\text{Al}} \cdots \text{H–}(C_6H_5^{+})\text{COR} \xrightarrow{\text{aromatization}} C_6H_5COR + AlCl_3$$

Friedel–Crafts *alkylation* reaction:

$$RCl + AlCl_3 \longrightarrow R\text{–}Cl^{+}\text{–}\overset{-}{A}lCl_3 \longrightarrow \overset{-}{A}lCl_4 + R^+ \xrightarrow{C_6H_6} C_6H_5R$$

alkyl cation

References

1. Friedel, P.; Crafts, J. M. *Compt. Rend.* **1877**, *84*, 1392.
2. Pearson, D. E.; Buehler, C. A. *Synthesis* **1972**, 533.
3. Gore, P. H. *Chem. Ind.* **1974**, 727.
4. Chevrier, B.; Weis, R. *Angew. Chem.* **1974**, *86*, 12.
5. Schriesheim, A.; Kirshenbaum, I. *Chemtech* **1978**, *8*, 310.

6. Ottoni, O.; Neder, A. V. F.; Dias, A. K. B.; Cruz, R. P. A.; Aquino, L. B. *Org. Lett.* **2000**, *3*, 1005.
7. Fleming, I. *Chemtracts* **2001**, *14*, 405.

Friedländer synthesis

Quinoline synthesis from the condensation of *o*-aminobenzaldehyde with aldehyde or ketone in the presence of NaOH.

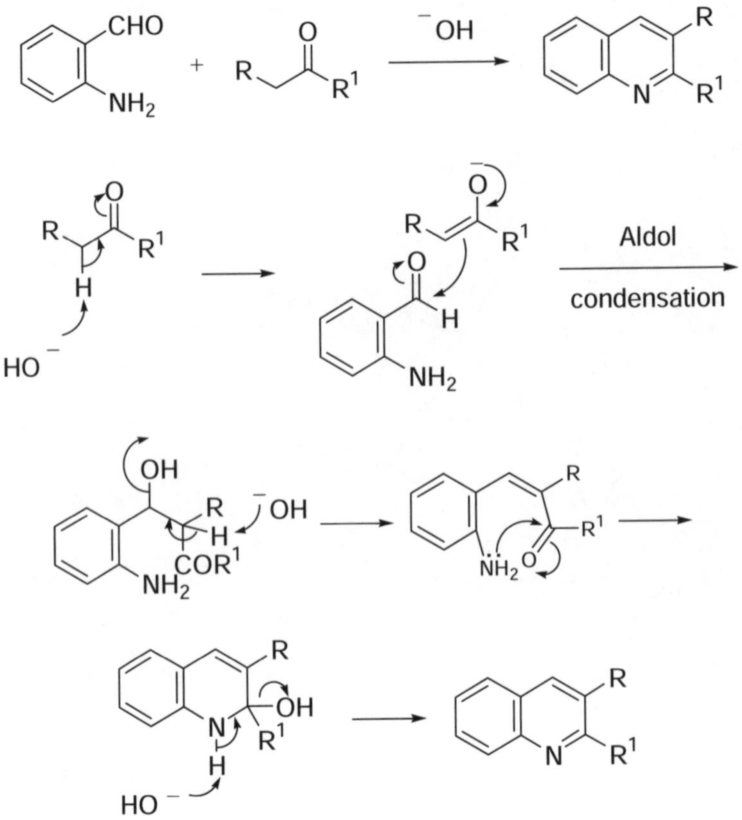

References

1. Friedländer, P. *Ber.* **1882**, *15*, 2572.
2. Cheng, C.-C.; Yan, S.-J. *Org. Recat.* **1982**, *28*, 37.
3. Thummel, R. P. *Synlett* **1992**, 1.
4. Riesgo, E. C.; Jin, X.; Thummel, R. P. *J. Org. Chem.* **1996**, *61*, 3017.
5. Mori, T.; Imafuku, K.; Piao, M.-Z.; Fujimori, K. *J. Heterocycl. Chem.* **1996**, *33*, 841.
6. Ubeda, J. I.; Villacampa, M.; Avendano, C. *Synthesis* **1998**, 1176.
7. Bu, X.; Deady, L. W. *Synth. Commun.* **1999**, *29*, 4223.
8. Strekowski, L.; Czarny, A.; Lee, H. *J. Fluorine Chem.* **2000**, *104*, 281.

9. Chen, J.; Deady, L. W.; Desneves, J.; Kaye, A. J.; Finlay, G. J.; Baguley, B. C.; Denny, W. A. *Bioorg. Med. Chem.* **2000**, *8*, 2461.
10. Gladiali, S.; Chelecci, G.; Mudadu, M. S.; Gastaut, M.-A.; Thummel, R. P. *J. Org. Chem.* **2001**, *66*, 400.

Fries rearrangement

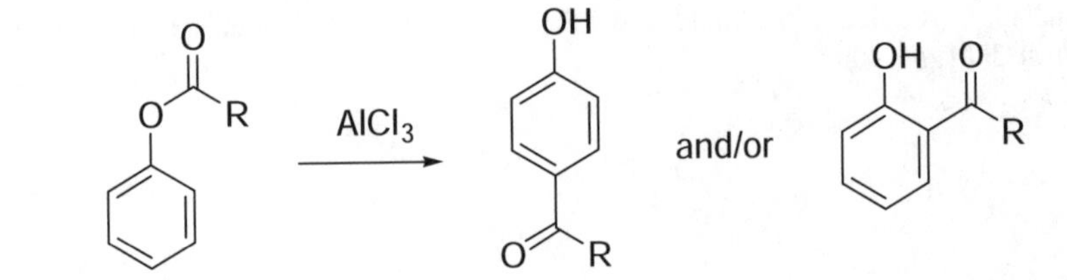

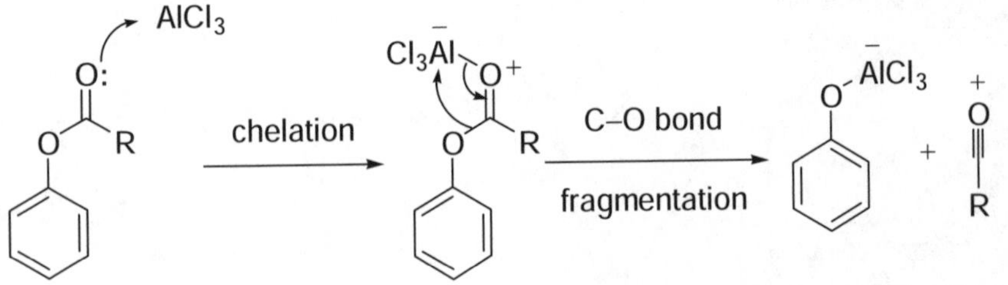

aluminum phenolate, acylium ion

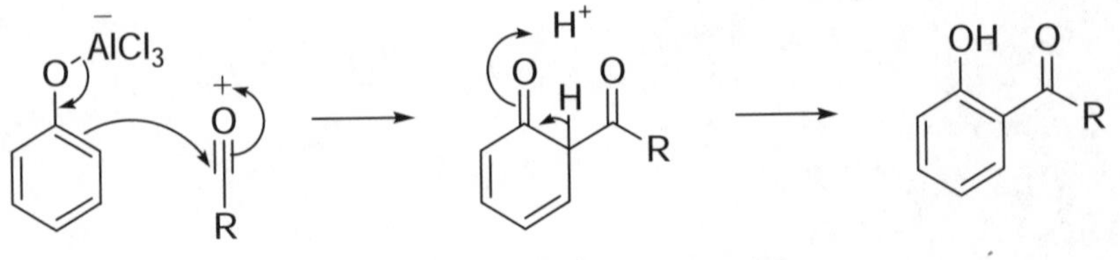

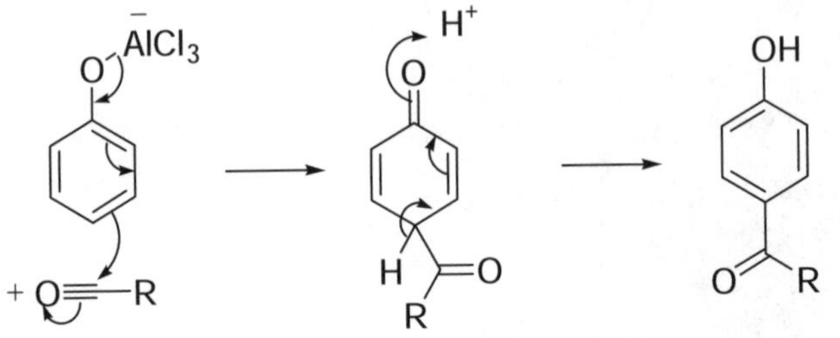

References

1. Fries, K.; Fink, G. *Ber.* **1908**, *41*, 4271.
2. Martin, R. *Org. Prep. Proced. Int.* **1992**, *24*, 369.
3. Trehan, I. R.; Brar, J. S.; Arora, A. K.; Kad, G. L. *J. Chem. Educ.* **1997**, *74*, 324.
4. Boyer, J. L.; Krum, J. E.; Myers, M. C.; Fazal, A. N.; Wigal, C. T. *J. Org. Chem.* **2000**, *65*, 4712.
5. Focken, T.; Hopf, H.; Snieckus, V.; Dix, I.; Jones, P. G. *Eur. J. Org. Chem.* **2001**, 2221.

Fritsch–Buttenberg–Wiechell rearrangement

R H R^1 X —Base→ R^1—≡—R

e.g. a variant:

Br Br Bu Bu —n-BuLi→ Bu—≡—≡—≡—Bu

Bu$^-$ Br Br Bu Bu —halogen-metal exchange→ Li Br Bu Bu

—α-elimination, – Br$^-$→ Bu Bu ≡ Bu Bu

alkylidene carbene/carbenoid

—1,2-shift→ Bu—≡—≡—≡—Bu

References

1. Fritsch, P. *Liebigs Ann. Chem.* **1894**, *272*, 319.
2. Koebrich, G.; Merkel, D. *Angew. Chem., Int. Ed. Engl.* **1970**, *9*, 243.
3. Fienemann, H.; Koebrich, G. *Chem. Ber.* **1974**, *104*, 2797.
4. Sket, B.; Zupan, M. *J. Chem. Soc., Perkin Trans. 1* **1979**, 752.
5. Creton, I.; Rezaei, H.; Marek, I.; Normant, J. F. *Tetrahedron Lett.* **1999**, *40*, 1899.
6. Rezaei, H.; Yamanoi, S.; Chemla, F.; Normant, J. F. *Org. Lett.* **2000**, *2*, 419.
7. Eisler, S.; Tykwinski, R. R. *J. Am. Chem. Soc.* **2000**, *122*, 10736.

Fujimoto–Belleau reaction

R MgX
O
O
O
R
nucleophilic addition
ring opening
isomerization
tautomerization
OH
Robinson annulation
HO
H
:B
E1cb

References

1. Fujimoto, C. I. *J. Am. Chem. Soc.* **1951**, *73*, 1856.
2. Weill-Raynal, J. *Synthesis* **1969**, 49.
3. Heys, J. R.; Senderoff, S. G. *J. Org. Chem.* **1989**, *54*, 4702.
4. Aloui, M.; Lygo, B.; Trabsa, H. *Synlett* **1994**, 115.
5. Revial, G.; Jabin, I.; Redolfi, M.; Pfau, M. *Tetrahedron: Asymmetry* **2001**, *12*, 1683.

Fukuyama amine synthesis

SO_2Cl NO$_2$ NO$_2$ — 1. R^1NH_2, pyr. 2. R^2OH, PPh_3, DEAD 3. $HSCH_2CO_2H$ → R^1-NH-R^2 + S-CH$_2$CO$_2$H, NO$_2$, NO$_2$ + SO_2

R^1NH_2, Cl, SO_2, NO$_2$, NO$_2$ —[$-H^+$]→ SO_2NHR^1, NO$_2$, NO$_2$ —[R^2OH, PPh_3, DEAD]→ $SO_2NR^1R^2$, NO$_2$, NO$_2$

See Mitsunobu reaction (page 238) for the mechanism.

H-S-CH$_2$CO$_2$H, $SO_2NR^1R^2$, NO$_2$, NO$_2$ —[S_NAr]→ HO_2C, S, SO_2, NR^1R^2, NO$_2$, H^+, NO$_2$

Meisenheimer complex

→ R^1-NH-R^2 + S-CH$_2$CO$_2$H, NO$_2$, NO$_2$ + SO_2

References

1. Fukuyama, T.; Jow, C.-K.; Cheung, M. *Tetrahedron Lett.* **1995**, *36*, 6373.
2. Fukuyama, T.; Cheung, M.; Jow, C.-K.; Hidai, Y.; Kan, T. *ibid.* **1997**, *38*, 5831.
3. Yang, L.; Chiu, K. *ibid.* **1997**, *38*, 7307.
4. Piscopio, A. D.; Miller, J. F.; Koch, K. *ibid.* **1998**, *39*, 2667.
5. Bolton, G. L.; Hodges, J. C. *J. Comb. Chem.* **1999**, *1*, 130.
6. Lin, X.; Dorr, H.; Nuss, J. M. *Tetrahedron Lett.* **2000**, *41*, 3309.

Gabriel synthesis

Synthesis of primary amines using potassium phthalimide and alkyl halides.

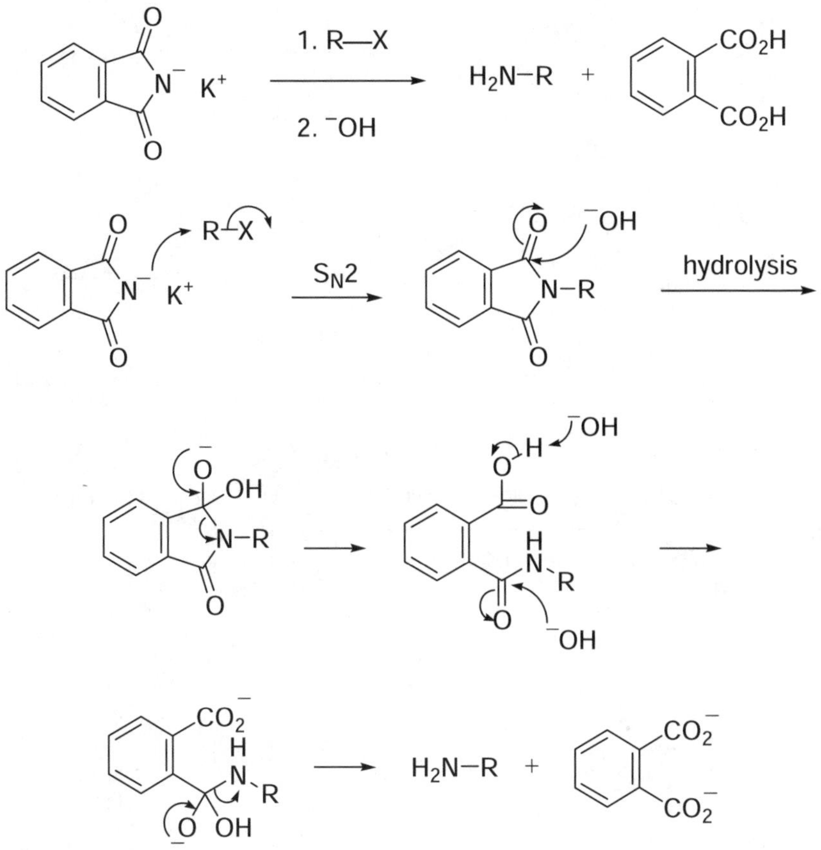

1. R—X

2. $^{-}$OH

$H_2N{-}R$ +

hydrolysis

S_N2

References

1. Gabriel, S. *Ber.* **1887**, *20*, 2224.
2. Press, J. B.; Haug, M. F.; Wright, W. B., Jr. *Synth. Commun.* **1985**, *15*, 837.
3. Slusarska, E.; Zwierzak, A. *Liebigs Ann. Chem.* **1986**, 402.
4. Han, Y.; Hu, H. *Synthesis* **1990**, 122.
5. Ragnarsson, U.; Grehn, L. *Acc. Chem. Res.* **1991**, *24*, 285.
6. Toda, F.; Soda, S.; Goldberg, I. *J. Chem. Soc., Perkin Trans. 1* **1993**, 2357.
7. Khan, M. N. *J. Org. Chem.* **1996**, *61*, 8063.
8. Mamedov, V. A.; Tsuboi, S.; Mustakimova, L. V.; Hamamoto, H.; Gubaidullin, A. T.; Litvinov, I. A.; Levin, Ya. A. *Chem. Heterocycl. Compd.* **2001**, *36*, 911.

Gassman indole synthesis

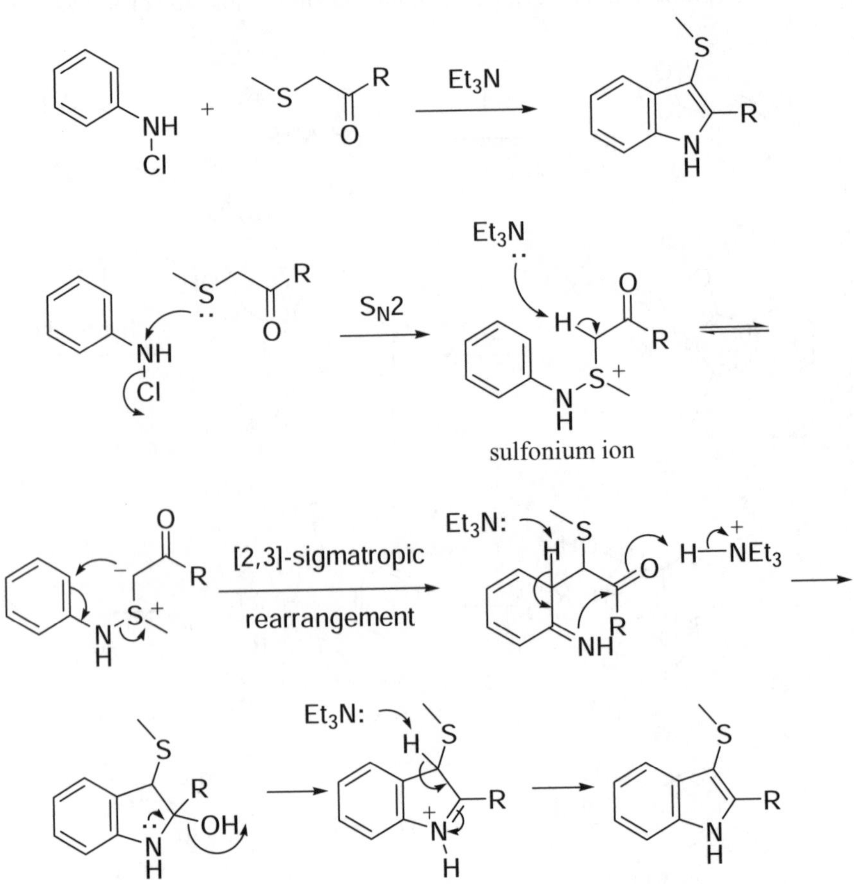

References

1. Gassman, P. G.; van Bergen, T. J.; Gilbert, D. P.; Cue, B. W. *J. Am. Chem. Soc.* **1974**, *96*, 5495.
2. Ishikawa, H.; Uno, T.; Miyamoto, H.; Ueda, H.; Tamaoka, H.; Tominaga, M.; Nakagawa, K. *Chem. Pharm. Bull.* **1990**, *38*, 2459.
3. Wierenga, W. *J. Am. Chem. Soc.* **1981**, *103*, 5621.
4. Smith, A. B., III; Sunazuka, T.; Leenay, T. L.; Kingery-Wood, J. *ibid.* **1990**, *112*, 8197.
5. Smith, A. B., III; Kingery-Wood, J.; Leenay, T. L.; Nolen, E. G.; Sunazuka, T. *ibid.* **1992**, *114*, 1438.

Gattermann–Koch reaction

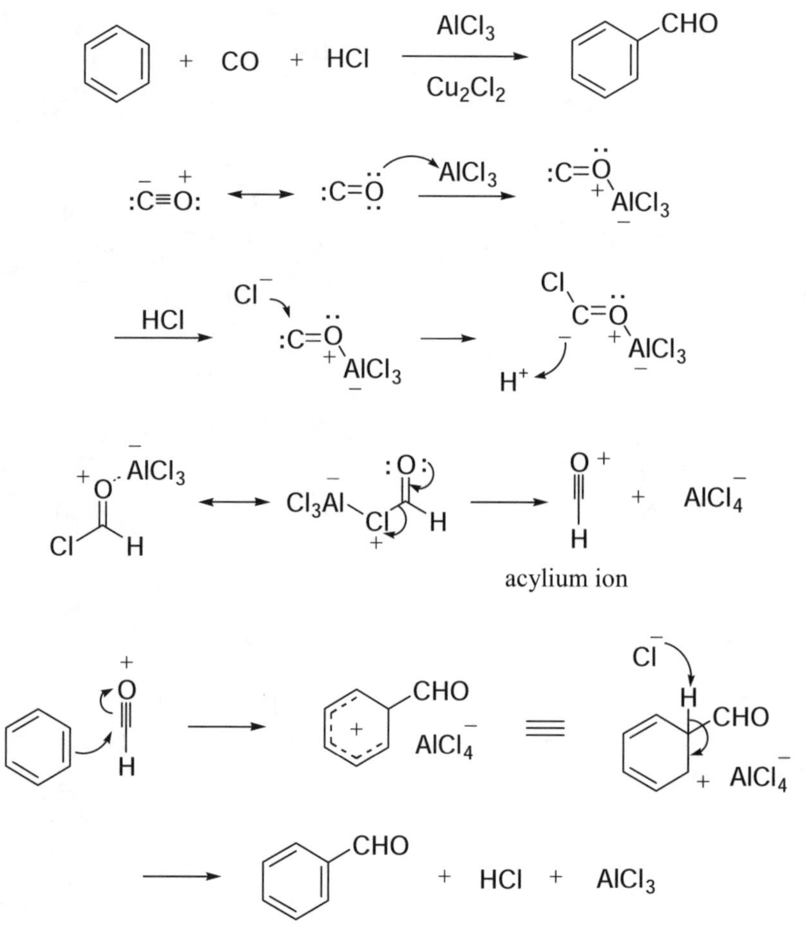

$$C_6H_6 + CO + HCl \xrightarrow[Cu_2Cl_2]{AlCl_3} C_6H_5CHO$$

acylium ion

References

1. Gattermann, L.; Koch, J. A. *Ber.* **1897**, *30*, 1622.
2. Tanaka, M.; Fujiwara, M.; Ando, H. *J. Org. Chem.* **1995**, *60*, 2106.
3. Tanaka, M.; Fujiwara, M.; Ando, H.; Souma, Y. *Chem. Commun.* **1996**, 159.
4. Tanaka, M.; Fujiwara, M.; Xu, Q.; Souma, Y.; Ando, H.; Laali, K. K. *J. Am. Chem. Soc.* **1997**, *119*, 5100.
5. Tanaka, M. *Trends Org. Chem.* **1998**, *7*, 45.
6. Tanaka, M.; Fujiwara, M.; Xu, Q.; Ando, H.; Raeker, T J. *J. Org. Chem.* **1998**, *63*, 4408.
7. Kantlehner, W.; Vettel, M.; Gissel, A; Haug, E.; Ziegler, G.; Ciesielski, M.; Scherr, O.; Haas, R. *J. Prakt. Chem.* **2000**, *342*, 297.

Gewald aminothiophene synthesis

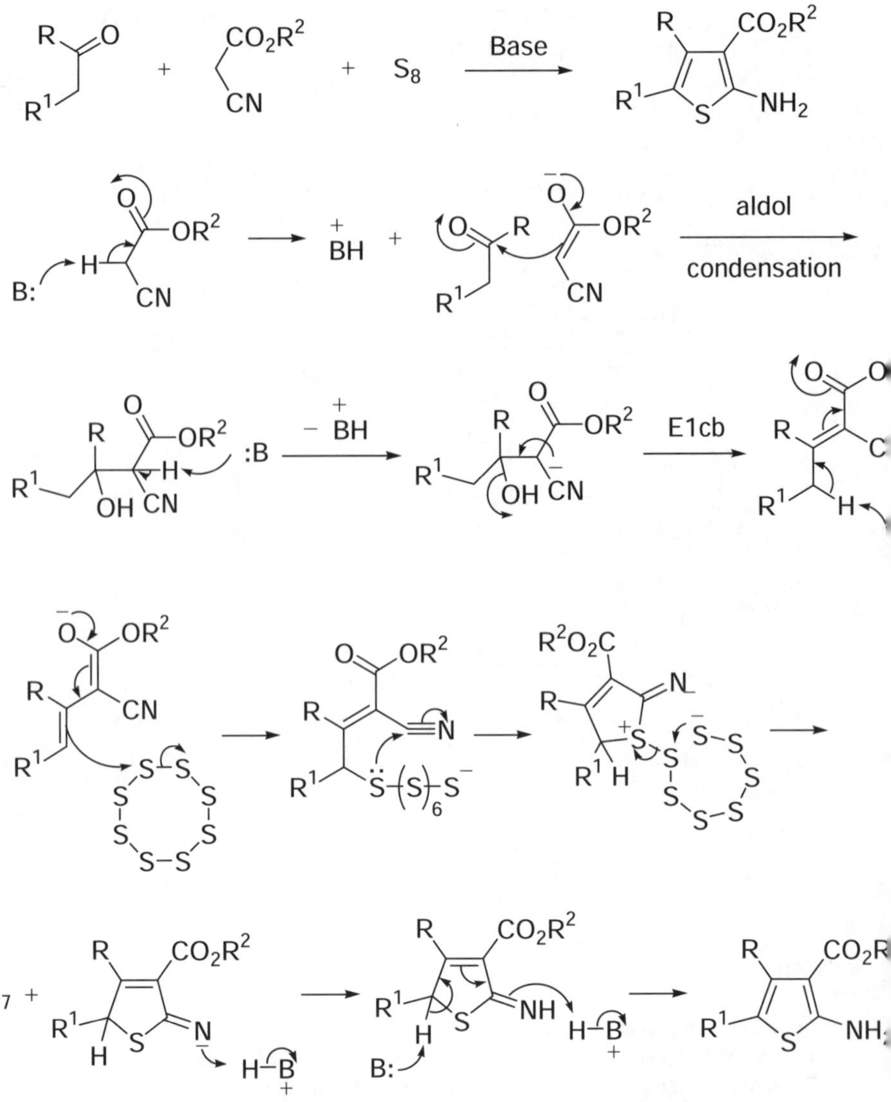

References

1. Peet, N. P.; Sunder, S.; Barbuch, R. J.; Vinogradoff, A. P. *J. Heterocycl. Chem.* **1986**, *23*, 129.
2. Guetschow, M.; Schroeter, H.; Kuhnle, G.; Eger, K. *Monatsh. Chem.* **1996**, *127*, 297.
3. Hallas, G.; Towns, A. D. *Dyes Pigm.* **1996**, *32*, 135.

4. Zhang, M.; Harper, R. W. *Bioorg. Med. Chem. Lett.* **1997**, *7*, 1629.
5. Sabnis, R. W.; Rangnekar, D. W.; Sonawane, N. D. *J. Heterocycl. Chem.* **1999**, *36*, 333.
6. Baraldi, P. G.; Zaid, A. Z.; Lampronti, I.; Fruttarolo, F. F.; Pavani, M. G.; Tabrizi, M. A.; Shryock, J. C. S.; Leung, E.; Romagnoli, R. *Bioorg. Med. Chem. Lett.* **2000**, *10*, 1953.
7. Pinto, I. L.; Jarvest, R. L.; Serafinowska, H. T. *Tetrahedron Lett.* **2000**, *41*, 1597.
8. Buchstaller, H.-P.; Siebert, C. D.; Lyssy, R. H.; Frank, I.; Duran, A.; Gottschlich, R.; Noe, C. R. *Monatsh. Chem.* **2001**, *132*, 279.

Glaser coupling

Oxidative homocoupling of terminal alkynes using copper catalyst.

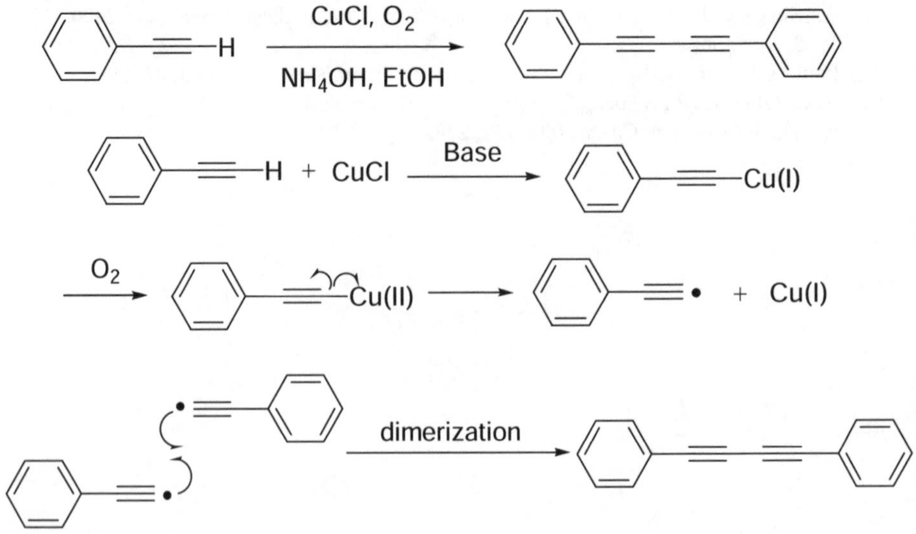

References

1. Glaser, C. *Ber.* **1869**, *2*, 422.
2. Hoeger, S.; Meckenstock, A.-D.; Pellen, H. *J. Org. Chem.* **1997**, *62*, 4556.
3. Li, J.; Jiang, H. *Chem. Commun.* **1999**, 2369.
4. Siemsen, P.; Livingston, R. C.; Diederich, F. *Angew. Chem., Int. Ed.* **2000**, *39*, 2632.
5. Setzer, W. N.; Gu, X.; Wells, E. B.; Setzer, M. C.; Moriarity, D. M. *Chem. Pharm. Bull.* **2001**, *48*, 1776.
6. Kabalka, G. W.; Wang, L.; Pagni, R. M. *Synlett* **2001**, 108.

Gomberg–Bachmann reaction

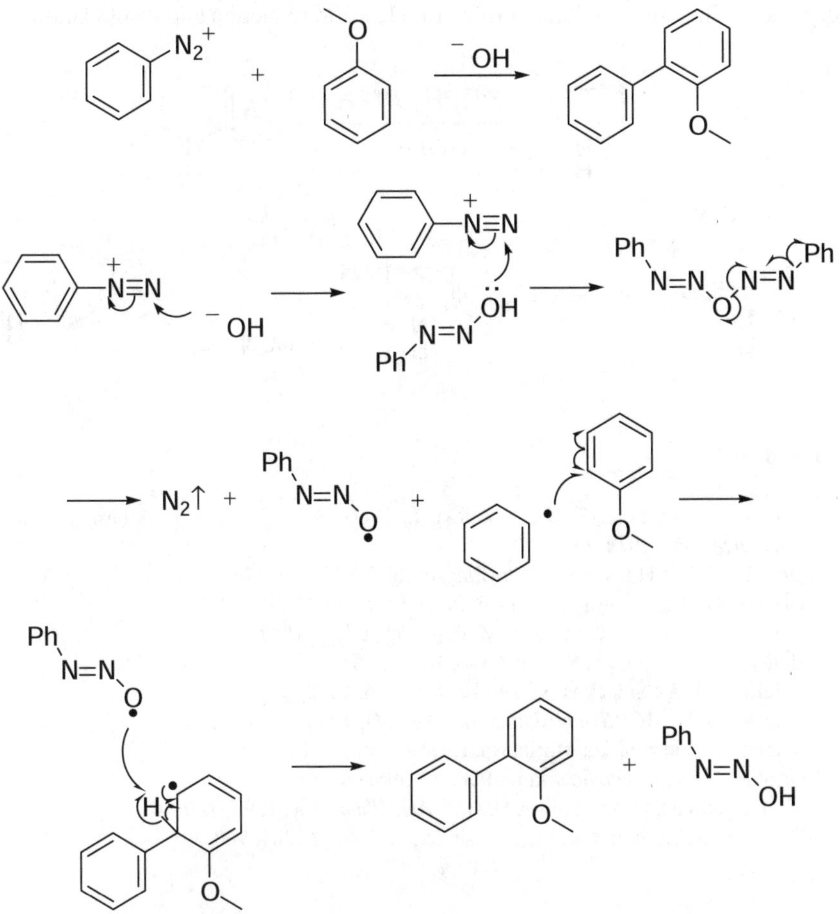

References

1. Gomberg, M.; Bachmann, W. E. *J. Am. Chem. Soc.* **1924**, *46*, 2339.
2. Beadle, J. R.; Korzeniowski, S. H.; Rosenberg, D. E.; Garcia-Slanga, B. J.; Gokel, G. W. *J. Org. Chem.* **1984**, *49*, 1594.
3. McKenzie, T. C.; Rolfes, S. M. *J. Heterocycl. Chem.* **1987**, *24*, 859.
4. Gurczynski, M.; Tomasik, P. *Org. Prep. Proced. Int.* **1991**, *23*, 438.
5. Hales, N. J.; Heaney, H.; Hollinshead, J. H.; Sharma, R. P. *Tetrahedron* **1995**, *51*, 7403.
6. Lai, Y.-H.; Jiang, J. *J. Org. Chem.* **1997**, *62*, 4412.

Gribble indole reduction

Reduction of the indole double bond using sodium cyanoborohydride in glacial acetic acid. The use of sodium borohydride leads to reduction and *N*-alkylation.

$NaBH_3CN$

HOAc

H^+

H–$\bar{B}H_2CN$

References

1. Gribble, G. W.; Lord, P. D.; Skotnicki, J.; Dietz, S.E.; Eaton, J. T.; Johnson, J. L. *J. Am. Chem. Soc.* **1974**, *96*, 7812.
2. Gribble, G. W.; Hoffman, J. H. *Synthesis* **1977**, 859.
3. Gribble, G. W.; Nutaitis, C. F. *Org. Prep. Proc. Int.* **1985**, *17*, 317.
4. Rawal, V.H.; Jones, R. J.; Cava, M. P. *J. Org. Chem.* **1987**, *52*, 19.
5. Boger, D. L.; Coleman, R.S.; Invergo, B. L. *J. Org. Chem.* **1987**, *52*, 1521.
6. Siddiqui, M. A.; Snieckus, V. *Tetrahedron Lett.* **1990**, *31*, 1523.
7. Gribble, G. W. *ACS Symposium Series No. 641*, **1996**, pp 167–200.
8. Somei, M.; Yamada, F.; Morikawa, H. *Heterocycles* **1997**, *46*, 91.
9. Gribble, G. W. *Chem. Soc. Rev.* **1998**, *27*, 395.
10. He, F.; Foxman, B. M.; Snider, B. B. *J. Am. Chem. Soc.* **1998**, *120*, 6417.
11. Nicolaou, K.C.; Safina, B. S.; Winssinger, N. *Synlett* **2001**, 900

Gribble reduction of diaryl ketones

Reduction of diaryl ketones and diarylmethanols to diarylmethanes using sodium borohydride in trifluoroacetic acid. Also applicable to diheteroaryl ketones and alcohols.

$$Ar^1C(=O)Ar^2 \xrightarrow[CF_3CO_2H]{NaBH_4} \left[Ar^1CH(OH)Ar^2 \right] \longrightarrow Ar^1CH_2Ar^2$$

$$Ar^1C(=O)Ar^2 \;\; (H^+,\; (CF_3CO_2)_3\overset{-}{B}\text{–}H) \longrightarrow Ar^1CH(OH)Ar^2 \xrightarrow{CF_3CO_2H}$$

$$Ar^1CH(\overset{+}{O}H_2)Ar^2 \xrightarrow{-H_2O} Ar^1\overset{+}{C}HAr^2 \xrightarrow{H\overset{-}{B}(OCOCF_3)_3} Ar_1CH_2Ar^2$$

References

1. Gribble, G. W.; Leese, R. M.; Evans, B. E. *Synthesis* **1977**, 172.
2. Gribble, G. W.; Kelly, W. J.; Emery, S. E. *ibid.* **1978**, 763.
3. Gribble, G. W.; Nutaitis, C. F. *Org. Prep. Proc. Int.* **1985**, *17*, 317.
4. Kabalka, G. W.; Kennedy, T. P. *ibid.* **1989**, *21*, 348.
5. Daich, A.; Decroix, B. *J. Heterocycl. Chem.* **1992**, *29*, 1789.
6. Gribble, G. W. *ACS Symposium Series No. 641*, **1996**, pp 167–200.
7. Gribble, G. W. *Chem. Soc. Rev.* **1998**, *27*, 395.
8. Sattelkau, T.; Qandil, A. M.; Nichols, D. E. *Synthesis* **2001**, 267.

Grob fragmentation

General scheme:

Y ⌒ X → Y= + // + X^-

$X = OH_2^+$, OTs, I, Br, Cl; $Y = O^-$, NR_2

e.g.:

MsO, OMOM, OH —NaH, 15-crown-5→ MsO, OMOM, O^- → OMOM, O

e.g.:

MsO, O, O, H, H–O, H^- —LAH, DME, reflux, 90%→ H, O, H, O, O —H^-→ HO, H, O, O

References

1. Grob, C. A.; Baumann, W. *Helv. Chim. Acta* **1955**, *38*, 594.
2. Grob, C. A.; Schiess, P. W. *Angew. Chem., Int. Ed. Engl.* **1967**, *6*, 1.
3. French, L. G.; Charlton, T. P. *Heterocycles* **1993**, *35*, 305.

4. Harmata, M.; Elahmad, S. *Tetrahedron Lett.* **1993**, *34*, 789.
5. Armesto, X. L.; Canle L., M.; Losada, M.; Santaballa, J. A. *J. Org. Chem.* **1994**, *59*, 4659.
6. Yoshimitsu, T.; Yanagiya, M.; Nagaoka, H. *Tetrahedron Lett.* **1999**, *40*, 5215.
7. Hu, W.-P.; Wang, J.-J.; Tsai, P.-C. *J. Org. Chem.* **2000**, *65*, 4208.
8. Molander, G. A.; Le Huerou, Y.; Brown, G. A. *ibid.* **2001**, *66*, 4511.

Guareschi–Thorpe condensation

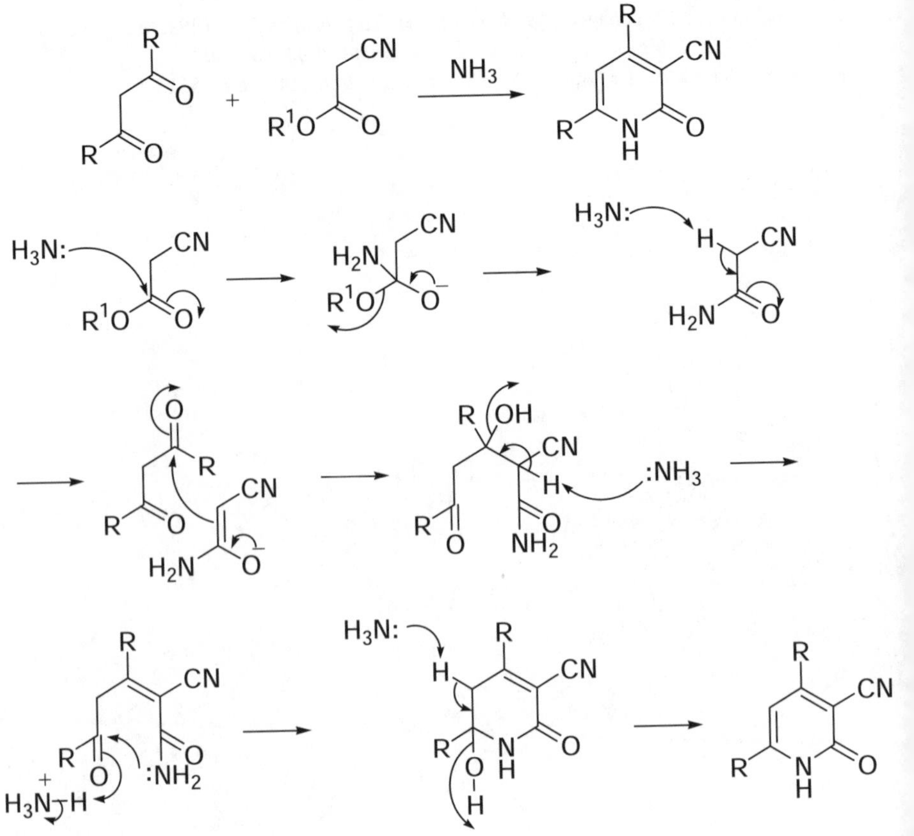

References

1. Baron, H.; Renfry, F. G. P.; Thorpe, J. F. *J. Chem. Soc.* **1904**, *85*, 1726.
2. Brunskill, J. S. A. *J. Chem. Soc. (C)* **1968**, 960.
3. Brunskill, J. S. A. *J. Chem. Soc., Perkin Trans. 1* **1972**, 2946.

Hajos–Wiechert reaction

Asymmetric Robinson annulation catalyzed by (*S*)-(−)-proline.

cat. HO_2C H N H

CH_3CN

Michael addition

enamine formation

catalytic asymmetric enamine aldol

H_2O:

hydrolysis of iminium salt

B:

E1cb

References

1. Hajos, Z. G.; Parrish, D. R. *J. Org. Chem.* **1974**, *39*, 1615.
2. Eder, U.; Sauer, G.; Wiechert, R. *Angew. Chem., Int. Ed. Engl.* **1971**, *10*, 496.
3. Brown, K. L.; Dann, L.; Duntz, J. D.; Eschenmoser, A.; Hobi, R.; Kratky, C. *Helv. Chim. Acta* **1978**, *61*, 3108.
4. Agami, C. *Bull. Soc. Chim. Fr.* **1988**, 499.
5. Nelson, S. G. *Tetrahedron: Asymmetry* **1998**, *9*, 357.
6. List, B.; Lerner, R. A.; Barbas, C. F., III. *J. Am. Chem. Soc.* **2000**, *122*, 2395.
7. List, B.; Pojarliev, P.; Castello, C. *Org. Lett.* **2001**, *3*, 573.

Haller–Bauer reaction

Base-induced cleavage of non-enolizable ketones leading to carboxylic acid derivative and a neutral fragment in which the carbonyl group is replaced by a hydrogen.

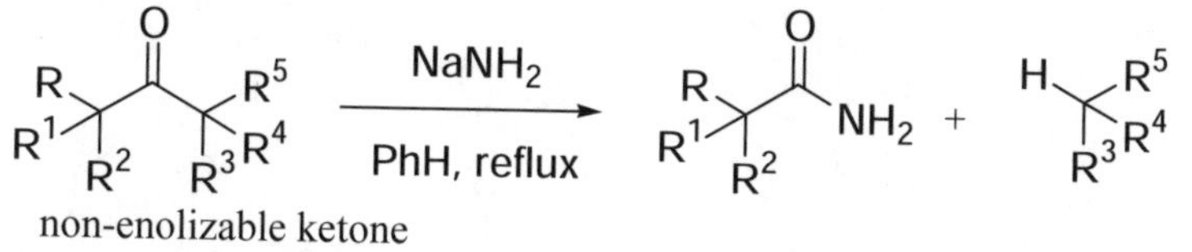

NaNH$_2$

PhH, reflux

non-enolizable ketone

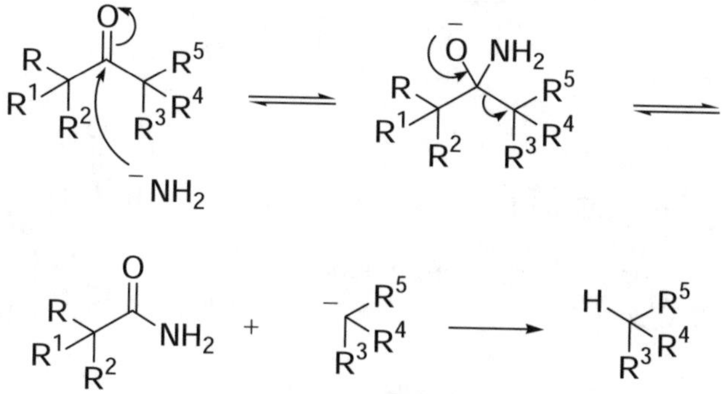

References

1. Haller, A.; Bauer, E. *Compt. Rend.* **1908**, *147*, 824.
2. Paquette, L. A.; Gilday, J. P.; Maynard, G. D. *J. Org. Chem.* **1989**, *54*, 5044.
3. Paquette, L. A.; Gilday, J. P. *Org. Prep. Proc. Int.* **1990**, *22*, 167.
4. Mehta, G.; Venkateswaran, R. V. *Tetrahedron* **2000**, *56*, 1399.
5. Arjona, O.; Medel, R.; Plumet, J. *Tetrahedron Lett.* **2001**, *42*, 1287.

Hantzsch pyridine synthesis

O
R H + CO_2Et
O R^1
NH_3
R
EtO_2C CO_2Et
R^1 N R^1
H

HNO_3
R
EtO_2C CO_2Et
R_1 N R^1

H_3N: H CO_2Et
O R^1
enolate formation
O
R H
CO_2Et
H_3N: H O R^1

aldol condensation
OH :NH_3
H
R CO_2Et
O R^1
OH
R CO_2Et
O R^1
E1cb
R
CO_2Et
O R^1

CO_2Et
:NH_3
O R^1
enamine formation
CO_2Et
HO R^1
H_2N:

$- H_2O$
H_3N: H CO_2Et
H_2N R^1
+

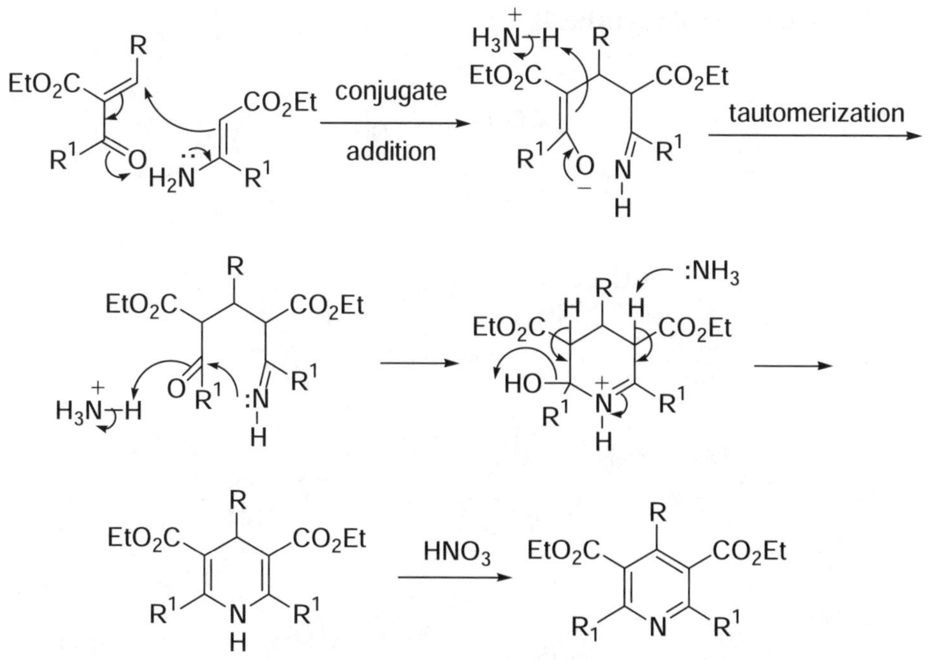

References

1. Hantzsch, A. *Ann.* **1882**, *215*, 1.
2. Balogh, M.; Hermecz, I.; Naray-Szabo, G.; Simon, K.; Meszaros, Z. *J. Chem. Soc., Perkin Trans. 1* **1986**, 753.
3. Katritzky, A. R.; Ostercamp, D. L.; Yousaf, T. I. *Tetrahedron* **1986**, *42*, 5729.
4. Shah, A. C.; Rehani, R.; Arya, V. P. *J. Chem. Res., (S)* **1994**, 106.
5. Menconi, I.; Angeles, E.; Martinez, L.; Posada, M. E.; Toscano, R. A.; Martinez, R. *J. Heterocycl. Chem.* **1995**, *32*, 831.
6. Muceniece, D.; Zandersons, A.; Lusis, V. *Bull. Soc. Chim. Belg.* **1997**, *106*, 467.
7. Goerlitzer, K.; Heinrici, C.; Ernst, L. *Pharmazie* **1999**, *54*, 35.
8. Raboin, J.-C.; Kirsch, G.; Beley, M. *J. Heterocycl. Chem.* **2000**, *37*, 1077.

Hantzsch pyrrole synthesis

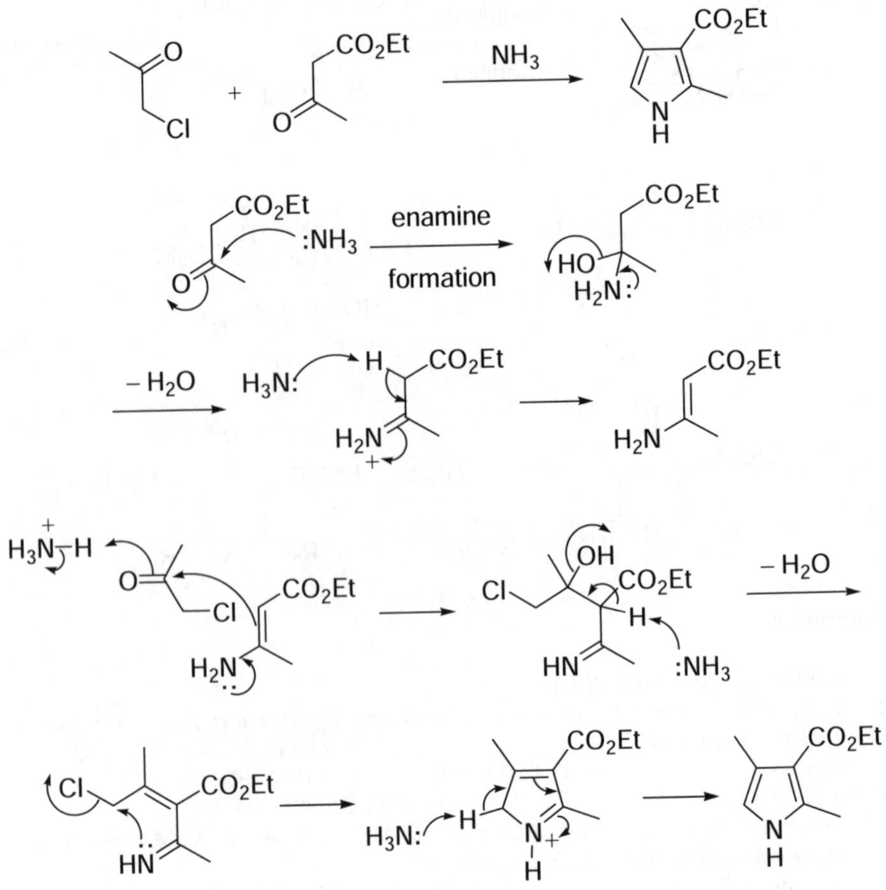

NH$_3$

CO$_2$Et

enamine formation

$-$ H$_2$O

References

1. Hantzsch, A. *Ber.* **1890**, *23*, 1474.
2. Roomi, M. W.; MacDonald, S. F. *Can. J. Chem.* **1970**, *48*, 1689.
3. Hort, E. V.; Anderson, L. R. *Kirk-Othmer Encycl. Chem. Technol.*, 3rd Ed. **1982**, *19*, 499.
4. Katritzky, A. R.; Ostercamp, D. L.; Yousaf, T. I. *Tetrahedron* **1987**, *43*, 5171.
5. Kirschke, K.; Costisella, B.; Ramm, M.; Schulz, B. *J. Prakt. Chem.* **1990**, *332*, 143.
6. Trautwein, A. W.; Süßmuth, R. D.; Jung, G. *Bioorg. Med. Chem. Lett.* **1998**, *8*, 2381.

Haworth reaction

$AlCl_3$ Zn HCl H_2SO_4

$AlCl_3$ $O-AlCl_3$ $AlCl_3$

Friedel–Crafts acylation OH Zn, HCl reduction OH

H_2SO_4 OH_2

acylium ion

Friedel–Crafts acylation H $-H^+$

References

1. Haworth, R. D. *J. Chem. Soc.* **1932**, 1125.
2. Agranat, I.; Shih, Y. *J. Chem. Educ.* **1976**, *53*. 488.
3. Silveira, A., Jr.; McWhorter, E. J. *J. Org. Chem.* **1972**, *37*. 3687.
4. Aichaoui, H.; Poupaert, J. H.; Lesieur, D.; Henichart, J. P. *Bull. Soc. Chim. Belg.* **1992**, *101*. 1053.

Hayashi rearrangement

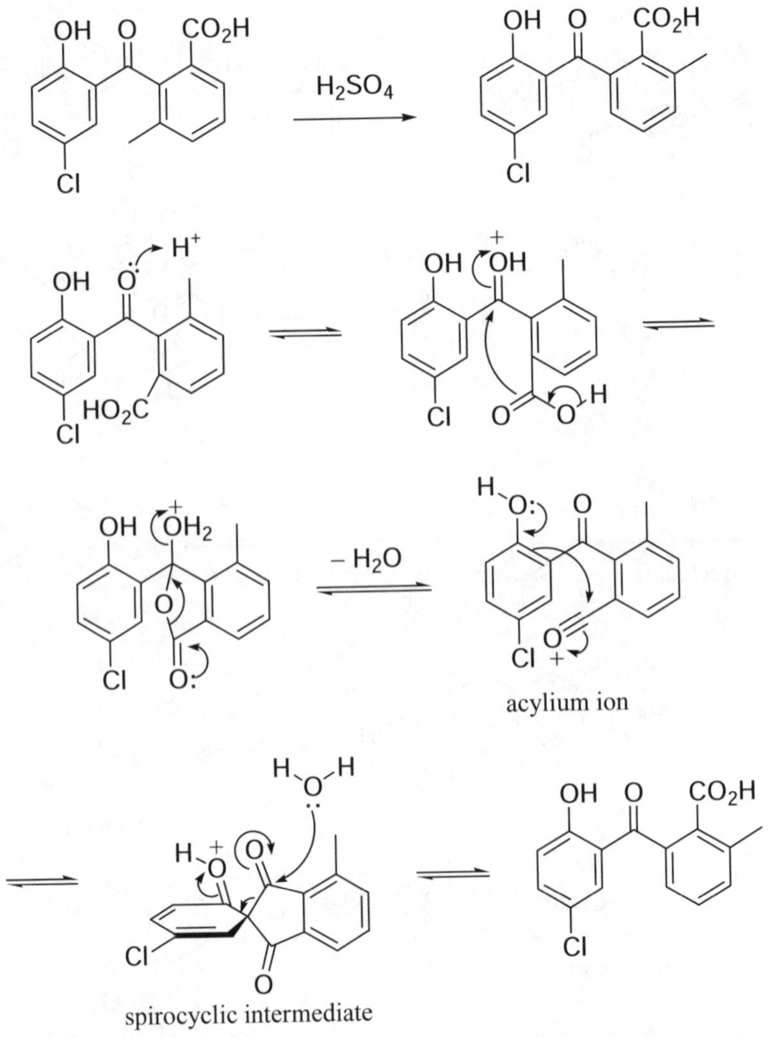

H_2SO_4

$- H_2O$

acylium ion

spirocyclic intermediate

References

1. Hayashi, M. *J. Chem. Soc.* **1927**, 2516.
2. Sandin, R. B.; Melby, R.; Crawford, R.; McGreer, D. G. *J. Am. Chem. Soc.* **1956**, *78*, 3817.
3. Newman, M. S.; Ihrman, K. G. *ibid.* **1958**, *80*, 3652.
4. Cristol, S. J.; Caspar, M. L. *J. Org. Chem.* **1968**, *33*, 2020.
5. Cadogan, J. I. G.; Kulik, S.; Tood, M. J. *J. Chem. Soc., Chem. Commun.* **1968**, 736.
6. Newmann, M. S. *Acc. Chem. Res.* **1972**, *5*, 354.

7. Cushman, M.; Choong, T.-C.; Valko, J. T.; Koleck, M. P. *J. Org. Chem.* **1980**, *45*, 5067.
8. Opitz, A.; Roemer, E.; Haas, W.; Gorls, H.; Werner, W.; Grafe, U. *Tetrahedron* **2000**, *56*, 5147.

Heck reaction

Palladium-catalyzed coupling between organohalides or triflates with olefins.

$Pd(Ph_3P)_4$, KOAc
DMA, 100 °C, 15 h
95%

Pd(0)
oxidative addition

Pd(II)–Cl

coordination

insertion

Pd(II)–Cl
Ph

internal rotation

syn-β-hydride elimination

H–Pd(II)–Cl

base

Pd(0) + $BH^+ Cl^-$

References

1. Heck, R. F.; Nolley, J. P., Jr. *J. Am. Chem. Soc.* **1968**, *90*, 5518.
2. Akita, Y.; Inoue, A.; Mori, Y.; Ohta, A. *Heterocycles* **1986**, *24*, 2093.
3. Beletskaya, I. P.; Cheprakov, A. V. *Chem. Rev.* **2000**, *100*, 3009.
4. Amatore, C.; Jutand, A. *Acc. Chem. Res.* **2000**, *33*, 314.
5. Franzen, R. *Can. J. Chem.* **2000**, *78*, 957.

6. Mayasundari, A.; Young, D. G. J. *Tetrahedron Lett.* **2001**, *42*, 203.
7. Haeberli, A.; Leumann, C. J. *Org. Lett.* **2001**, *3*, 489.
8. Gilbertson, S. R.; Fu, Z.; Xie, D. *Tetrahedron Lett.* **2001**, *42*, 365.

Hegedus indole synthesis

Stoichiometric Pd(II)-mediated oxidative cyclization of alkenyl anilines to indoles. *Cf.* Wacker oxidation.

MeO_2C NH_2 — $PdCl_2(CH_3CN)_2$, THF; then, Et_3N, 84% → MeO_2C N H

MeO_2C NH_2 — $PdCl_2(CH_3CN)_2$ palladation → MeO_2C NH_2 Pd Cl Cl precipitate

Et_3N → Cl Cl Et_3N-Pd MeO_2C NH_2 →

Et_3N H Pd Cl Cl MeO_2C N H_2 — – HCl, – "PdH" → MeO_2C N H "Pd

H PdL_n CH_3 MeO_2C N H — β-elimination → CH_3 MeO_2C N H

References

1. Hegedus, L. S.; Allen, G. F.; Waterman, E. L. *J. Am. Chem. Soc.* **1976**, *98,* 2674.
2. Hegedus, L. S.; Allen, G. F.; Bozell, J. J.; Waterman, E. L. *ibid.* **1978**, *100,* 5800.
3. Hegedus, L. S. *Angew. Chem., Int. Ed. Engl.* **1988**, *27,* 1113.

Hell–Volhardt–Zelinsky reaction

α-Bromination of carboxylic acids using Br_2/PBr_3.

References

1. Hell, C. *Ber.* **1881**, *14*, 891.
2. Little, J. C.; Sexton, A. R.; Tong, Y.-L. C.; Zurawic, T. E. *J. Am. Chem. Soc.* **1969**, *91*, 7098.
3. Chatterjee, N. R. *Indian J. Chem., Sect. B* **1978**, *16B*, 730.

Henry reaction (nitroaldol reaction)

References

1. Henry, L. *Compt. Rend.* **1895**, *120*, 1265.
2. Matsumoto, K. *Angew. Chem.* **1984**, *96*, 599.
3. Sakanaka, O.; Ohmori, T.; Kozaki, S.; Suami, T.; Ishii, T.; Ohba, S.; Saito, Y. *Bull. Chem. Soc. Jpn.* **1986**, *59*, 1753.
4. Rosini, G. In *Comprehensive Organic Synthesis* Trost, B. M.; Fleming, I., Eds, Pergamon, **1991**, *2*, 321–340.
5. Barrett, A. G. M.; Robyr, C.; Spilling, C. D. *J. Org. Chem.* **1989**, *54*, 1233.
6. Ballini, R.; Bosica, G. *ibid.* **1994**, *59*, 5466.
7. Ballini, R.; Bosica, G. *ibid,* **1997**, *62*, 425.
8. Kisanga, P. B.; Verkade, J. G. *ibid.* **1999**, *64*, 4298.
9. Simoni, D.; Rondanin, R.; Morini, M.; Baruchello, R.; Invidiata, F. P. *Tetrahedron Lett.* **2000**, *41*, 1607.
10. Bandgar, B. P.; Uppalla, L. S. *Synth. Commun.* **2000**, *30*, 2071.
11. Luzzio, F. A. *Tetrahedron* **2001**, *57*, 915.

Herz reaction

References

1. Herz, R. Ger. Pat. 360,690, **1914**.
2. Ried, W.; Valentin, J. *Ann.* **1966**, 183.
3. Schneller, S. W. *Int. J. Sulfur Chem. B* **1972**, *7,* 155.
4. Schneller, S. W. *ibid,* **1976**, *8,* 579.
5. Chenard, B. L. *J. Org. Chem.* **1984**, *49*, 1224.
6. Belica, P. S.; Manchand, P. S. *Synthesis* **1990**, 539.
7. Grandolini, G.; Perioli, L.; Ambrogi, V. *Gazz. Chim. Ital.* **1997**, *127*, 411.

Heteroaryl Heck reaction

Intermolecular or intramolecular Heck reaction that occurs onto a heteroaryl recipient.

$Pd(Ph_3P)_4$, Ph_3P, CuI

Cs_2CO_3, DMF, 140 °C

Pd(0)

oxidative addition

Pd(II)I

insertion

B:

H

Pd(II)

I

Cs_2CO_3

+ Pd(0) + CsI + $CsHCO_3$

References

1. Ohta, A.; Akita, Y.; Ohkuwa, T.; Chiba, M.; Fukunaka, R.; Miyafuji, A.; Nakata, T.; Tani, N. Aoyagi, Y. *Heterocycles* **1990**, *31*, 1951.
2. Aoyagi, Y.; Inoue, A.; Koizumi, I.; Hashimoto, R.; Tokunaga, K.; Gohma, K.; Komatsu, J.; Sekine, K.; Miyafuji, A.; Konoh, J. Honma, R. Akita, Y.; Ohta, A. *ibid.* **1992**, *33*, 257.
3. Proudfoot, J. R. *et al. J. Med. Chem.* **1995**, *38*, 4930.
4. Pivsa-Art, S.; Satoh, T.; Kawamura, Y.; Miura, M.; Nomura, M. *Bull. Chem. Soc. Jpn.* **1998**, *71*, 467.
5. Li, J. J.; Gribble, G. W. In *Palladium in Heterocyclic Chemistry* **2000**, Pergamon: Oxford, p16.

Hiyama cross-coupling reaction

Palladium-catalyzed cross-coupling reaction of organosilicons with organic halides, triflates, *etc.* in the presence of an activating agent such as fluoride or hydroxide (transmetallation is reluctant to occur without the effect of an activating agent). For the catalytic cycle, see the Kumada coupling on page 208.

MeO_2C ... Br + F_3Si ...

$Pd(Ph_3P)_4$, TBAF / THF → MeO_2C ...

F_3Si ... + Bu_4N^+ F^- —nucleophilic attack→ [F_4Si ...]$^-$ Bu_4N^+ → F_4Si^- ... Bu_4N^+

MeO_2C ... Br + $Pd(Ph_3P)_4$ —oxidative addition→ MeO_2C ... Pd—Br

F_4Si^- ... —trans"metal"lation→ MeO_2C ... Pd ... CO_2Me

—reductive elimination→ MeO_2C ... + $Pd(Ph_3P)_4$

References

1. Hiyama, T.; Hatanaka, Y. *Pure Appl. Chem.* **1994**, *66*, 1471.

2. Matsuhashi, H.; Kuroboshi, M.; Hatanaka, Y.; Hiyama, T. *Tetrahedron Lett.* **1994**, *35*, 6507.
3. Mateo, C.; Fernandez-Rivas, C.; Echavarren, A. M.; Cardenas, D. J. *Organometallics* **1997**, *16*, 1997.
4. Hiyama, T. In *Metal-Catalyzed Cross-Coupling Reactions* **1998**, Diederich, F.; Stang, P. J., Eds.; Wiley–VCH Verlag GmbH: Weinheim, Germany, 421–53.
5. Denmark, S. E.; Wang, Z. *J. Organomet. Chem.* **2001**, *624*, 372.

Hodges–Vedejs metallation of oxazoles

Metallation of an oxazole followed by treatment with benzaldehyde results in a 4-substituted oxazole as the major product [1]:

LDA

PhCHO

2-lithiooxazole

H^+

However, the ring-opening process can be prevented by addition of boranes [3]:

$H_3B{\cdot}THF$

complexation

LTMP

metalation

LTMP = lithium tetramethylpiperidine

1. PhCHO

2. H^+

References

1. Hodges, J. C.; Patt, W. C.; Connolly, C. J. *J. Org. Chem.* **1991**, *56*, 449.
2. Iddon, B. *Heterocycles* **1994**, *37*, 1321.
3. Vedejs, E.; Monahan, S. D. *J. Org. Chem.* **1996**, *61*, 5192.
4. Vedejs, E.; Luchetta, L. M. *ibid.* **1999**, *64*, 1011.

Hofmann rearrangement (Hofmann degradation reaction)

$$R{-}C({=}O){-}NH_2 \xrightarrow[\text{NaOH}]{\text{Br}_2} R{-}N{=}C{=}O \xrightarrow{\text{H}_2\text{O}} R{-}NH_2 + CO_2\uparrow$$

$$\longrightarrow R{-}N{=}C{=}O \;(^{-}OH) \longrightarrow \longrightarrow R{-}NH_2 + CO_2\uparrow$$

isocyanate intermediate

References

1. Hofmann, A. W. *Ber.* **1881**, *14*, 2725.
2. Grillot, G. F. *Mech. Mol. Migr.* **1971**, 237.
3. Jew, S.-s.; Kang, M.-h. *Arch. Pharmacal Res.* **1994**, *17*, 490.
4. Huang, X.; Seid, M.; Keillor, J. W. *J. Org. Chem.* **1997**, *62*, 7495.
5. Monk, K. A.; Mohan, R. S. *J. Chem. Educ.* **1999**, *76*, 1717.
6. Togo, H.; Nabana, T.; Yamaguchi, K. *J. Org. Chem.* **2000**, *65*, 8391.
7. Yu, C.; Jiang, Y.; Liu, B.; Hu, L. *Tetrahedron Lett.* **2001**, *42*, 1449.

Hofmann–Löffler–Freytag reaction

References

1. Hofmann, A. W. *Ber.* **1883**, *16*, 558.
2. Furstoss, R.; Teissier, P.; Waegell, B. *Tetrahedron Lett.* **1970**, 1263.
3. Deshpande, R. P.; Nayak, U. R. *Indian J. Chem., Sect. B* **1979**, *17B*, 310.
4. Hammerum, S. *Tetrahedron Lett.* **1981**, *22*, 157.
5. Uskokovic, M. R.; Henderson, T.s; Reese, C.; Lee, H. L.; Grethe, G.; Gutzwiller, J. *J. Am. Chem. Soc.* **1978**, *100*, 571.
6. Madsen, J.; Viuf, C.; Bols, M. *Chem. Eur. J.* **2000**, *6*, 1140.
7. Togo, H.; Katohgi, M. *Synlett* **2001**, 565.

Hofmann–Martius reaction

HCl

S_N2

+ CH_3Cl

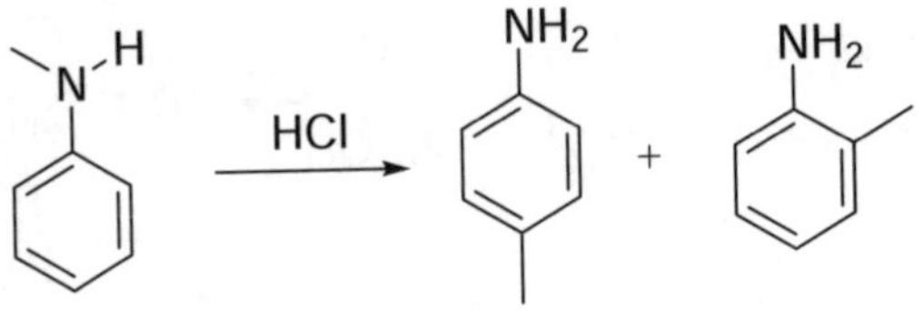

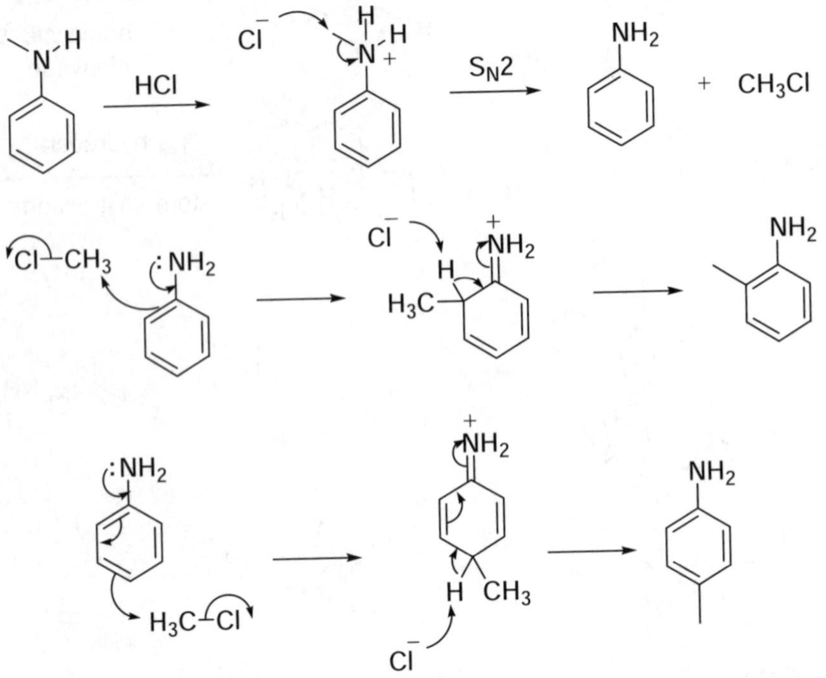

Reilly–Hickinbottom rearrangement is a variation of the Hofmann–Martius reaction in which a Lewis acid is used instead of a protic acid. The reaction follows an analogous pathway:

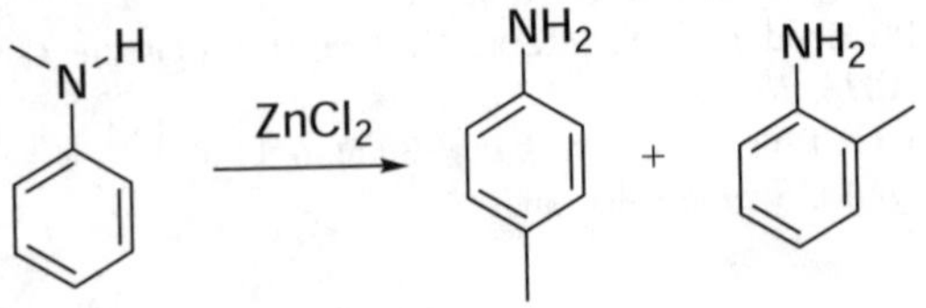

$ZnCl_2$

References

1. Hofmann, A. W.; Martius, C. A. *Ber.* **1964**, *20*, 2717.
2. Ogata, Y.; *et al. Tetrahedron* **1964**, 1263.
3. Ogata, Y.; *et al. J. Org. Chem.* **1970**, *35*, 1642.
4. Grillot, G. F. *Mech. Mol. Migr.* **1971**, *3*, 237.
5. Giumanini, A. G.; Roveri, S.; Del Mazza, D. *J. Org. Chem.* **1975**, *40*, 1677.
6. Hori, M.; Kataoka, T.; Shimizu, H.; Hsu, C. F.; Hasegawa, Y.; Eyama, N. *J. Chem. Soc., Perkin Trans. 1* **1988**, 2271.
7. Siskos, M. G.; Tzerpos, N.; Zarkadis, A. *Bull. Soc. Chim. Belg.* **1996**, *105*, 759.

Hooker oxidation

lapachol

$KMnO_4$

NaOH

dihydroxylation

alcohol oxidation

enolization

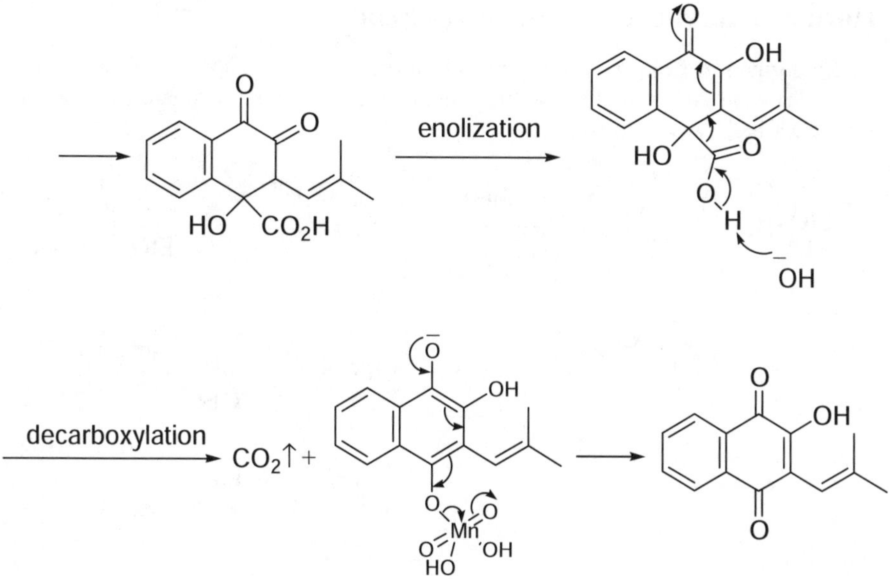

enolization

decarboxylation

CO_2↑ +

References

1. Hooker, S. C. *J. Am. Chem. Soc.* **1936**, *58,* 1174.
2. Fieser, L. F.; Sachs, D. H. *ibid.* **1968**, *90,* 4129.
3. Lee, K. Hee; Moore, H. W. *Tetrahedron Lett.* **1993**, *34,* 235.
4. Lee, K.; Turnbull, P.; Moore, H. W. *J. Org. Chem.* **1995**, *60,* 461.

Horner–Wadsworth–Emmons reaction

Olefin formation from aldehydes and phosphonates. Workup is more advantageous than the corresponding Wittig reaction because the phosphate by-product can be washed away with water.

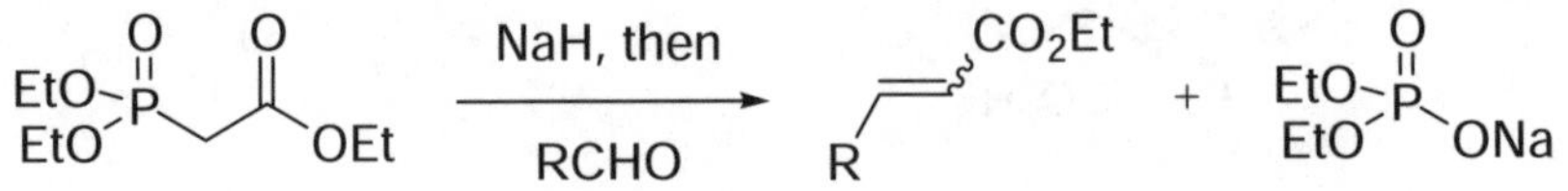

NaH, then RCHO

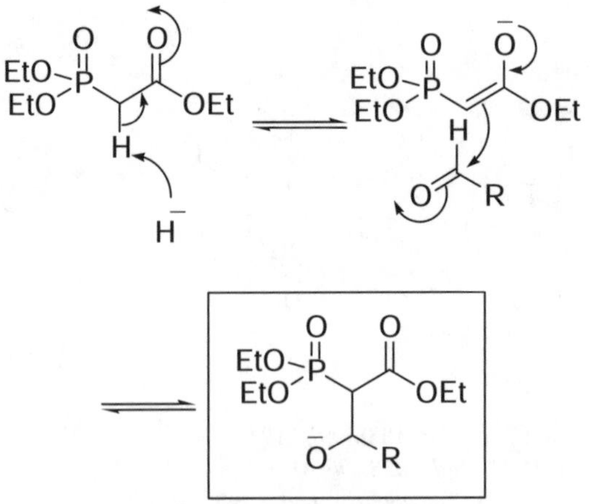

erythro (kinetic) or *threo* (thermodynamic)

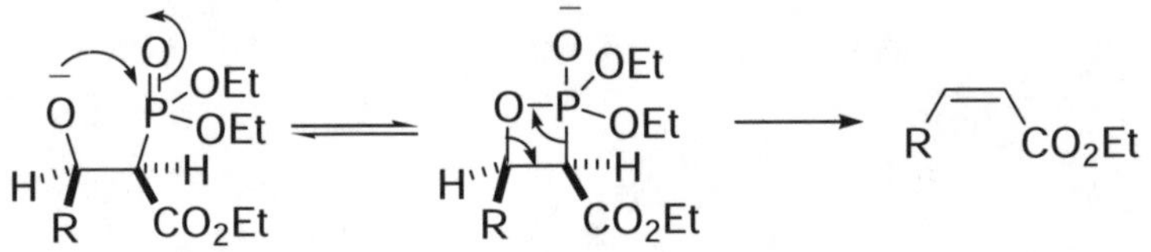

erythro, kinetic adduct

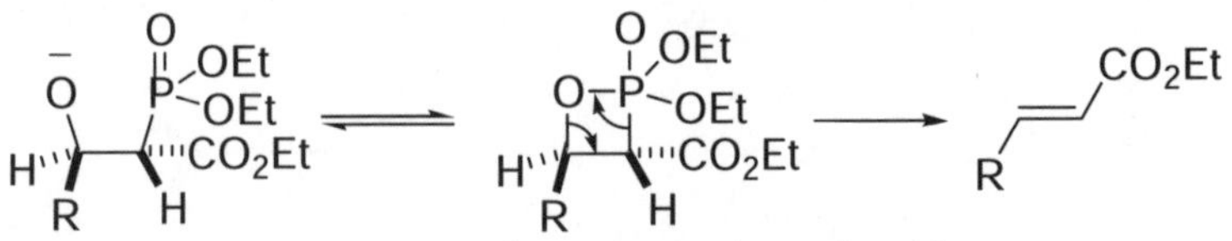

threo, thermodynamic adduct

References

1. Horner, L.; Hoffmann, H.; Wippel, H. G.; Klahre, G. *Chem. Ber.* **1959**, *92*, 2499.
2. Wadsworth, W. S., Jr.; Emmons, W. D. *J. Am. Chem. Soc.* **1961**, *62*, 1733.

3. Wadsworth, D. H.; Schupp, O. E.; Seus, E. J.; Ford, J. A., Jr. *J. Org. Chem.* **1965**, *30*, 680.
4. Maryanoff, B. E.; Reitz, A. B. *Chem. Rev.* **1989**, *89*, 863.
5. Ando, K. *J. Org. Chem.* **1997**, *62*, 1934.
6. Ando, K. *ibid.* **1999**, *64*, 6815.
7. Simoni, D.; Rossi, M.; Rondanin, R.; Mazzali, A.; Baruchello, R.; Malagutti, C.; Roberti, M.; Invidiata, F. P. *Org. Lett.* **2000**, *2*, 3765.
8. Mawaziny, S.; Lakany, A. M. *Phosphorus, Sulfur Silicon Relat. Elem.* **2000**, *163*, 99.
9. Reiser, U.; Jauch, J. *Synlett* **2001**, 90.
10. Comins, D. L.; Ollinger, C. G. *Tetrahedron Lett.* **2001**, *42*, 4115.

Houben–Hoesch reaction

Acid-catalyzed acylation of phenols using nitriles.

RCN

HCl, $AlCl_3$

OH OH NH•HCl R

OH OH R O

$AlCl_3$ chelation R—≡N:

:OH OH R—≡N−$AlCl_3$ + −

electrophilic substitution

H^+ O OH H Cl^- R NH

OH H_2O: OH R NH•HCl

hydrolysis

OH OH R NH_2 O H Cl^- H^+

OH OH R O

References

1. Hoesch, K. *Ber.* **1915**, *48,* 1122.
2. Amer, M. I.; Booth, B. L.; Noori, G. F. M.; Proenca, M. F. J. R. P. *J. Chem. Soc., Perkin Trans. 1* **1983**, 1075.

3. Yato, M.; Ohwada, T.; Shudo, K. *J. Am. Chem. Soc.* **1991**, *113*, 691.
4. Sato, Y.; Yato, M.; Ohwada, T.; Saito, S.; Shudo, K. *ibid.* **1995**, *117*, 3037.
5. Kawecki, R.; Mazurek, A. P.; Kozerski, L.; Maurin, J. K. *Synthesis* **1999**, 751.

Hunsdiecker reaction

$$RCO_2Ag \xrightarrow{X_2} R{-}X + CO_2\uparrow + AgX$$

$$RCO_2Ag + X{-}X \longrightarrow AgX + RCO_2X \xrightarrow[\text{cleavage}]{\text{homolytic}}$$

$$X\bullet + RCO_2\bullet \longrightarrow CO_2\uparrow + R\bullet \xrightarrow{RCO_2X} R{-}X + RCO_2\bullet$$

References

1. Hunsdiecker, H.; Hunsdiecker, C. *Ber.* **1942**, *75,* 291.
2. Naskar, D.; Chowdhury, S.; Roy, S. *Tetrahedron Lett.* **1998**, *39*, 699.
3. Camps, P.; Lukach, A. E.; Pujol, X.; Vazquez, S. *Tetrahedron* **2000**, *56*, 2703.
4. De Luca, L.; Giacomelli, G.; Porcu, G.; Taddei, M. *Org. Lett.* **2001**, *3*, 855.

Ing–Manske procedure

A variant of Gabriel amine synthesis where hydrazine is used to release the amine from the corresponding phthalimide:

1. RX
2. NH_2NH_2

$H_2N{-}R$ +

S_N2

$:NH_2NH_2$

NHNH$_2$

N–R

References

1. Ing, H. R.; Manske, R. H. F. *J. Chem. Soc.* **1926**, 2348.
2. Khan, M. N. *J. Org. Chem.* **1995**, *60*, 4536.
3. Hearn, M. J.; Lucas, L. E. *J. Heterocycl. Chem.* **1984**, *21*, 615.
4. Khan, M. N. *J. Org. Chem.* **1996**, *61*, 8063.

Jacobsen–Katsuki epoxidation

Mangnese(III)-catalyzed asymmetric epoxidation of (*Z*)-olefins.

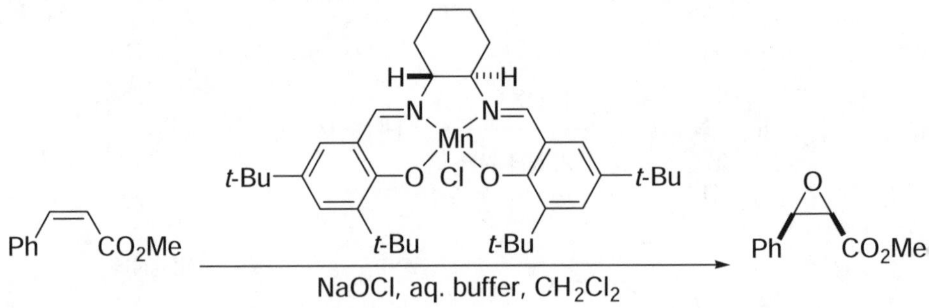

1. Concerted oxygen transfer (*cis*-epoxide):

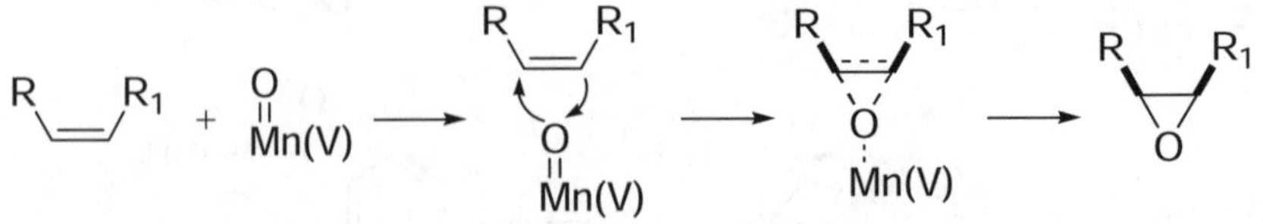

2. Oxygen transfer *via* radical intermediate (*trans*-epoxide):

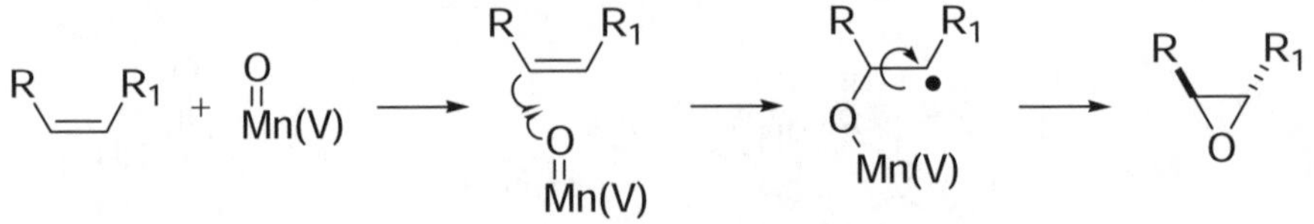

3. Oxygen transfer *via* manganaoxetane intermediate (*cis*-epoxide):

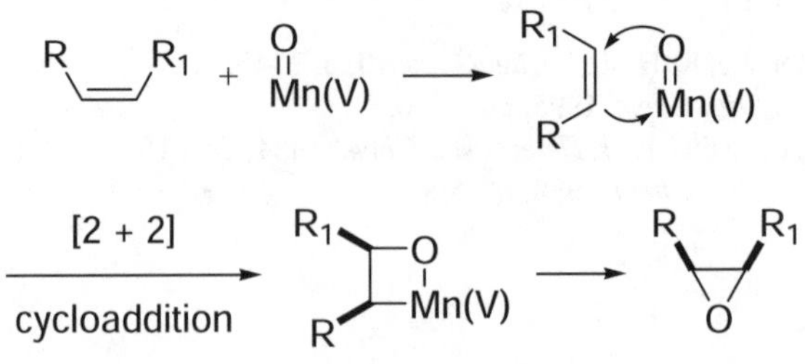

References

1. Zhang, W.; Loebach, J. L.; Wilson, S. R.; Jacobsen, E. N. *J. Am. Chem. Soc.* **1990**, *112,* 2801.
2. Irie, R.; Noda, K.; Ito, Y.; Katsuki, T. *Tetrahedron Lett.* **1991**, *32*, 1055.
3. Zhang, W.; Jacobsen, E. N. *J. Org. Chem.* **1991**, *56*, 2296.
4. Schurig, V.; Betschinger, F. *Chem. Rev.* **1992**, *92*, 873.
5. Jacobsen, E. N. In *Catalytic Asymmetric Synthesis* Ojima, I., Ed., VCH: Weinheim, New York, **1993**, Ch. 4.2.
6. Palucki, M.; McCormick, G. J.; Jacobsen, E. N. *Tetrahedron Lett.* **1995**, *36*, 5457.
7. Linker, T. *Angew. Chem., Int. Ed. Engl.* **1997**, *36*, 2060.
8. Katsuki, T. In *Catalytic Asymmetric Synthesis* 2nd ed., Ojima, I., ed.; Wiley-VCH: New York, **2000**, 287.
9. El-Bahraoui, J.; Wiest, O.; Feichtinger, D.; Plattner, D. A. *Angew. Chem., Int. Ed.* **2001**, *40*, 2073.

Jacobsen rearrangement

conc. H_2SO_4

SO_3H

Mechanism 1:

HO–S(=O)(=O)–OH

electrophilic substitution

H SO_3H

re-aromatization

SO_3H H^+

protonation

SO_3H H

1,2-alkyl migration

SO_3H H ↔ SO_3H H

1,2-alkyl migration

SO_3H

Mechanism 2:

HO–S(=O)(=O)–OH

electrophilic substitution

SO_3H

1,2-alkyl migration

SO_3H ↔ SO_3H

1,2-alkyl migration

SO

Mechanism 3:

SO_3H SO_3H^+ migration SO_3H SO_3H

1,2-alkyl migration SO_3H 1,2-alkyl migration SO_3H

References

1. Jacobsen, O. *Ber.* **1952**, *578*, 122.
2. Shine, H. J. *Aromatic Rearrangement* Elsevier: New York, **1967**, pp 23–32, 48–55.
3. Hart, H.; Janssen, J. F. *J. Org. Chem.* **1970**, *35*, 3637.
4. Marvell, E. N.; Graybill, B. M. *ibid.* **1965**, *30*, 4014.
5. Kilpatrick, M.; Meyer, M. *J. Phys. Chem.* **1961**, *65*, 1312.
6. Hart, H.; Janssen, J. F. *J. Org. Chem.* **1970**, *35*, 3637.
7. Suzuki, H.; Sugiyama, T. *Bull. Chem. Soc. Jpn.* **1973**, *46*, 586.
8. Norula, J. L.; Gupta, R. P. *Chem. Era* **1974**, *10*, 7.
9. Solari, E.; Musso, F.; Ferguson, R.; Floriani, C.; Chiesi-Villa, A.; Rizzoli, C. *Angew. Chem., Int. Ed. Engl.* **1995**, *35*, 1510.
10. Dotrong, M.; Lovejoy, S. M.; Wolfe, J. F.; Evers, R. C. *J. Heterocycl. Chem.* **1997**, *34*, 817.

Japp–Klingemann hydrazone synthesis

$ArN_2^+\ Cl^-$ + α-keto-ester $\xrightarrow{KOH}$ hydrazone + CH_3CO_2H

Diazonium salt α-keto-ester hydrazone

deprotonation

coupling

References

1. Japp, F. R.; Klingemann, F. *Liebigs Ann. Chem.* **1888**, *247*, 190.
2. Laduree, D.; Florentin, D.; Robba, M. *J. Heterocycl. Chem.* **1980**, *17*, 1189.
3. Loubinoux, B.; Sinnes, J.-L.; O'Sullivan, A. C.; Winkler, T. *J. Org. Chem.* **1995**, *60*, 953.
4. Saha, C., Miss; Chakraborty, A.; Chowdhury, B. K. *Indian J. Chem.* **1996**, *35B*, 677.
5. Pete, B.; Bitter, I.; Harsanyi, K.; Toke, L. *Heterocycles* **2000**, *53*, 665.
6. Atlan, V.; Kaim, L. E.; Supiot, C. *Chem. Commun.* **2000**, 1385.

Julia–Lythgoe olefination

(*E*)-Olefins from sulfones and aldehydes.

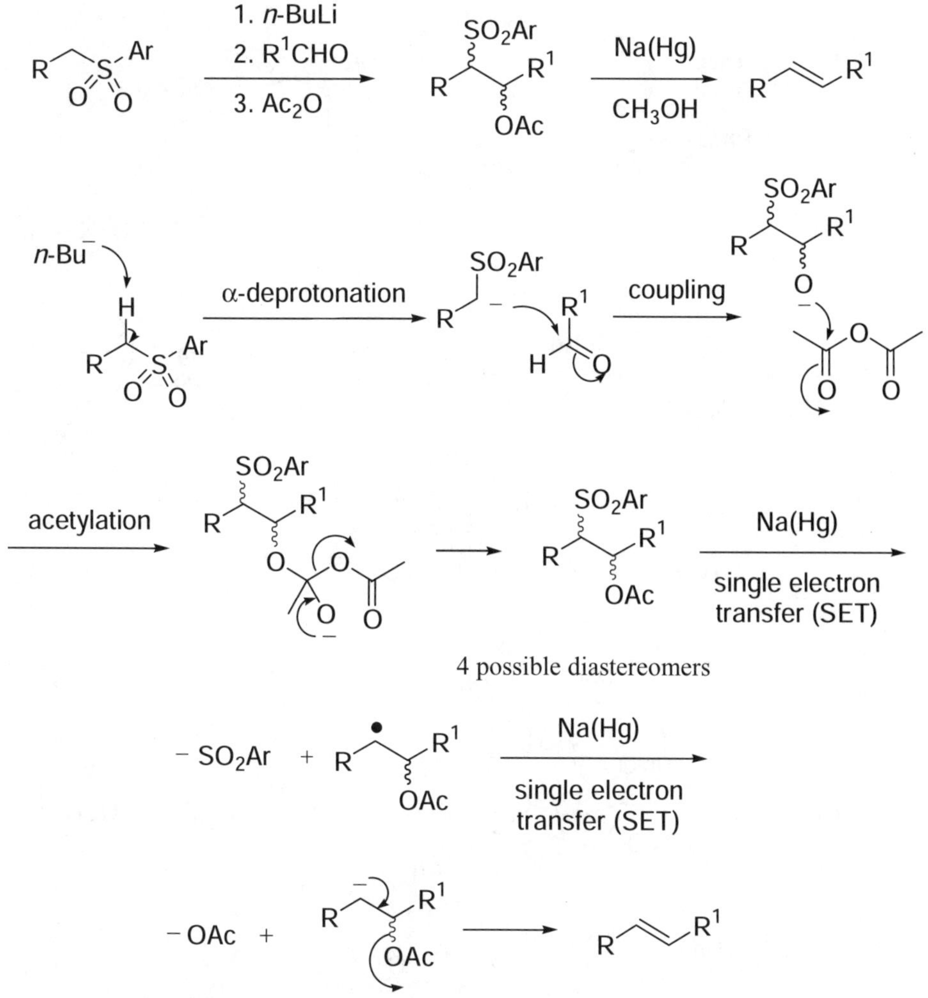

1. *n*-BuLi
2. R^1CHO
3. Ac_2O

Na(Hg)

CH_3OH

α-deprotonation

coupling

acetylation

Na(Hg)

single electron transfer (SET)

4 possible diastereomers

Na(Hg)

single electron transfer (SET)

References

1. Julia, M.; Paris, J. M. *Tetrahedron. Lett.* **1973**, 4833.
2. Keck, G. E.; Savin, K. A.; Weglarz, M. A. *J. Org. Chem.* **1995**, *60*, 3194.
3. Marko, I. E.; Murphy, F.; Dolan, S. *Tetrahedron Lett.* **1996**, *37*, 2089.
4. Satoh, T.; Yamada, N.; Asano, T. *ibid.* **1998**, *39*, 6935.
5. Satoh, T.; Hanaki, N.; Yamada, N.; Asano, T. *Tetrahedron* **2000**, *56*, 6223.
6. Charette, A. B.; Berthelette, C.; St-Martib, D. *Tetrahedron Lett.* **2001**, *42*, 5149.

Kahne glycosidation

Diastereoselective glycosidation of a sulfoxide at the anomeric center as the glycosyl acceptor. The sulfoxide activation is achieved using Tf_2O.

PivO OPiv O PivO PivO O S Ph + Ph O O O HO N3 SPh

CH3 t-Bu N t-Bu

Tf_2O, CH_2Cl_2, –78 to –30 °C

oxonium ion

PivO OPiv Ph O O O O PivO PivO O N3 SPh

References

1. Yan, L.; Taylor, C. M.; Goodnow, R., Jr.; Kahne, D. *J. Am. Chem. Soc.* **1994**, *116*, 6953.
2. Yan, L.; Kahne, D. *ibid.* **1996**, *118*, 9239.
3. Crich, D.; Li, H. *J. Org. Chem.* **2000**, *65*, 801.
4. Berkowitz, D. B.; Choi, S.; Bhuniya, D.; Shoemaker, R. K. *Org. Lett.* **2000**, *2*, 1149.

Keck stereoselective allylation

Asymmetric allylation of aldehydes with allylstannane in the presence of a Lewis acid and catalytic chiral BINAP (or other chiral ligands).

cat.
OH
OH
O
H
SnBu3
$Ti(Oi\text{-}Pr)_4$, 4Å MS, CH_2Cl_2
OH

$Ti(Oi\text{-}Pr)_2$(binap)*
:Ö:
H
chelation
$Ti(Oi\text{-}Pr)_2$(binap)*
+O
H
SnBu3
nucleophilic
addition

$Ti(Oi\text{-}Pr)_2$(binap)*
O
H_2O
workup
OH

The enantioselectivity is imparted by the steric bias of the chiral ligands which displace *iso*-propoxide of titanium *iso*-propoxide. Therefore, the chiral Lewis acid becomes $Ti(Oi\text{-}Pr)_2$(binap), which is substitutionally labile:

O
Ti
O
O*i*-Pr
O*i*-Pr

References

1. Keck, G. E.; Tarbet, K. H.; Geraci, L. S. *J. Am. Chem. Soc.* **1993**, *115*, 8467.
2. Keck, G. E.; Geraci, L. S. *Tetrahedron Lett.* **1993**, *34*, 7827.
3. Keck, G. E.; Krishnamurthy, D.; Grier, M. C. *J. Org. Chem.* **1993**, *58*, 6543.

4. Roe, B. A.; Boojamra, C. G.; Griggs, J. L.; Bertozzi, C. R. *ibid.* **1996**, *61*, 6442.
5. Fürstner, A.; Langemann, K. *J. Am. Chem. Soc.* **1997**, *119*, 9130.
6. Marshall, J. A.; Palovich, M. R. *J. Org. Chem.* **1998**, *63*, 4381.
7. Evans, P. A.; Manangan, T. *ibid.* **2000**, *65*, 4523.
8. Keck, G. E.; Wager, C. A.; Wager, T. T.; Savin, K. A.; Covel, J. A.; McLaws, M. D.; Krishnamurthy, D.; Cee, V. J. *Angew. Chem., Int. Ed.* **2001**, *40*, 231.

Keck macrolactonization

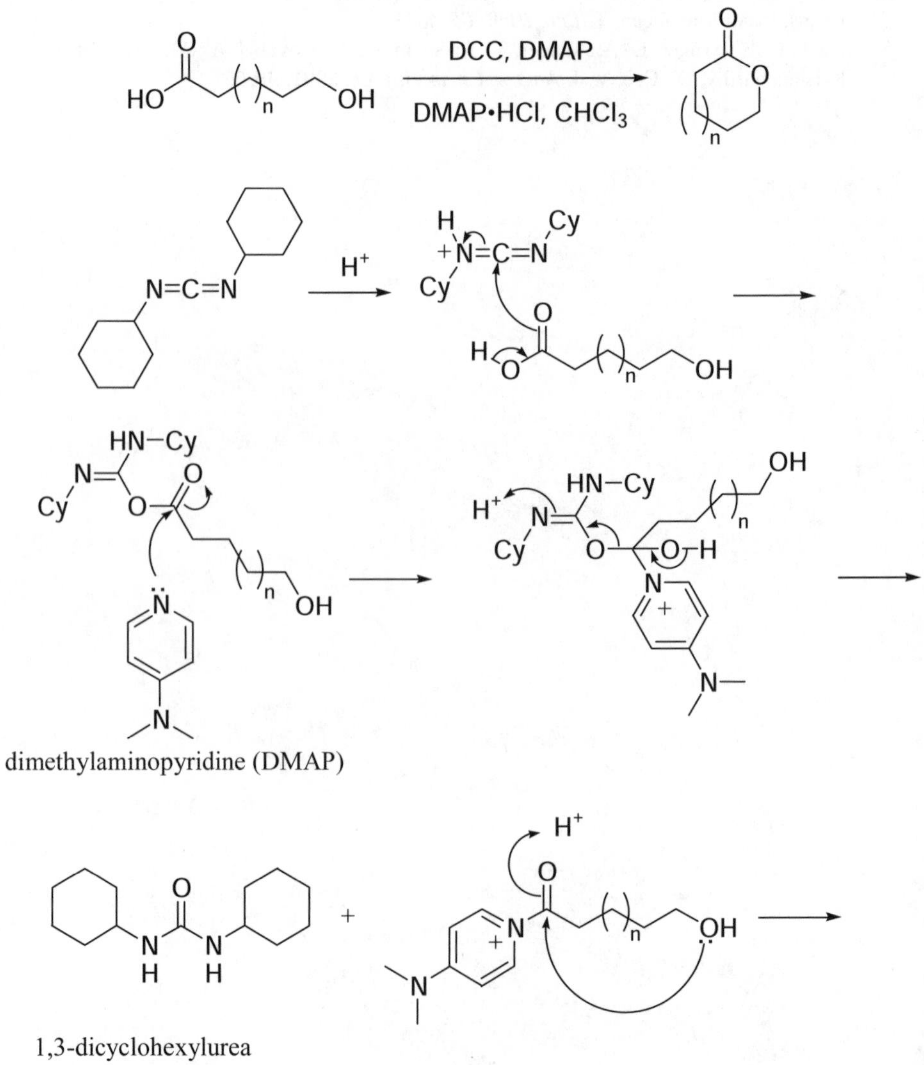

DCC, DMAP

DMAP•HCl, $CHCl_3$

dimethylaminopyridine (DMAP)

1,3-dicyclohexylurea

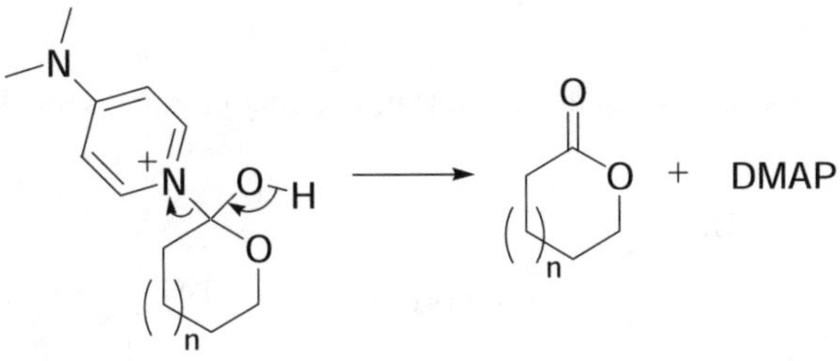

References

1. Boden, E. P.; Keck, G. E. *J. Org. Chem.* **1985**, *50*, 2394.
2. Keck, G. E.; Sanchez, C.; Wager, C. A. *Tetrahedron Lett.* **2000**, *41*, 8673.
3. Tsai, C.-Y.; Huang, X.; Wong, C.-H. *ibid.* **2000**, *41*, 9499.

Kemp elimination

Treatment of benzisoxazole results in the ring-opening product, salicylonitrile.

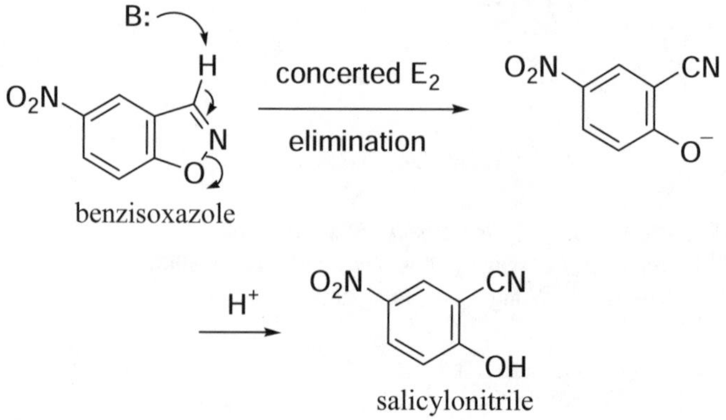

References

1. Casey, M. L.; Kemp, D. S.; Paul, K. G.; Cox, D. D. *J. Org. Chem.* **1973**, *38*, 2294.
2. Kemp, D. S.; Casey, M. L. *J. Am. Chem. Soc.* **1973**, *95*, 6670.
3. Kemp, D. S.; Cox, D. D.; Paul, K. G. *ibid.* **1975**, *97*, 7312.
4. Shulman, H.; Keinan, E. *Org. Lett.* **2000**, *2*, 3747.
5. Hollfelder, F.; Kirky, A. J.; Tawfik, D. S. *J. Org. Chem.* **2001**, *66*, 5866.

Kennedy oxidative cyclization

Asymmetric synthesis of tetrahydrofuran by treatment of a δ-hydroxyolefin with Re_2O_7.

1. Re_2O_7, 2,6-lutidine
2. NaOOH, 40%

trans:cis > 12:1

[2 + 2]
cycloaddition

perrhenate ester

reductive
elimination

Re(V), Re(VII)
exchange

NaOOH

References

1. Kennedy, R. M.; Tang, S. *Tetrahedron Lett.* **1992**, *33*, 3729.
2. Tang, S.; Kennedy, R. M. *ibid.* **1992**, *33*, 5299.
3. Tang, S.; Kennedy, R. M. *ibid.* **1992**, *33*, 5303.
4. Tang, S.; Kennedy, R. M. *ibid.* **1992**, *33*, 7823.
5. Boyce, R. S.; Kennedy, R. M. *ibid.* **1994**, *35*, 5133.
6. Sinha, S. C.; Sinha, A.; Santosh, C.; Keinan, E. *J. Am. Chem. Soc.* **1997**, *119*, 12014.
7. Avedissian, H.; Sinha, S. C.; Yazbak, A.; Sinha, A.; Neogi, P.; Sinha, S. C.; Keinan, E. *J. Org. Chem.* **2000**, *65*, 6035.

Kharasch addition reaction

Transition metal-catalyzed radical addition of $CXCl_3$ to olefins.

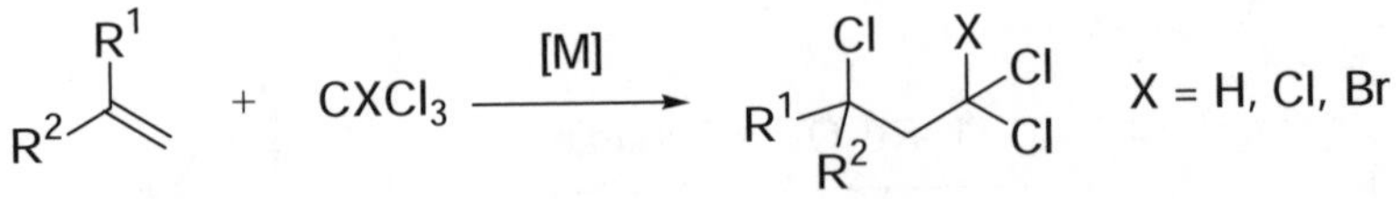

X = H, Cl, Br

M = organometallic reagent containing Ru, Re, Mo, W, Fe, Al, B, Cr, Sm, *etc.*

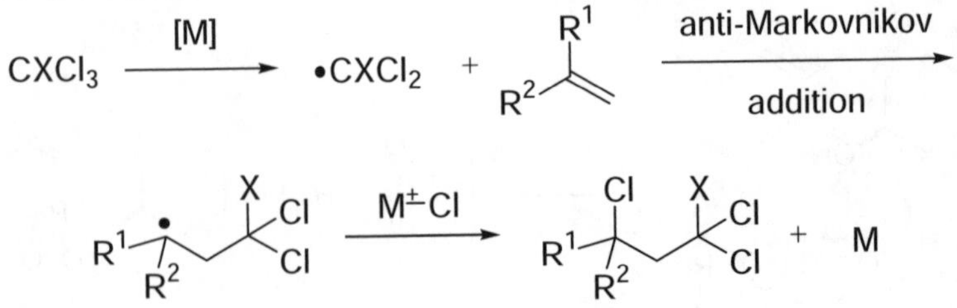

References

1. Kharasch, M. S.; Jensen, E. V.; Urry, W. H. *Science* **1945**, *102*, 2640.
2. Gossage, R. A.; van de Kuil, L. A.; van Koten, G. *Acc. Chem. Res.* **1998**, *31*, 423.
3. Simal, F.; Wlodarczak, L.; Demonceau, A.; Noels, A. F. *Tetrahedron Lett.* **2000**, *41*, 6071.

Knoevenagel condensation

R–CHO + $CH_2(CO_2R^1)_2$ → (pyrrolidine, N–H) → R–CH=C(CO_2R^1)$_2$ → ($^-$OH) → R–CH=CH–CO_2H

pyrrolidine (NH) + H–$CH(CO_2R^1)_2$ ⇌ (deprotonation) pyrrolidinium ($\overset{+}{N}H_2$) + $^-CH(CO_2R^1)_2$

pyrrolidine (NH) + R–CHO ⇌ (iminium ion formation) R–CH(OH)–N(pyrrolidine) ⇌ R–CH=N^+(pyrrolidine) + $(CO_2R^1)_2\overset{-}{C}H$

⇌ (H^+) R–CH(H–N^+ pyrrolidine)–CH(CO_2R^1)$_2$ + pyrrolidine (NH) ⇌ R–CH=C(CO_2R^1)(CO_2R_1) + $^-$OH, $^-$OH ⇌ (hydrolysis)

R–CH=C[C(HO)(O^-)–OR^1][C(O^-)(HO)–OR^1] ⇌ R–CH=C(CO_2^-)$_2$ → (H^+ workup) R–CH=C(CO_2H)(C(=O)–OH)

→ (decarboxylation) CO_2↑ + R–CH=C=C(OH)–OH (H^+) → R–CH=CH–CO_2H

References

1. Knoevenagel, E. *Ber.* **1898**, *31*, 2596.
2. Jones, G. *Org. React.* **1967**, *15*, 204.
3. Van der Baan, J. L.; Bickelhaupt, F. *Tetrahedron* **1974**, *30*, 2088.
4. Green, B.; Khaidem, I. S.; Crane, R. I.; Newaz, S. S. *ibid.* **1975**, *32*, 2997.
5. Angeletti, E.; Canepa, C.; Martinetti, G.; Venturello, P. *J. Chem. Soc., Perkin Trans. 1* **1989**, 105.
6. Paquette, L. A.; Kern, B. E.; Mendez-Andino, J. *Tetrahedron Lett.* **1999**, *40*, 4129.
7. Balalaie, S.; Nemati, N. *Synth. Commun.* **2000**, *30*, 869.
8. Kim, P.; Olmstead, M. M.; Nantz, M. H.; Kurth, M. *Tetrahedron Lett.* **2000**, *41*, 4029.
9. Siebenhaar, B.; Casagrande, B.; Studer, M.; Blaser, H.-U. *Can. J. Chem.* **2001**, *79*, 566.

Knorr pyrrole synthesis

R, O, CO_2Et, R^1, NH_2, O, R^2, +, R, CO_2Et, R^1, N, H, R^2

B:, H, :B, HO, OH

References

1. Knorr, L. *Ber.* **1884**, *17*, 1635.
2. Hort, E. V.; Anderson, L. R. *Kirk-Othmer Encycl. Chem. Technol.*, 3rd Ed. **1982**, *19*, 499.
3. Jones, R. A.; Rustidge, D. C.; Cushman, S. M. *Synth. Commun.* **1984**, *14*, 575.
4. Fabiano, E.; Golding, B. T. *J. Chem. Soc., Perkin Trans. 1* **1991**, 3371.
5. Hamby, J. M.; Hodges, J. C. *Heterocycles* **1993**, *35*, 843.
6. Alberola, A.; Ortega, A. G.; Sadaba, M. L.; Sanudo, C. *Tetrahedron* **1999**, *55*, 6555.
7. Braun, R. U.; Zeitler, K.; Mueller, Th. J. J. *Org. Lett.* **2001**, *3*, 3297.

Koch carbonylation reaction (Koch–Haaf carbonylation reaction)

Strong acid-catalyzed tertiary carboxylic acid formation from alcohols or olefins and CO.

R, OH, CO, H_2O, H^+, R, CO_2H

H, R, O:, H^+, protonation, R, H, O+, H, E1

R, +, alkyl migration, R, C≡O, +, R, +, O

the tertiary carbocation is thermodynamically favored

R, O+, :OH_2, R, O, $\overset{+}{O}H_2$, – H^+, R, O, OH

References

1. Koch, H.; Haaf, W. *Liebigs Ann. Chem.* **1958**, *618*, 251.
2. Kell, D. R.; McQuillin, F. J. *J. Chem. Soc., Perkin Trans. 1* **1972**, 2096.
3. Norell, J. R. *J. Org. Chem.* **1972**, *37*, 1971.
4. Nambudiry, M. E. N.; Rao, G. S. Krishna. *Tetrahedron Lett.* **1972**, 4707.
5. Booth, B. L.; El-Fekky, T. A. *J. Chem. Soc., Perkin Trans. 1* **1979**, 2441.
6. Langhals, H.; Mergelsberg, I.; Ruechardt, C. *Tetrahedron Lett.* **1981**, *22*, 2365.
7. Farooq, O.; Marcelli, M.; Prakash, G. K. S.; Olah, G. A. *J. Am. Chem. Soc.* **1988**, *110*, 864.
8. Stepanov, A. G.; Luzgin, M. V.; Romannikov, V. N.; Zamaraev, K. I. *ibid.* **1995**, *117*, 3615.
9. Mori, H.; Wada, A.; Xu, Q.; Souma, Y. *Chem. Lett.* **2000**, 136.

10. Olah, G. A.; Prakash, G. K. S.; Mathew, T.; Marinez, E. R. *Angew. Chem., Int. Ed.* **2000**, *39*, 2547.
11. Xu, Q.; Inoue, S.; Tsumori, N.; Mori, H.; Kameda, M.; Tanaka, M.; Fujiwara, M.; Souma, Y. *J. Mol. Catal. A: Chem.* **2001**, *170*, 147.

Koenig–Knorr glycosidation

Formation of the β-glycoside from α-halocarbohydrate under the influence of silver salt.

ROH

Ag_2CO_3

+ AgBr + $CO_2\uparrow$ + H_2O

oxonium ion

β-anomer is favored

β-anomer

References

1. Koenig, W.; Knorr, E. *Ber.* **1901**, *34*, 957.
2. Schmidt, R. R. *Angew. Chem.* **1986**, *98*, 213.
3. Greiner, J.; Milius, A.; Riess, J. G. *Tetrahedron Lett.* **1988**, *29*, 2193.
4. Smith, A. B., III; Rivero, R. A.; Hale, K. J.; Vaccaro, H. A. *J. Am. Chem. Soc.* **1991**, *113*, 2092.
5. Li, H.; Li, Q.; Cai, M.-S.; Li, Z.-J. *Carbohydr. Res.* **2000**, *328*, 611.
6. Fürstner, A.; Radkowski, K.; Grabowski, J.; Wirtz, C.; Mynott, R. *J. Org. Chem.* **2000**, *65*, 8758.
7. Josien-Lefebvre, D.; Desmares, G.; Le Drian, C. *Helv. Chim. Acta* **2001**, *84*, 890.

Kolbe–Schmitt reaction

$O^- Na^+$

CO_2

pressure

OH

$CO_2^- Na^+$

H^+

OH

CO_2H

$O^- Na^+$

CO_2

complexation

Na^+

O

O

O

nucleophilic

addition

O

$O^- Na^+$

O

H

aromatization

OH

$CO_2^- Na^+$

H^+

workup

OH

CO_2H

References

1. Kolbe, H. *Liebigs Ann. Chem.* **1860**, *113*, 1125.
2. Schmitt, R. *J. Prakt. Chem.* **1885**, *31*, 397.
3. Lindsey, A. S.; Jeskey, H. *Chem. Rev.* **1957**, *57*, 583.
4. Kunert, M.; Dinjus, E.; Nauck, M.; Sieler, J. *Ber.* **1997**, *130*, 1461.
5. Kosugi, Y.; Takahashi, K. *Stud. Surf. Sci. Catal.* **1998**, *114*, 487.
6. Kosugi, Y.; Rahim, M. A.; Takahashi, K.; Imaoka, Y.; Kitayama, M. *Appl. Organomet. Chem.* **2000**, *14*, 841.

Kostanecki reaction

1 → 2 + 3

Also known as **Kostanecki–Robinson reaction**. Transformation **1**→**2** represents an **Allan–Robinson reaction** (see page 3), whereas **1**→**3** is a **Kostanecki (acylation) reaction**:

acyl transfer

enolization

6-*exo-trig* ring closure

$- H_2O$

References

1. von Kostanecki, S.; Rozycki, A. *Ber.* **1901**, *34*, 102.
2. Cook, D.; McIntyre, J. S. *J. Org. Chem.* **1968**, *33*, 1746.
3. Szell, T.; Dozsai, L.; Zarandy, M.; Menyharth, K. *Tetrahedron* **1969**, *25*, 715.
4. Pardanani, N. H.; Trivedi, K. N. *J. Indian Chem. Soc.* **1972**, *49*, 599.
5. Ahluwalia, V. K. *Indian J. Chem., Sect. B* **1976**, *14B*, 682.

6. Looker, J. H.; McMechan, J. H.; Mader, J. W. *J. Org. Chem.* **1978**, *43*, 2344.
7. Iyer, P. R.; Iyer, C. S. R.; Prasad, K. J. R. *B* **1983**, *22B*, 1055.
8. Flavin, M. T.; Rizzo, J. D.; Khilevich, A.; Kucherenko, A.; Sheinkman, A. K.; Vilaychack, V.; Lin, L.; Chen, W.; Mata, E.; Greenwood, E. M.; Pengsuparp, T.; Pezzuto, J. M.; Hughes, S. H.; Flavin, T. M.; Cibulski, M.; Boulanger, W. A.; Shone, R. L.; Xu, Z-Q. *J. Med. Chem.* **1996**, *39*, 1303.

Krapcho decarboxylation

Nucleophilic decarboxylation of β-ketoesters, malonate esters, α-cyanoesters, and α-sulfonylesters.

NaCl, DMSO; + $CH_3Cl\uparrow$ + $CO_2\uparrow$

$\longrightarrow$ $CH_3Cl\uparrow$ + $CO_2\uparrow$ + ; ^{-}Cl; H^+

References

1. Krapcho, A. P.; Glynn, G. A.; Grenon, B. J. *Tetrahedron Lett.* **1967**, 215.
2. Flynn, D. L.; Becker, D. P.; Nosal, R.; Zabrowski, D. L. *ibid.* **1992**, *33*, 7283.
3. Martin, C. J.; Rawson, D. J.; Williams, J. M. J. *Tetrahedron: Asymmetry* **1998**, *9*, 3723.

Kröhnke reaction (pyridine synthesis)

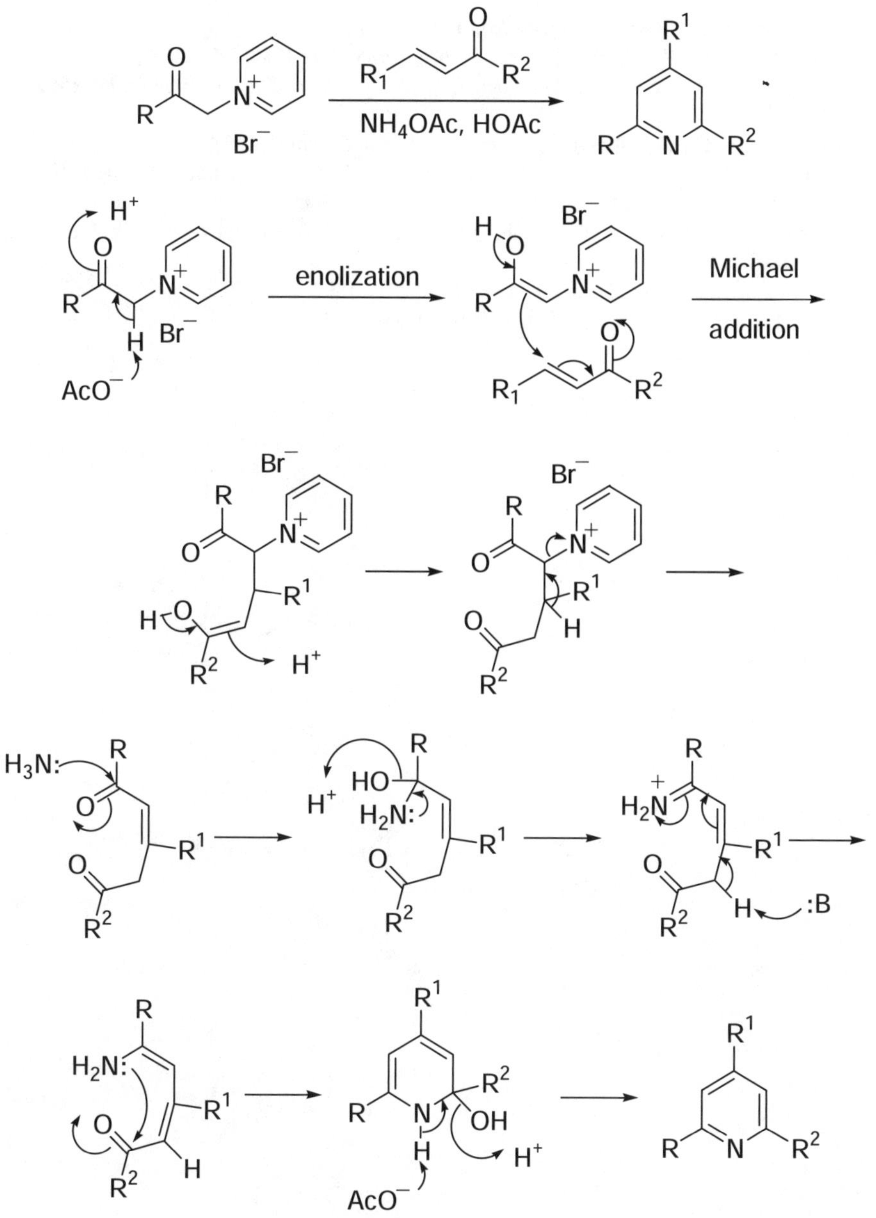

NH_4OAc, HOAc

enolization

Michael addition

References

1. Zecher, W.; Kröhnke, F. *Ber.* **1961**, *94*, 690.
2. Kröhnke, F. *Synthesis* **1976**, 1.
3. Constable, E. C.; Lewis, J. *Tetrahedron* **1982**, *1*, 303.
4. Constable, E. C.; Ward, M. D.; Corr, J. *Inorg. Chim. Acta* **1988**, *141*, 201.
5. Constable, E. C.; Ward, M. D.; Tocher, D. A. *J. Chem. Soc., Dalton Trans.* **1991**, 1675.
6. Constable, E. C.; Chotalia, R. *J. Chem. Soc., Chem. Commun.* **1992**, 65.
7. Markovac, A.; Ash, A. B.; Stevens, C. L.; Hackley, B. E., Jr.; Steinberg, G. M. *J. Heterocycl. Chem.* **1977**, *14*, 19.
8. Chatterjea, J. N.; Shaw, S. C.; Singh, J. N.; Singh, S. N. *Indian J. Chem., Sect. B* **1977**, *15B*, 430.
9. Kelly, T. R.; Lee, Y.-J.; Mears, R. J. *J. Org. Chem.* **1997**, *62*, 2774.
10. Bark, T.; Von Zelewsky, A. *Chimia* **2000**, *54*, 589.

Kumada cross-coupling reaction

$$\text{R–X} + \text{R}^1\text{–MgX} \xrightarrow{\text{Pd(0)}} \text{R–R}^1 + \text{MgX}_2$$

$$\text{R–X} + \text{L}_2\text{Pd(0)} \xrightarrow[\text{addition}]{\text{oxidative}} \text{R(L)Pd(L)X} \xrightarrow[\text{transmetallation, isomerization}]{\text{R}^1\text{–MgX}}$$

$$\text{MgX}_2 + \text{L}_2\text{Pd(R)(R}^1) \xrightarrow[\text{elimination}]{\text{reductive}} \text{R–R}^1 + \text{L}_2\text{Pd(0)}$$

The Kumada cross-coupling reaction (also occasionally known as the Kharasch cross-coupling reaction) is a nickel- or palladium-catalyzed cross-coupling reaction of a Grignard reagent with an organic halide, triflate, *etc.* Along with Negishi, Stille, Hiyama, and Suzuki cross-coupling reactions, they belong to the same category of Pd-catalyzed cross-coupling reactions of organic halides, triflates and other electrophiles with organometallic reagents. These reactions follow a general mechanistic cycle as shown on the next page. There are slight variations for the Hiyama and Suzuki reactions, for which an additional activation step is required for the transmetallation to occur.

The catalytic cycle:

$$\text{L}_n\text{Pd(II)} + \text{R}^1\text{M} \xrightarrow{\text{transmetallation}} \text{L}_n\text{Pd(II)(R}^1)_2$$

$$\xrightarrow[\text{elimination}]{\text{reductive}} \text{R}^1\text{–R}^1 + \text{L}_n\text{Pd(0)}$$

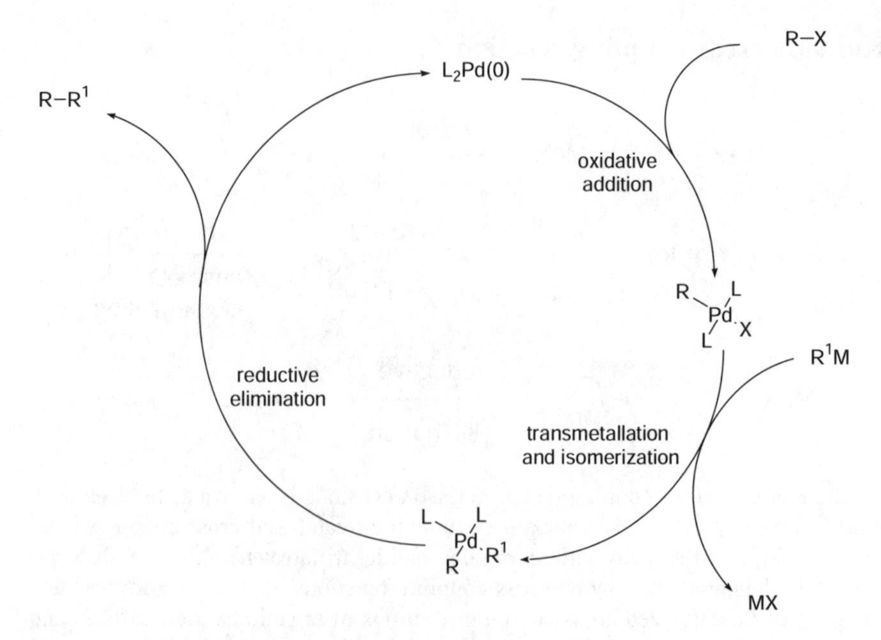

References

1. Tamao, K.; Sumitani, K.; Kiso, Y.; Zembayashi, M.; Fujioka, A.; Kodma, S.-i.; Nakajima, I.; Minato, A.; Kumada, M. *Bull. Chem. Soc. Jpn.* **1976**, *49*, 1958.
2. Wright, M. E.; Jin, M. J. *J. Organomet. Chem.* **1990**, *387*, 373.
3. Kalinin, V. N. *Synthesis* **1992**, 413.
4. Stanforth, S. P. *Tetrahedron* **1998**, *54*, 263.
5. Park, M.; Buck, J. R.; Rizzo, C. J. *Tetrahedron* **1998**, *54*, 12707.
6. Huang, J.; Nolan, S. P. *J. Am. Chem. Soc.* **1999**, *121*, 9889.
7. Lipshutz, B. H.; Tomioka, T.; Blomgren, P. A.; Sclafani, J. A. *Inorg. Chim. Acta* **1999**, *296*, 164.
8. Uenishi, J.; Matsui, K. *Tetrahedron Lett.* **2001**, *42*, 4353.

Larock indole synthesis

Indole synthesis using the palladium-catalyzed coupling reaction of an *o*-iodoaniline with a propargyl alcohol.

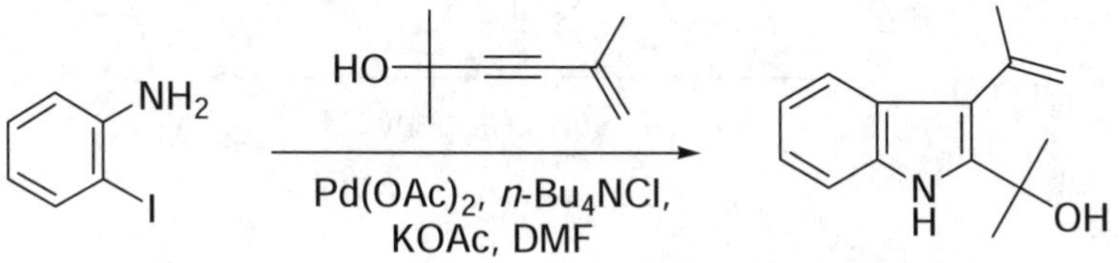

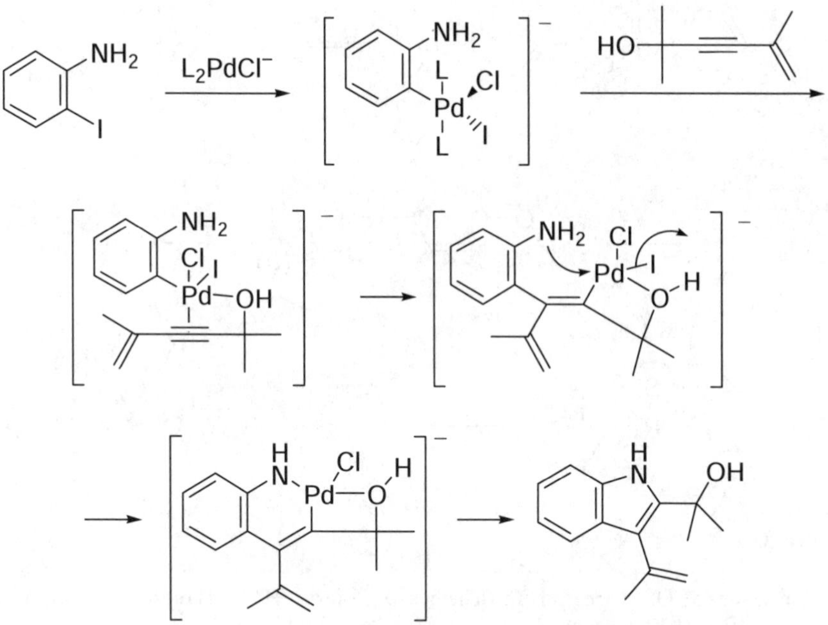

References

1. Larock, R. C.; Yum, E. K. *J. Am. Chem. Soc.* **1991**, *113,* 6689.
2. Larock, R. C.; Yum, E. K.; Refvik, M. D. *J. Org. Chem.* **1998**, *63,* 7652.
3. Larock, R. C. *J. Organomet. Chem.* **1999**, *576,* 111.

Lawesson's reagent

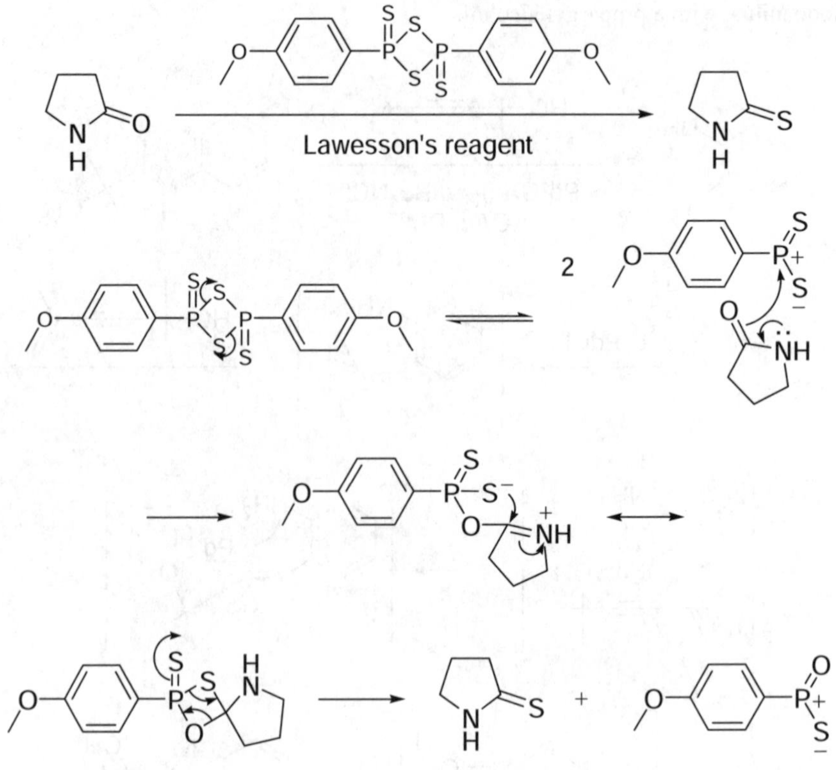

Lawesson's reagent

2

References

1. Lawesson, S. O.; Perregaad, J.; Scheibye, S.; Meyer, H. J.; Thomsen, I. *Bull. Soc. Chim. Belg.* **1977**, *86,* 679.
2. Navech, J.; Majoral, J. P.; Kraemer, R. *Tetrahedron Lett.* **1983**, *24*, 5885.
3. Cava, M. P.; Levinson, M. I. *Tetrahedron* **1985**, *41*, 5061.
4. Luheshi, A. B. N.; Smalley, R. K.; Kennewell, P. D.; Westwood, R. *Tetrahedron Lett.* **1990**, *31*, 123.
5. Luo, Y.; He, L.; Ding, M.; Yang, G.; Luo, A.; Liu, X.; Wu, T. *Heterocycl. Commun.* **2001**, *7*, 37.

Leuckart–Wallach reaction

$$R^1R^2C{=}O + HNR^3R^4 \xrightarrow{HCO_2H} R^1R^2CH{-}NR^3R^4 + CO_2\uparrow + H_2O$$

formic acid is the reducing reagent here

$- H_2O$

HCO_2H

iminium ion intermediate

$- CO_2$

$- H^+$

reduction

References

1. Leuckart, R. *Ber.* **1885**, *18*, 2341.
2. Wallach, O. *Liebigs Ann. Chem.* **1892**, *272*, 99.
3. Doorenbos, N. J.; Solomons, W. E. *Chem. Ind.* **1970**, 1322.
4. Ito, K.; Oba, H.; Sekiya, M. *Bull. Chem. Soc. Jpn.* **1976**, *49*, 2485.
5. Musumarra, G.; Sergi, C. *Heterocycles* **1994**, *37*, 1033.

Lieben haloform reaction

Iodoform, a yellow precipitate in water, is often used for detection of methyl ketones.

$$RCOCH_3 \xrightarrow[NaOH]{X_2} RCOO^- \, Na^+ + CHX_3 + NaX \qquad X = Br, I$$

References

1. Lieben, A. *Liebigs Ann. Chem.* **1870**, *Suppl. 7*, 218.
2. Rothenberg, G.; Sasson, Y. *Tetrahedron* **1996**, *52*, 13641.
3. Tietze, L. F.; Voss, E.; Hartfiel, U. *Org. Synth.* **1990**, *69*, 238.

Liebeskind–Srogl coupling

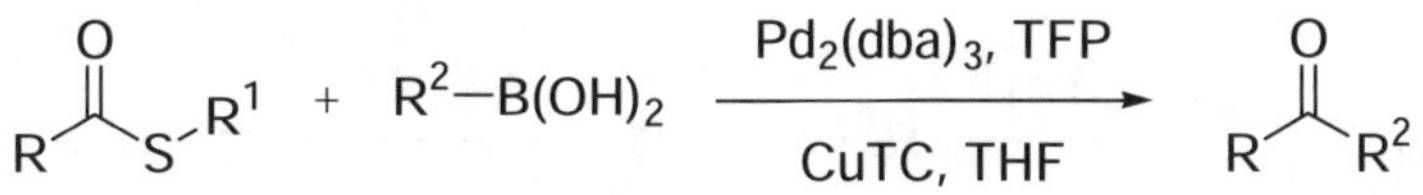

$Pd_2(dba)_3$, TFP

CuTC, THF

TFP = tris(2-furyl)phosphine, CuTC = copper(I) thiophene-2-carboxylate

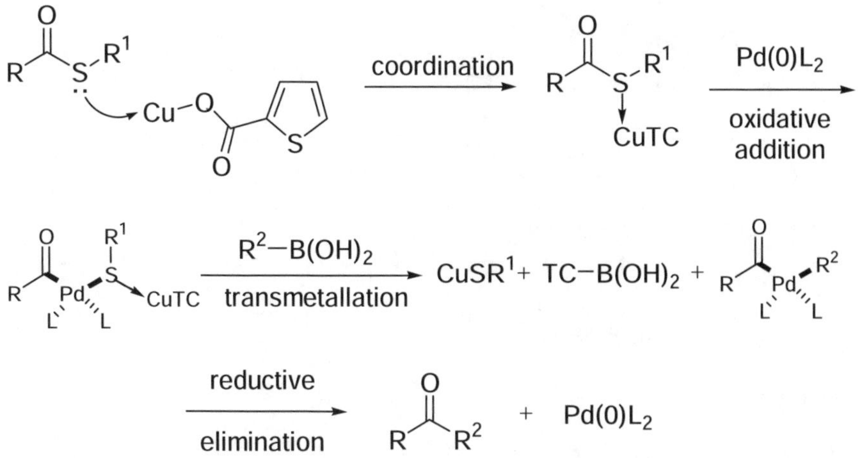

coordination

$Pd(0)L_2$

oxidative addition

R^2–$B(OH)_2$

transmetallation

$CuSR^1$ + TC–$B(OH)_2$ +

reductive elimination

+ $Pd(0)L_2$

Reference

Liebeskind, L. S.; Srogl, J. *J. Am. Chem. Soc.* **2000**, *122,* 11260.

Lossen rearrangement

$$R^1C(O)NHOC(O)R^2 \xrightarrow{^-OH} R^1-N=C=O \xrightarrow{H_2O} R^1-NH_2 + CO_2\uparrow$$

$$R^1C(O)N(H)OC(O)R^2 + {}^-OH \longrightarrow R^1C(O)\bar{N}OC(O)R^2 \longrightarrow$$

$$R^2CO_2^- + R^1-N=C=O \xrightarrow{:OH_2} R^1NHC(O)OH \xrightarrow{:B} R^1-NH_2 + CO_2\uparrow$$

isocyanate intermediate

References

1. Lossen, W. *Ann.* **1872**, *161*, 347.
2. Bauer, L.; Exner, O. *Angew. Chem.* **1974**, *86*, 419.
3. Lipczynska-Kochany, E. *Wiad. Chem.* **1982**, *36*, 735.
4. Casteel, D. A.; Gephart, R. S.; Morgan, T. *Heterocycles* **1993**, *36*, 485.
5. Zalipsky, S. *Chem. Commun.* **1998**, 69.
6. Anilkumar, R.; Chandrasekhar, S.; Sridhar, M. *Tetrahedron Lett.* **2000**, *41*, 5291.
7. Needs, P. W.; Rigby, N. M.; Ring, S. G.; MacDougall, A. *Carbohydr. Res.* **2001**, *333*, 47.

Luche reduction

1,2-Reduction of enones using $NaBH_4$–$CeCl_3$.

O $\xrightarrow[CeCl_3]{NaBH_4}$ OH

$NaBH_4$ + $CeCl_3$ ⟶ $HCeCl_2$ ⟶ O

O $CeCl_2$ H $\xrightarrow[\text{addition}]{\text{1,2-hydride}}$ O $CeCl_2$ $\xrightarrow{\text{workup}}$ OH

References

1. Li, K.; Hamann, L. G.; Koreeda, M. *Tetrahedron Lett.* **1992**, *33*, 6569.
2. Cook, G. P.; Greenberg, M. M. *J. Org. Chem.* **1994**, *59*, 4704.
3. Hutton, G.; Jolliff, T.; Mitchell, H.; Warren, S. *Tetrahedron Lett.* **1995**, *36*, 7905.
4. Moreno-Dorado, F. J.; Guerra, F. M.; Aladro, F. J.; Bustamante, J. M.; Jorge, Z. D.; Massanet, G. M. *Tetrahedron* **1999**, *55*, 6997.
5. Barluenga, J.; Fananas, F. J.; Sanz, R.; Garcia, F.; Garcia, N. *Tetrahedron Lett.* **1999**, *40*, 4735.
6. Haukaas, M. H.; O'Doherty, G. A. *Org. Lett.* **2001**, *3*, 401.

McFadyen–Stevens reduction

Treatment of acylbenzenesulfonylhydrazines with base delivers the corresponding aldehydes.

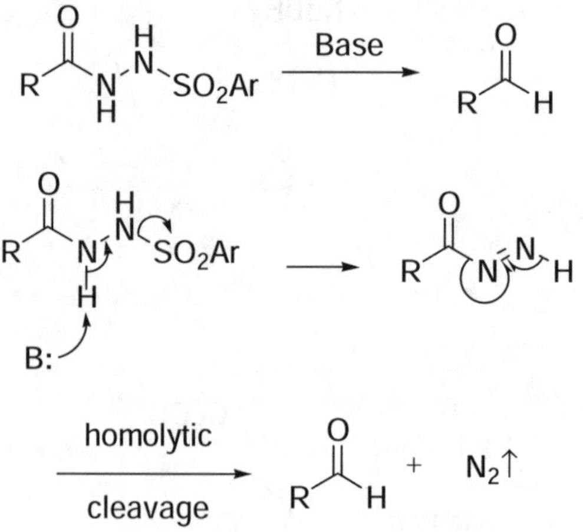

References

1. Babad, H.; Herbert, W.; Stiles, A. W. *Tetrahedron Lett.* **1966**, 2927.
2. Graboyes, H.; Anderson, E. L.; Levinson, S. H.; Resnick, T. M. *J. Heterocycl. Chem.* **1975**, *12*, 1225.
3. Eichler, E.; Rooney, C. S.; Williams, H. W. R. *ibid.* **1976**, *13*, 841.
4. Nair, M.; Shechter, H. *J. Chem. Soc., Chem. Commun.* **1978**, 793.
5. Dudman, C. C.; Grice, P.; Reese, C. B. *Tetrahedron Lett.* **1980**, *21*, 4645.
6. Manna, R. K.; Jaisankar, P.; Giri, Venkatachalam S. *Synth. Commun.* **1998**, *28*, 9.

McLafferty fragmentation

Intramolecular fragmentation of carbonyls in mass spectra.

1,5-hydrogen atom abstraction

Norrish type II cleavage

References

1. McLafferty, F. W. *Anal. Chem.* **1956**, *28*, 306.
2. Gilpin, J. A.; McLafferty, F. W. *Anal. Chem.* **1957**, *29*, 990.
3. Zollinger, M.; Seibl, J. *Org. Mass Spectrom.* **1985**, *20*, 649.
4. Kingston, D. G. I.; Bursey, J. T.; Bursey, M. M. *Chem. Rev.* **1974**, *74*, 215.
5. Budzikiewicz, H.; Bold, P. *Org. Mass Spectrom.* **1991**, *26*, 709.
6. Stringer, M. B.; Underwood, D. J.; Bowie, J. H.; Allison, C. E.; Donchi, K. F.; Derrick, P. J. *Org. Mass Spectrom.* **1991**, *27*, 270.
7. Lightner, D. A.; Steinberg, F. S.; Huestis, L. D. *J. Mass Spectrom. Soc. Jpn.* **1998**, *46*, 11.
8. Alvarez, R. M.; Fernandez, A. H.; Chioua, M.; Perez, P. R.; Vilchez, N. V.; Torres, F. G. *Rapid Commun. Mass Spectrom.* **1999**, *13*, 2480.
9. Rychlik, M. *J. Mass Spectrom.* **2001**, *36*, 555.

McMurry coupling

Olefination of carbonyls with $TiCl_3/LiAlH_4$.

$$R^1R^2C{=}O \xrightarrow[\text{2. } H_2O]{\text{1. } TiCl_3,\ LiAlH_4} R^1R^2C{=}CR^1R^2 + TiO$$

$$Ti(III)Cl_3 + LiAlH_4 \longrightarrow Ti(0)$$

single electron transfer

homocoupling

radical anion intermediate

oxide-coated titanium surface

References

1. McMurry, J. E.; Fleming, M. P. *J. Am. Chem. Soc.* **1974**, *96*, 4708.
2. McMurry, J. E. *Chem. Rev.* **1989**, *89*, 1513.
3. Ephritikhine, M. *Chem. Commun.* **1998**, 2549.
4. Hirao, T. *Synlett* **1999**, 175.
5. Yamato, T.; Fujita, K.; Tsuzuki, H. *J. Chem. Soc., Perkin Trans. 1* **2001**, 2089.

Madelung indole synthesis

Indoles from the cyclization of 2-(acylamino)-toluene using strong bases.

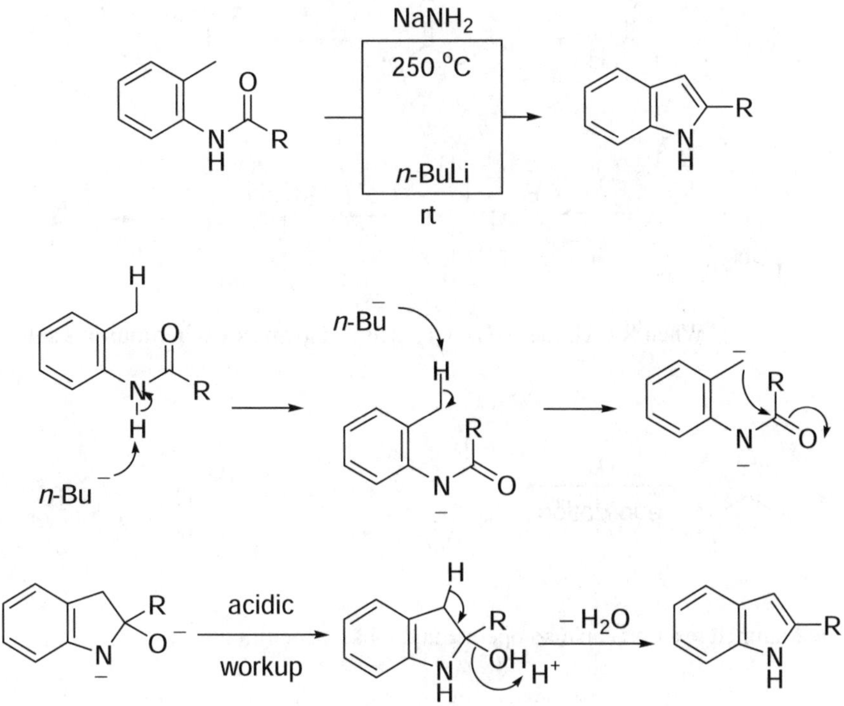

References

1. Madelung, W. *Ber.* **1912**, *45*, 1128.
2. Houlihan, W. J.; Parrino, V. A.; Uike, Y. *J. Org. Chem.* **1981**, *46*, 4511.
3. Houlihan, W. J.; Uike, Y.; Parrino, V. A. *ibid.* **1981**, *46*, 4515.
4. Orlemans, E. O. M.; Schreuder, A. H.; Conti, P. G. M.; Verboom, W.; Reinhoudt, D. N. *Tetrahedron* **1987**, *43*, 3817.
5. Smith, A. B., III; Haseltine, J. N.; Visnick, M. *ibid.* **1989**, *45*, 2431.

Mannich reaction

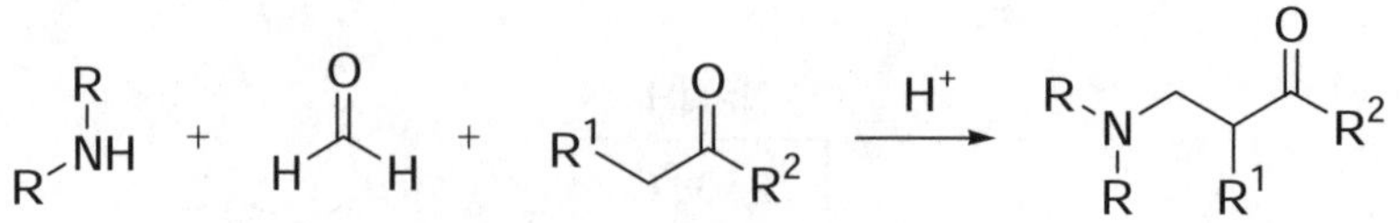

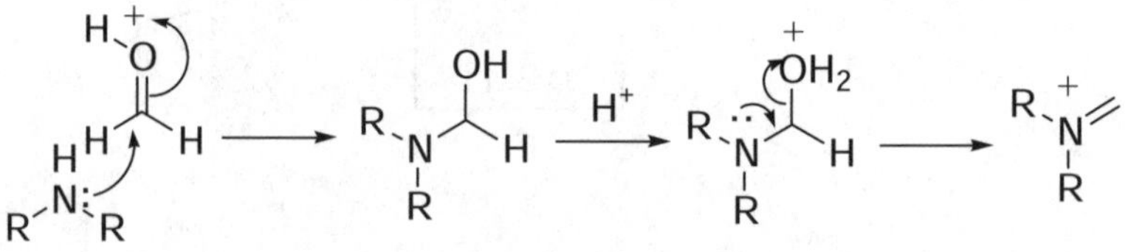

When R = H, the $^{+}NH_2{=}CH_2$ salt is known as **Eschenmoser's salt**

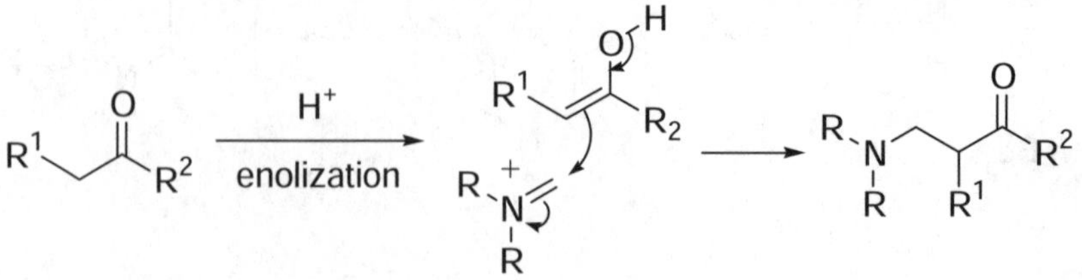

The Mannich reaction can also operate under basic conditions:

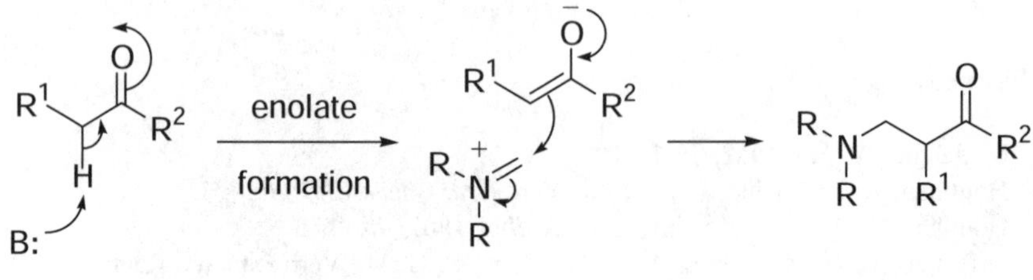

References

1. Mannich, C.; Krosche, W. *Arch. Pharm.* **1912**, *250*, 647.
2. Thompson, B. B. *J. Pharm. Sci.* **1968**, *57*, 715.
3. Bordunov, A. V.; Bradshaw, J. S.; Pastushok, V. N.; Izatt, R. M. *Synlett* **1996**, 933.
4. Arend, M.; Westermann, B.; Risch, N. *Angew. Chem., Int. Ed.* **1998**, *37*, 1045.
5. Padwa, A.; Waterson, A. G. *J. Org. Chem.* **2000**, *65*, 235.
6. List, B. *J. Am. Chem. Soc.* **2000**, *122*, 9336.
7. Schlienger, N.; Bryce, M. R.; Hansen, T. K. *Tetrahedron* **2000**, *56*, 10023.
8. Vicario, J. L.; Badía, D.; Carrillo, L. *Org. Lett.* **2001**, *3*, 773.

Marshall boronate fragmentation

Cf. Grob fragmentation. In fact, Marshall boronate fragmentation belongs to the Grob fragmentation category.

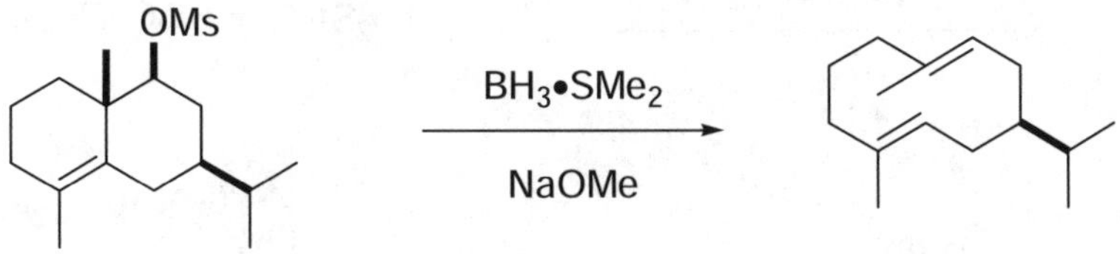

OMs

BH_3•SMe_2

NaOMe

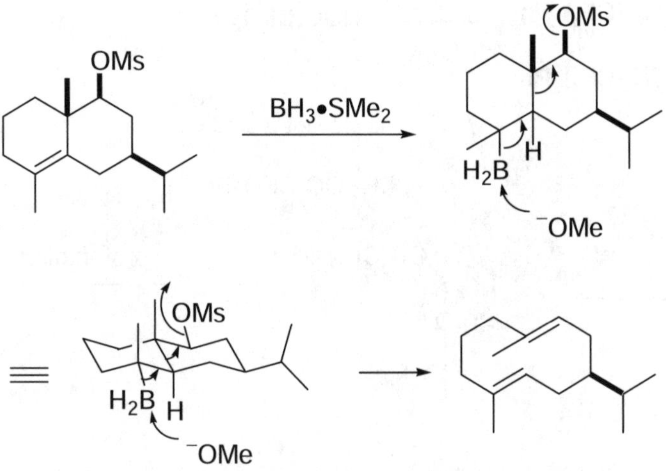

OMs

BH_3•SMe_2

OMs

H_2B H

$^-$OMe

≡

OMs

H_2B H

$^-$OMe

References

1. Marshall, J. A. *Synthesis* **1971**, 229.
2. Minnard, A. J.; Stork, G. A.; Wijinberg, J. B. P. A.; de Groot, A. *J. Org. Chem.* **1997**, *62*, 2344.

Martin's sulfurane dehydrating reagent

Cf. Burgess dehydrating reagent.

OH　$Ph_2S[OC(CF_3)_2Ph]_2$ → + $Ph_2S{=}O$ + $HOC(CF_3)_2Ph$

H–O*t*Bu
$OC(CF_3)_2Ph$
Ph_2S
$OC(CF_3)_2Ph$ ⇌ $HOC(CF_3)_2Ph$ + $Ph_2S^{+}{=}OC(CF_3)_2Ph$, ^{-}O

The alcohol is acidic

→ H–$OC(CF_3)_2Ph$, Ph_2S–$OC(CF_3)_2Ph$, O　protonation →

$^{-}OC(CF_3)_2Ph$ + H, Ph_2S–$O^{+}C(CF_3)_2Ph$, O → Ph_2S^{+}–O, H, $^{-}OC(CF_3$

β-elimination → E1cb　+ $Ph_2S{=}O$ + $HOC(CF_3)_2Ph$

References

1. Martin, J. C.; Arhart, R. J. *J. Am. Chem. Soc.* **1971**, *93*, 2339, 2341.
2. Martin, J. C.; Arhart, R. J. *ibid.* **1971**, *93*, 4327.
3. Tse, B.; Kishi, Y. *J. Org. Chem.* **1994**, *59*, 7807.

Masamune–Roush conditions

Applicable to base-sensitive aldehydes and phosphonates for the Horner–Wadsworth–Emmons reaction

α-keto or α-alkoxycarbonyl phosphonate required

LiCl, CH_3CN

1,8-diazabicyclo[5.4.0]undec-7-ene (DBU)

deprotonation

LiCl

chelation

References

1. Blanchette, M. A.; Choy, W.; Davis, J. T.; Essenfeld, A. P.; Masamune, S.; Roush, W. R.; Sakai, T. *Tetrahedron Lett.* **1984**, *25*, 2183.
2. Rychnovsky, S. D.; Khire, U. R.; Yang, G. *J. Am. Chem. Soc.* **1997**, *119*, 2058.
3. Dixon, D. J.; Foster, A. C.; Ley, S. V. *Org. Lett.* **2000**, *2*, 123.

Meerwein arylation

Arylation of unsaturated compounds by diazonium salts.

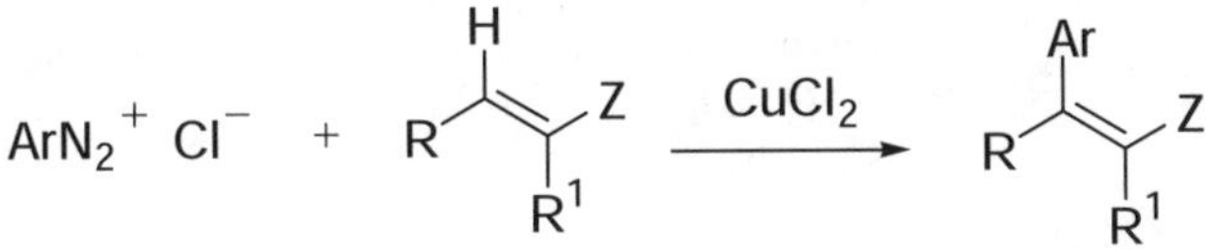

Z = Ar, C=C, C=O, CN, H

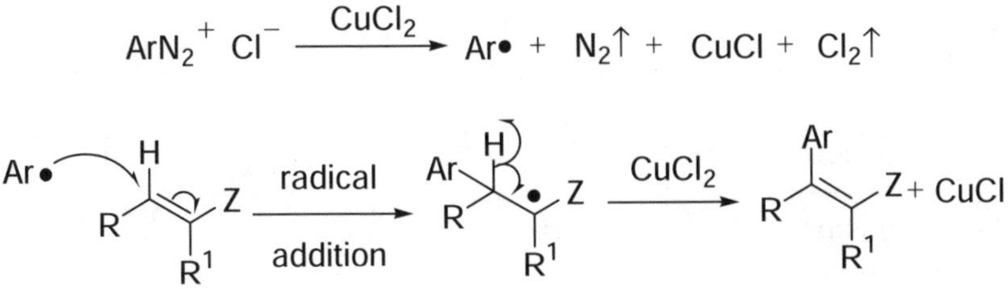

$$ArN_2^+\ Cl^- \xrightarrow{CuCl_2} Ar\bullet + N_2\uparrow + CuCl + Cl_2\uparrow$$

References

1. Meerwein, H.; Buchner, E.; van Emster, K. *J. Prakt. Chem.* **1939**, *152*, 237.
2. Rondestvedt, C. S., Jr. *Org. React.* **1976**, *24*, 225.
3. Raucher, S.; Koolpe, G. A. *J. Org. Chem.* **1983**, *48*, 2066.
4. Sutter, P.; Weis, C. D. *J. Heterocycl. Chem.* **1987**, *24*, 69.
5. Schmidt, A. H.; Schmitt, G.; Diedrich, H. *Synthesis* **1990**, 579.
6. Nock, H.; Schottenberger, H. *J. Org. Chem.* **1993**, *58*, 7045.
7. Takahashi, I.; Muramatsu, O.; Fukuhara, J.; Hosokawa, Y.; Takeyama, N.; Morita, T.; Kitajima, H. *Chem. Lett.* **1994**, 465.
8. Brunner, H.; Bluchel, C.; Doyle, M. P. *J. Organomet. Chem.* **1997**, *541*, 89.
9. Mella, M.; Coppo, P.; Guizzardi, B.; Fagnoni, M.; Freccero, M.; Albini, A. *J. Org. Chem.* **2001**, *66*, 6344.

Meerwein–Ponndorf–Verley reduction

Reduction of ketones to the corresponding alcohols using $Al(O\textit{i}\text{-}Pr)_3$ in isopropanol.

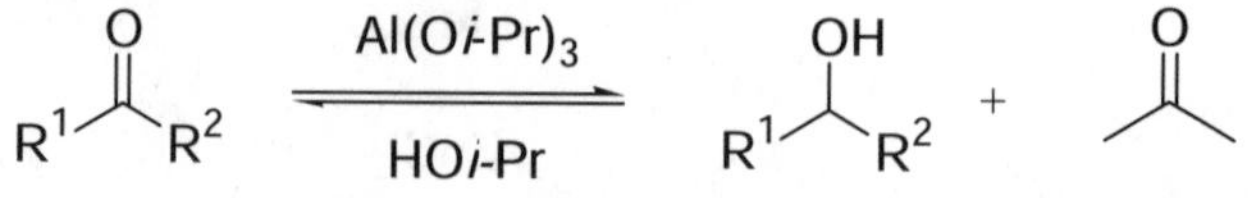

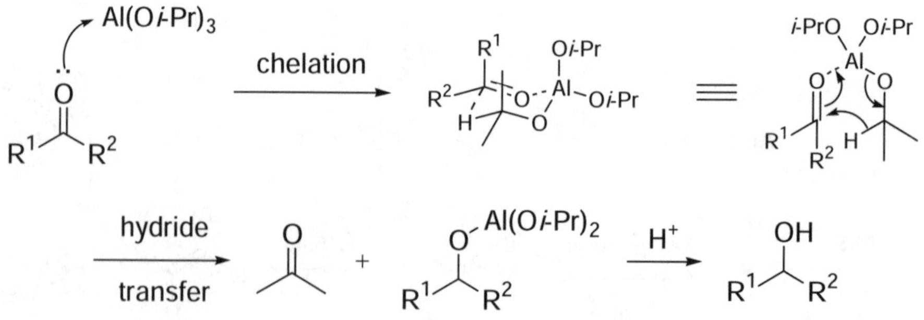

References

1. Meerwein, H.; Schmidt, R. *Liebigs Ann. Chem.* **1925**, *444*, 221.
2. Ashby, E. C. *Acc. Chem. Res.* **1988**, *21*, 414.
3. de Graauw, C. F.; Peters, J. A.; van Bekkum, H.; Huskens, J. *Synthesis* **1994**, 1007.
4. Aremo, N.; Hase, T. *Org. React.* **2001**, *42*, 3637.

Meinwald rearrangement

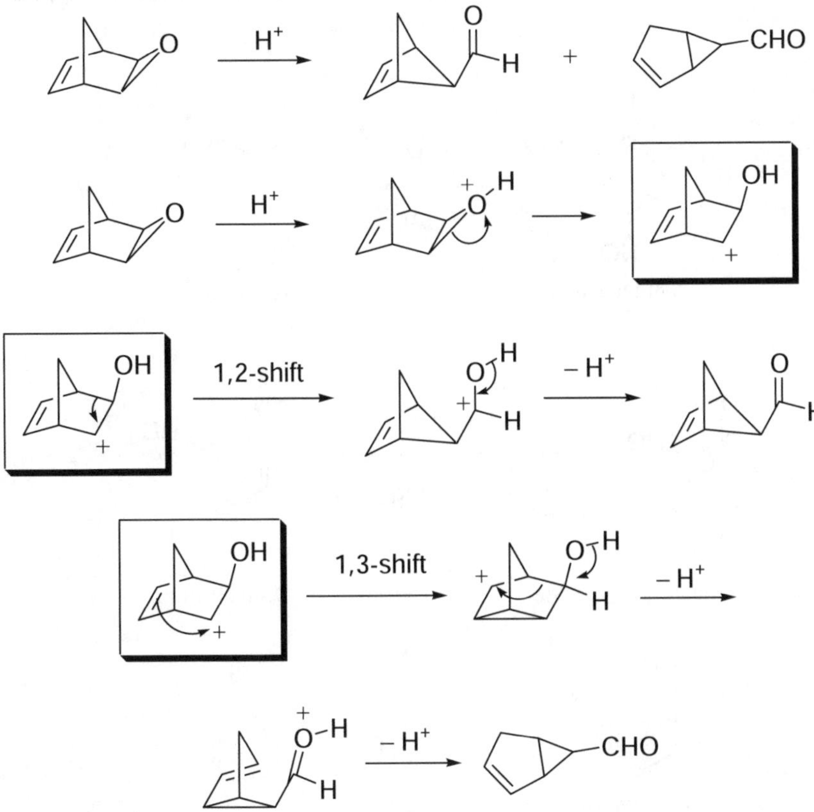

H⁺

1,2-shift

− H⁺

1,3-shift

References

1. Meinwald, J.; Labana, S. S.; Chadha, M. S. *J. Am. Chem. Soc.* **1962**, *85*, 582.
2. Meinwald, J.; Labana, S. S.; Labana, L. L.; Wahl, G. H. Jr. *Tetrahedron Lett.* **1965**, *23*, 1789.
3. Niwayama, S.; Noguchi, H.; Ohno, M.; Kobayashi, S. *ibid.* **1993**, *34*, 665.
4. Niwayama, S.; Kobayashi, S.; Ohno, M. *J. Am. Chem. Soc.* **1994**, *116*, 3290.
5. Kim, W.; Kim, H.; Rhee, H. *Heterocycles* . **2000**, *53*, 219.
6. Rhee, H.; Yoon, D.-O.; Jung, M. E. *Nucleosides, Nucleotides Nucleic Acids* **2000**, *19*, 619.
7. Sun, H.; Yang, J.; Amaral, K. E.; Horenstein, B. A. *Tetrahedron Lett.* **2001**, *42*, 2451.

Meisenheimer complex

Also known as Meisenheimer–Jackson salt, the stable intermediate for certain S_NAr reactions.

RNH_2, S_NAr

Sanger's reagent

S_NAr, slow, rate-determining step

ipso attack

ipso substitution

fast

Meisenheimer complex (**Meisenheimer–Jackson salt**)

The reaction using Sanger's reagent is faster than using the corresponding chloro-, bromo-, and iodo-dinitrobenzene — the fluoro-Meisenheimer complex is the most stabilized because F is the most electron-withdrawing. The reaction rate does not depend upon the leaving ability of the leaving group.

References

1. Meisenheimer, J. *Liebigs Ann. Chem.* **1902**, *323*, 205.
2. Strauss, M. J. *Acc. Chem. Res.* **1974**, *7*, 181.
3. Bernasconi, C. F. *Acc. Chem. Res.* **1978**, *11*, 147.
4. Terrier, F. *Chem. Rev.* **1982**, *82*, 77.
5. Buncel, E.; Dust, J. M.; Manderville, R. A. *J. Am. Chem. Soc.* **1996**, *118*, 6072.

6. Sepulcri, P.; Goumont, R.; Halle, J.-C.; Buncel, E.; Terrier, F. *Chem. Commun.* **1997**, 789.
7. Weiss, R.; Schwab, O.; Hampel, F. *Chem.—Eur. J.* **1999**, *5*, 968.
8. Hoshino, K.; Ozawa, N.; Kokado, H.; Seki, H.; Tokunaga, T.; Ishikawa, T. *J. Org. Chem.* **1999**, *64*, 4572.
9. Adam, W.; Makosza, M.; Zhao, C.-G.; Surowiec, M. *ibid.* **2000**, *65*, 1099.

Meisenheimer rearrangement

[1,2]-sigmatropic rearrangement:

$$R_1R_2N^+(R)O^- \xrightarrow{\Delta} R_1R_2N\text{-}O\text{-}R$$

[2,3]-sigmatropic rearrangement:

$$CH_2{=}CH\text{-}CH_2\text{-}N^+(R)_2O^- \xrightarrow{\Delta} CH_2{=}CH\text{-}CH_2\text{-}O\text{-}NR_2$$

References

1. Meisenheimer, J. *Ber.* **1919**, *52*, 1667.
2. [1,2]-sigmatropic rearrangement, Castagnoli, N. Jr.; Craig, J. C.; Melikian, A. P.; Roy, S. K. *Tetrahedron* **1970**, *26*, 4319.
3. [2,3]-sigmatropic rearrangement, Yamamoto, Y.; Oda, J.; Inouye, Y. *J. Org. Chem.* **1976**, *41*, 303.
4. Johnstone, R. A. W. *Mech. Mol. Migr.* **1969**, *2*, 249.
5. Kurihara, T.; Sakamoto, Y.; Matsumoto, H.; Kawabata, N.; Harusawa, S.; Yoneda, R. *Chem. Pharm. Bull.* **1994**, *42*, 475.
6. Molina, J. M.; El-Bergmi, R.; Dobado, J. A.; Portal, D. *J. Org. Chem.* **2000**, *65*, 8574.
7. Blanchet, J.; Bonin, M.; Micouin, L.; Husson, H.-P. *Tetrahedron Lett.* **2000**, *41*, 8279.

Meyer–Schuster rearrangement

The isomerization of secondary and tertiary α-acetylenic alcohols to α,β-unsaturated carbonyl groups *via* a 1,3-shift. When the acetylenic group is terminal, the products are aldehydes, whereas the internal acetylenes give ketones.
Cf. Rupe rearrangement

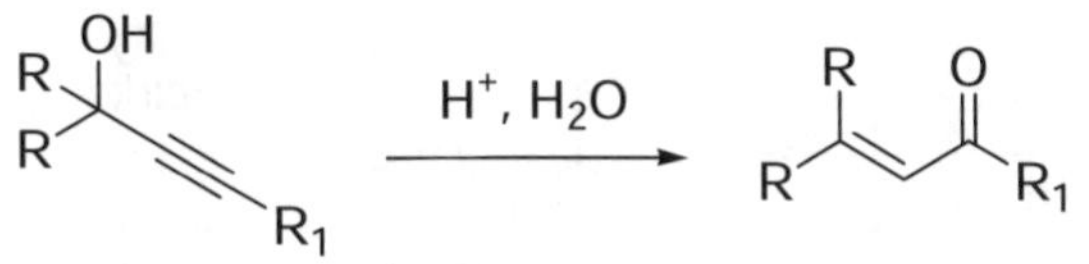

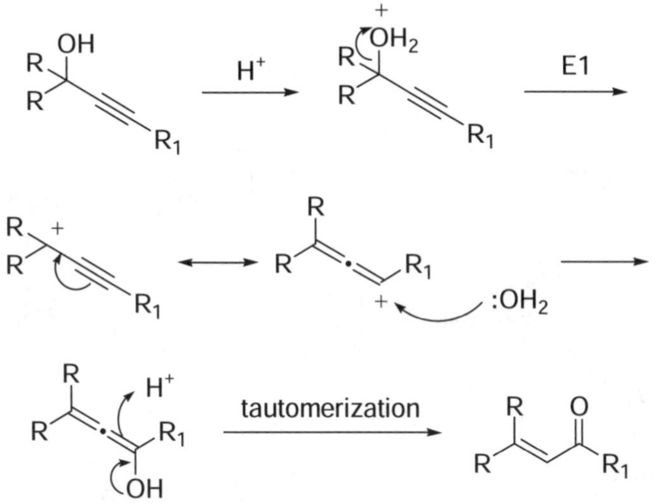

References

1. Swaminathan, S.; Narayanan, K. V. *Chem. Rev.* **1971**, *71*, 429.
2. Edens, M.; Boerner, D.; Chase, C. R.; Nass, D.; Schiavelli, M. D. *J. Org. Chem.* **1977**, *42*, 3403.
3. Cachia, P.; Darby, N.; Mak, T. C. W.; Money, T.; Trotter, J. *Can. J. Chem.* **1980**, *58*, 1172.
4. Andres, J.; Cardenas, R.; Silla, E.; Tapia, O. *J. Am. Chem. Soc.* **1988**, *110*, 666.
5. Tapia, O.; Lluch, J. M.; Cardenas, R.; Andres, J. *ibid.* **1989**, *111*, 829.
6. Omar, E. A.; Tu, C.; Wigal, C. T.; Braun, L. L. *J. Heterocycl. Chem.* **1992**, *29*, 947.
7. Yoshimatsu, M.; Naito, M.; Kawahigashi, M.; Shimizu, H.; Kataoka, T. *J. Org. Chem.* **1995**, *60*, 4798.
8. Lorber, C. Y.; Osborn, J. A. *Tetrahedron Lett.* **1996**, *37*, 853.
9. Chihab-Eddine, A.; Daich, A.; Jilale, A.; Decroix, B. *J. Heterocycl. Chem.* **2000**, *37*, 1543.

Michael addition

Conjugate addition of a carbon-nucleophile to an α,β-unsaturated system.

e.g.:

O
NaCH(CH$_3$O$_2$Et)$_2$
ONa
CO$_2$Et
CO$_2$Et
acidic
workup
O
CO$_2$Et
CO$_2$Et

e.g.:

O
R$_2$CuX
O
R

O
R
Cu
R
conjugate
addition
O
H$^+$
R
workup
O
R

References

1. Michael, A. *J. Prakt. Chem.* **1887**, *35*, 349.
2. Hunt, D. A. *Org. Prep. Proced. Int.* **1989**, *21*, 705.
3. D'Angelo, J.; Desmaele, D.; Dumas, F.; Guingant, Ae. *Tetrahedron: Asymmetry* **1992**, *3*, 459.
4. Hoz, S. *Acc. Chem. Res.* **1993**, *26*, 69.
5. Ihara, M.; Fukumoto, K. *Angew. Chem., Int. Ed. Engl.*, **1993**, *32*, 1010.
6. Itoh, T.; Shirakami, S. *Heterocycles* **2001**, *55*, 37.
7. Cai, C.; Soloshonok, V. A.; Hruby, V. J. *J. Org. Chem.* **2001**, *66*, 1339.
8. Sundararajan, G.; Prabagaran, N. *Org. Lett.* **2001**, *3*, 389.

Michaelis–Arbuzov phosphonate synthesis

General scheme:

$$(R^1O)_3P + R_2{-}X \xrightarrow{\Delta} R_2{-}P(=O)(OR^1){-}OR^1 + R^1{-}X$$

R^1 = alkyl, *etc.*; R_2 = alkyl, acyl, *etc.*; X = Cl, Br, I

e.g.:

$$(CH_3O)_3P + BrCH_2CO_2CH_3 \xrightarrow{\Delta} (CH_3O)_2P(=O)CH_2CO_2CH_3 + CH_3Br\uparrow$$

$$(CH_3O)_3P{:} + BrCH_2CO_2CH_3 \xrightarrow{S_N2} Br^- + (CH_3O)_3P^+CH_2CO_2CH_3$$

$$\xrightarrow{S_N2} (CH_3O)_2P(=O)CH_2CO_2CH_3 + CH_3Br\uparrow$$

References

1. Swaminathan, S.; Narayanan, K. V. *Chem. Rev.* **1971**, *71*, 429.
2. Gellespie, P.; Ramirez, F.; Ugi, I.; Marquarding, D. *Angew. Chem., Int. Ed. Engl.* **1973**, *12*, 91.
3. Bhattacharya, A. K.; Thyagarajan, G. *Chem. Rev.* **1981**, *81*, 415.
4. Waschbűsch, R.; Carran, J.; Marinetti, A.; Savignac, P. *Synthesis* **1997**, 672.
5. Kato, T.; Tejima, M.; Ebiike, H.; Achiwa, K. *Chem. Pharm. Bull.* **1996**, *44*, 1132.
6. Winum, J.-Y.; Kamal, M.; Agnaniet, H.; Leydet, A.; Montero, J.-L. *Phosphorus, Sulfur Silicon Relat. Elem.* **1997**, *129*, 83.
7. Griffith, J. A.; McCauley, D. J.; Barrans, R. E., Jr.; Herlinger, A. W. *Synth. Commun.* **1998**, *28*, 4317.
8. Kiddle, J. J.; Gurley, A. F. *Phosphorus, Sulfur Silicon Relat. Elem.* **2000**, *160*, 195.
9. Bhattacharya, A. K.; Stolz, F.; Schmidt, R. R. *Tetrahedron Lett.* **2001**, *42*, 5393.

Midland reduction

Asymmetric reduction of ketones using Alpine-borane®.
Alpine-borane® = *B*-isopinocampheyl-9-borabicyclo[3.3.1]nonane.

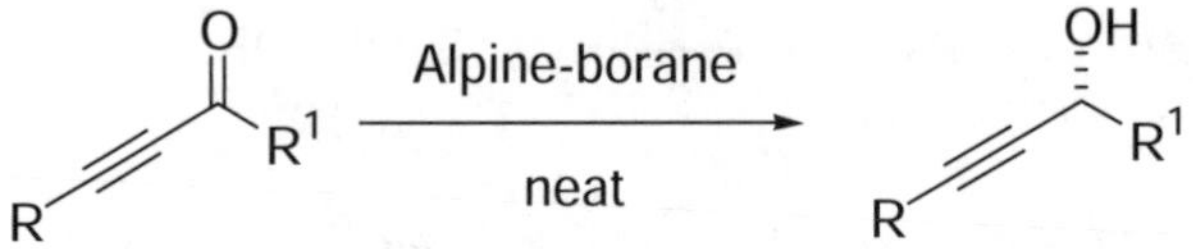

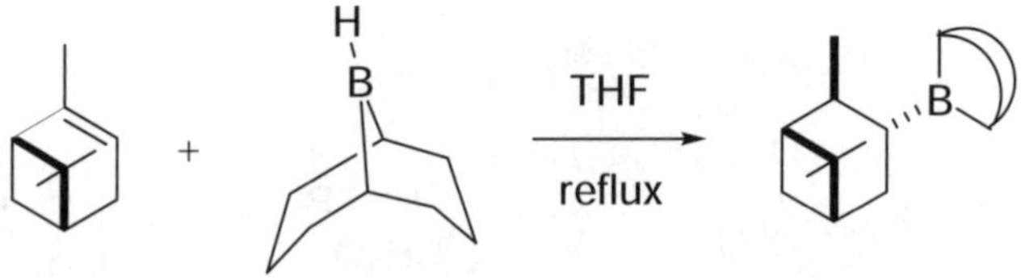

(1*R*)-(+)-α-pinene 9-BBN (*R*)-Alpine-borane

9-BBN = 9-borabicyclo[3.3.1]nonane

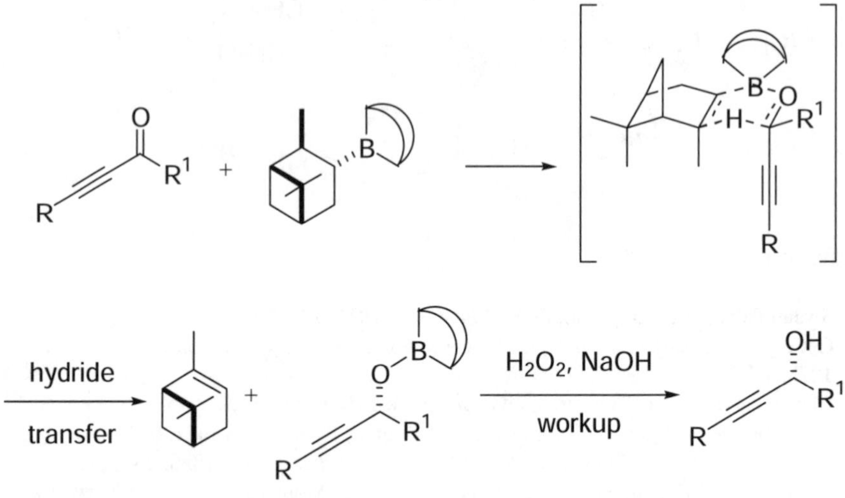

References

1. Midland, M. M.; Tramontano, A.; Zederic, S. A. *J. Am. Chem. Soc.* **1979**, *101*, 2352.
2. Midland, M. M.; McDowell, D. C.; Hatch, R. L.; Tramontano, A. *ibid.* **1980**, *102*, 867.
3. Brown, H. C.; Pai, G. G.; Jadhav, P. K. *ibid.* **1984**, *106*, 1531.
4. Brown, H. C.; Pai, G. G. *J. Org. Chem.* **1982**, *47*, 1606.
5. Midland, M. M.; Tramontano, A.; Kazubski, A.; Graham, R. S. Tsai, D. J. S.; Cardin, D. B. *Tetrahedron* **1984**, *40*, 1371.
6. Singh, V. K. *Synthesis* **1992**, 605.

Miller–Snyder aryl cyanide synthesis

Benzonitriles from *p*-nitrobenzaldehyde and *p*-nitrobenzonitrile.

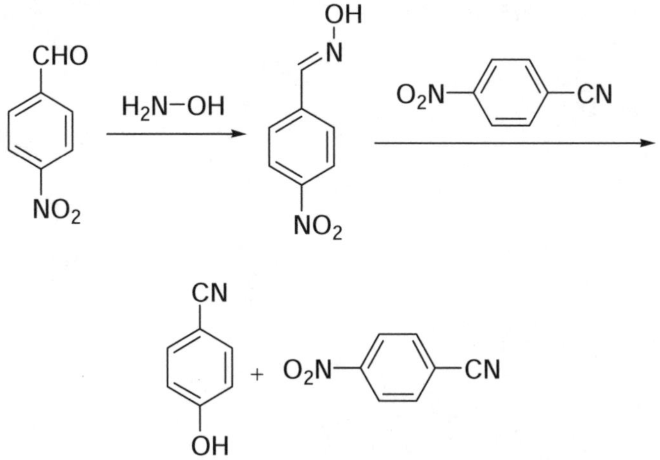

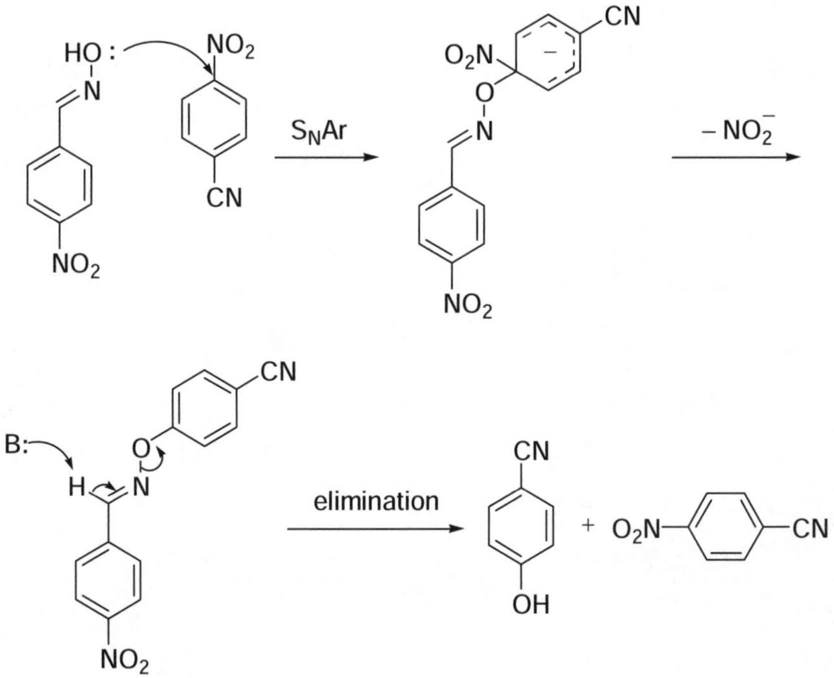

$H_2N{-}OH$

$O_2N{-}C_6H_4{-}CN$

+

S_NAr

$-\ NO_2^-$

elimination

+

References

1. Snyder, M. R. *J. Org. Chem.* **1974**, *39*, 3343.
2. Miller, M. J.; Loudon, G. M. *ibid.* **1975**, *40*, 126.
3. Snyder, M. R. *ibid.* **1975**, *40*, 2879.

Mislow–Evans rearrangement

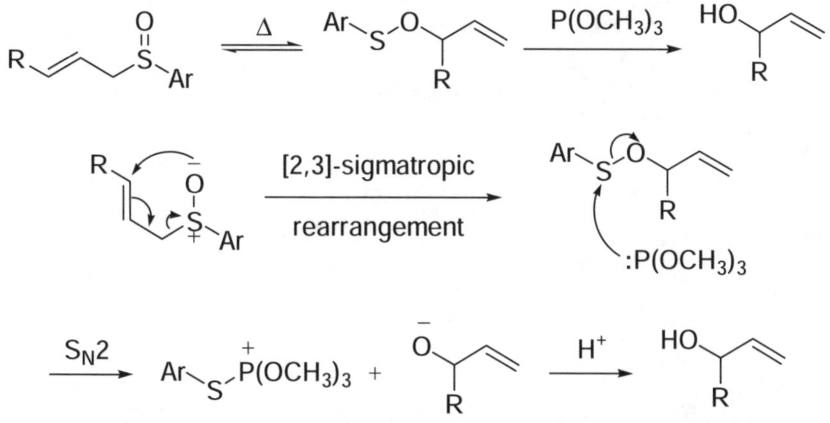

References

1. Tang, R.; Mislow, K. *J. Am. Chem. Soc.* **1970**, *92*, 2100.
2. Evans, D. A.; Andrews, G. C.; Sims, C. L. *ibid.* **1971**, *93*, 4956.
3. Evans, D. A.; Andrews, G. C. *ibid.* **1972**, *94*, 3672.
4. Evans, D. A.; Andrews, G. C. *Acc. Chem. Res.* **1974**, *7*, 147.
5. Masaki, Y.; Sakuma, K.; Kaji, K. *Chem. Pharm. Bull.* **1985**, *33*, 2531.
6. Jones-Hertzog, D. K.; Jorgensen, W. L. *J. Am. Chem. Soc.* **1995**, *117*, 9077.
7. Jones-Hertzog, D. K.; Jorgensen, W. L. *J. Org. Chem.* **1995**, *60*, 6682.
8. Mapp, A. K.; Heathcock, C. H. *ibid.* **1999**, *64*, 23.
9. Zhou, Z. S.; Flohr, A.; Hilvert, D. *ibid.* **1999**, *64*, 8334.

Mitsunobu reaction

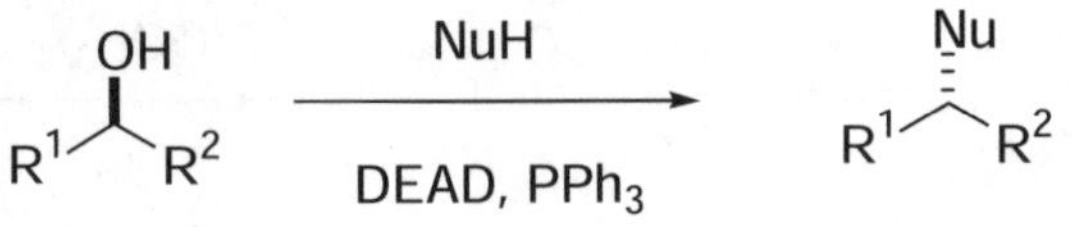

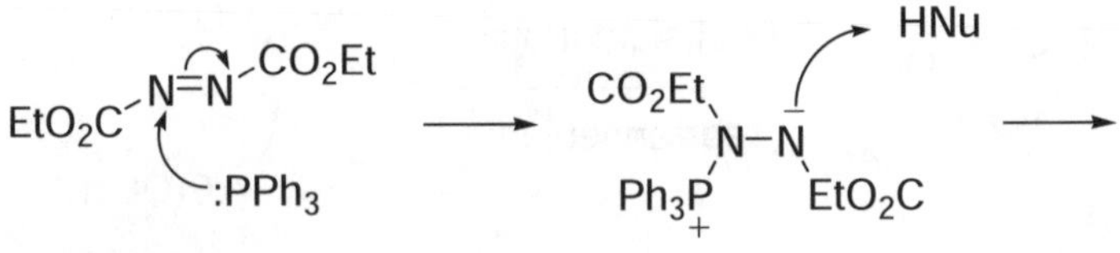

DEAD, diethyl azodicarboxylate

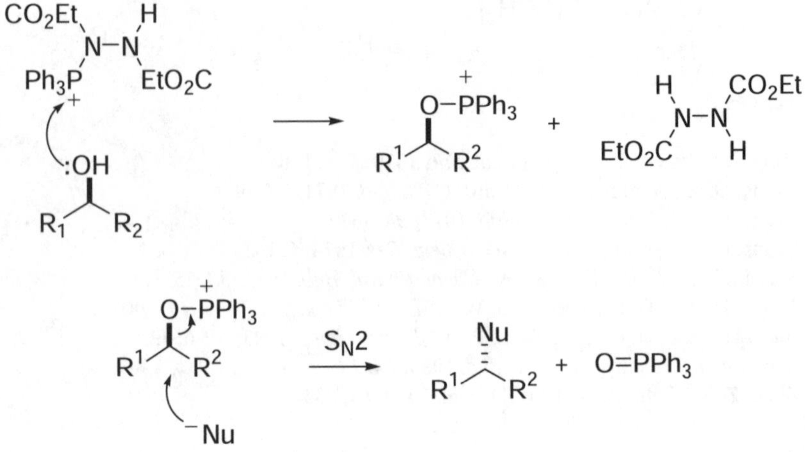

References

1. Mitsunobu, O.; Yamada, M. *Bull. Chem. Soc., Jpn.* **1967**, *40*, 2380.
2. Mitsunobu, O. *Synthesis* **1981**, 1.
3. Hughes, D. L. *Org. Prep. Proc. Int.* **1996**, *28*, 127.
4. Flynn, D. L.; Becker, D. P.; Nosal, R.; Zabrowski, D. L. *Tetrahedron Lett.* **2000**, *41*, 1959.
5. Barrett, A. G. M.; Roberts, R. S.; Schroeder, J. *Org. Lett.* **2000**, *2*, 2999.
6. Racero, J. C.; Macias-Sanchez, A. J.; Hernandez-Galan, R.; Hitchcock, P. B.; Hanson, J. R.; Collado, I. G. *J. Org. Chem.* **2000**, *65*, 7786.
7. Langlois, N.; Calvez, O. *Tetrahedron Lett.* **2000**, *41*, 8285.
8. Charette, A. B.; Janes, M. K.; Boezio, A. A. *J. Org. Chem.* **2001**, *66*, 2178.

Miyaura boration reaction

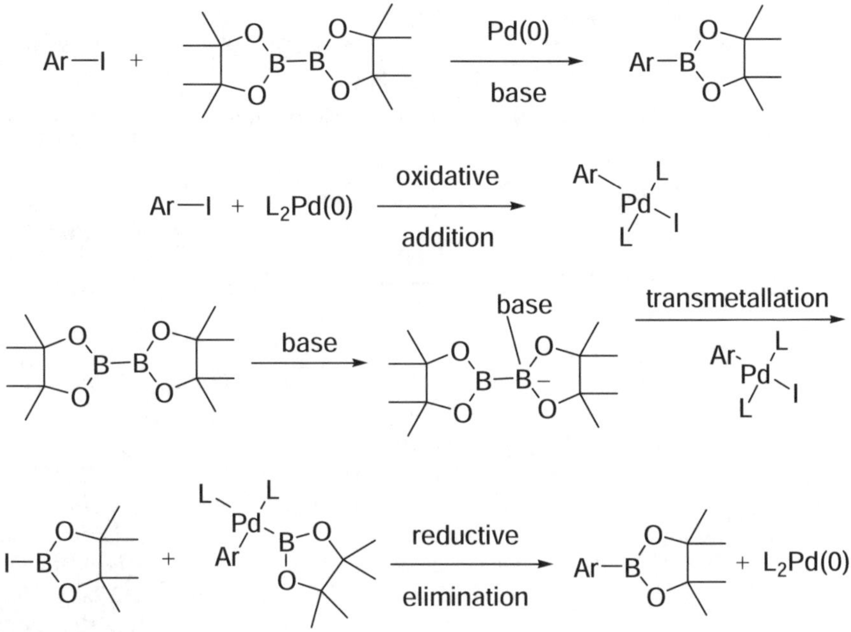

References

1. Ishiyama, T.; Murata, M.; Miyaura, N.; Suzuki, A. *J. Am. Chem. Soc.* **1993**, *115*, 11018.
2. Ishiyama, T.; Murata, M.; Miyaura, N. *J. Org. Chem.* **1995**, *60*, 7508.
3. Carbonnelle, A.-C.; Zhu, J. *Org. Lett.* **2000**, *2*, 3477.
4. Willis, D. M.; Strongin, R. M. *Tetrahedron Lett.* **2000**, *41*, 8683.
5. Takahashi, K.; Takagi, J.; Ishiyama, T.; Miyaura, N. *Chem. Lett.* **2000**, 126.

Moffatt oxidation

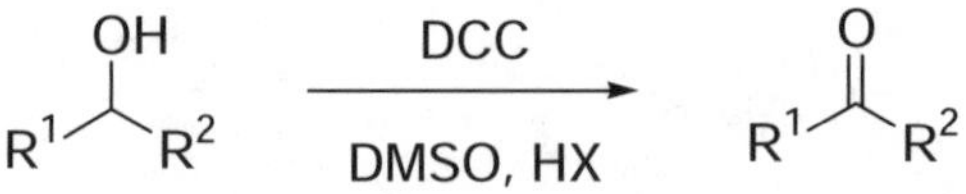

DCC
DMSO, HX

DCC, 1,3-dicyclohexylcarbodiimide

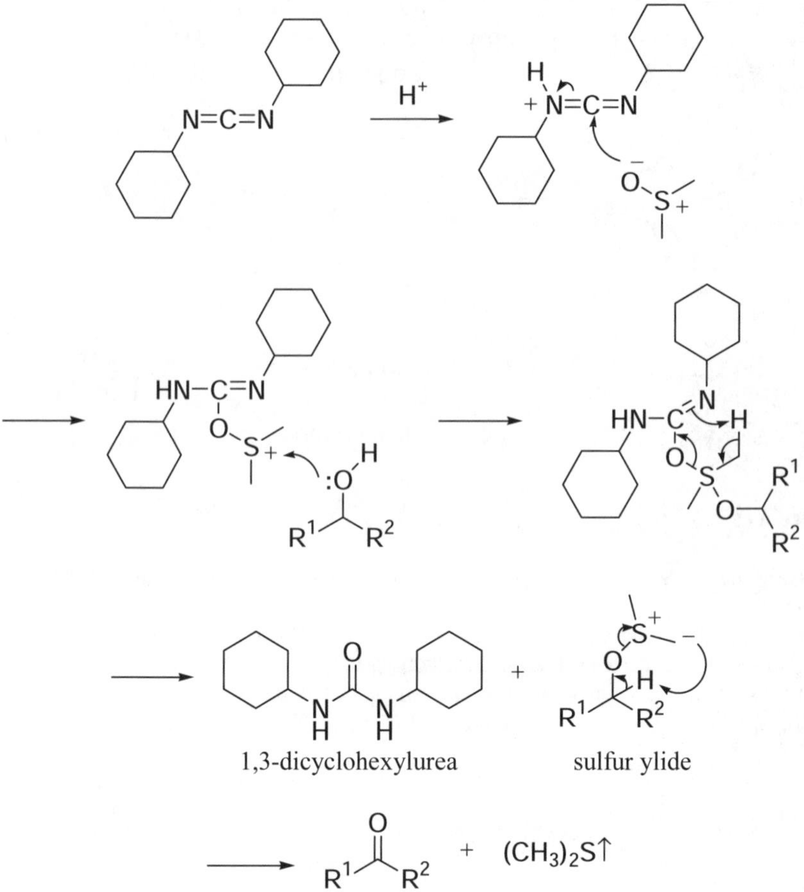

H^+

1,3-dicyclohexylurea

sulfur ylide

$(CH_3)_2S\uparrow$

References

1. Pfitzinger, K. E.; Moffatt, J. G. *J. Am. Chem. Soc.* **1963**, *85*, 3027.
2. Tidwell, T. T. *Org. React.* **1990**, *39*, 297.
3. Krysan, D. J.; Haight, A. R.; Lallaman, J. E.; Langridge, D. C.; Menzia, J. A.; Narayanan, B. A.; Pariza, R. J.; Reno, D. S.; Rockway, T. W.; *et al. Org. Prep. Proced. Int.* **1993**, *25*, 437.

Morgan–Walls reaction (Pictet–Hubert reaction)

Morgan–Walls reaction

R NH O $POCl_3$ R N R^1

O Cl P Cl Cl O NH R R^1 $POCl_2$ O NH R R^1

O_2PCl_2 R NH H R^1 R N R^1

Pictet–Hubert reaction

R NH O $ZnCl_2$ R N R^1

References

1. Pictet, A.; Hubert, A. *Ber.* **1896**, *29*, 1182.
2. Morgan, C. T.; Walls, L. P. *J. Chem. Soc.* **1931**, 2447.

Mori–Ban indole synthesis

$Pd(OAc)_2$, Ph_3P

$NaHCO_3$, DMF, 130 °C,

Reduction of $Pd(OAc)_2$ to Pd(0):

Mori–Ban indole synthesis:

Pd(0)

oxidative addition

insertion

β-hydride elimination

addition of PdH ⇌ β-hydride elimination ⇌

CO_2Me, $PdBrL_n$, N, Ac, H

Regeneration of Pd(0):

$$\text{H–PdBrL}_n + \text{NaHCO}_3 \longrightarrow \text{Pd(0)} + \text{NaBr} + \text{H}_2\text{O} + \text{CO}_2\uparrow$$

References

1. Reduction of $Pd(OAc)_2$ to Pd(0), (a) Amatore C.; Carre, E.; Jutand, A.; M'Barki, M. A.; Meyer, G. *Organometallics* **1995**, *14*, 5605; (b) Amatore C.; Carre, E.; M'Barki, M. A. *ibid.* **1995**, *14*, 1818; (c) Amatore C.; Jutand, A.; M'Barki, M. A. *ibid.* **1992**, *11*, 3009; (d) Amatore C.; Azzabi, M; Jutand, A. *J. Am. Chem. Soc.* **1991**, *113,* 8375.
2. Mori–Ban indole synthesis, (a) Mori, M.; Chiba, K.; Ban, Y. *Tetrahedron Lett.* **1977**, *12*, 1037.; (b) Ban, Y.; Wakamatsu, T.; Mori, M. *Heterocycles* **1977**, *6,* 1711.

Morin rearrangement

Acid-catalyzed conversion of penicillin sulfoxides to cephalosporins. The rearrangement seems to be general for a variety of other heterocyclic sulfoxides as well.

cat. TsOH

xylene, Δ

protonation

ring-opening

sulfenic acid

$- H^+$

References

1. Morin, R. B.; Jackson, B. G.; Mueller, R. A.; Lavagnino, E. R.; Scanlon, W. B.; Andrews, S. L. *J. Am. Chem. Soc.* **1963**, *85*, 1896.
2. Morin, R. B.; Jackson, B. G.; Mueller, R. A.; Lavagnino, E. R.; Scanlon, W. B.; Andrews, S. L. *ibid.* **1969**, *91*, 1401.
3. Morin, R. B.; Spry, D. O. *J. Chem. Soc., Chem. Commun.* **1970**, 335.
4. Gottstein, W. J.; Misco, P. F.; Cheney, L. C. *J. Org. Chem.* **1972**, *37*, 2765.
5. Chen, C. H. *Tetrahedron Lett.* **1976**, 17, 25.

6. Mah, H.; Nam, K. D.; Hahn, H.-G. *J. Heterocycl. Chem.* **1989**, *26,* 1447.
7. Farina, V.; Kant, J. *Synlett* **1994**, 565.
8. Hart, D. J.; Magomedov, N. A. *J. Org. Chem.* **1999**, *64*, 2990.
9. Freed, J. D.; Hart, D. J.; Magomedov, N. A. *ibid.* **2001**, *66*, 839.

Mukaiyama aldol reaction

R–CHO + R^1 (enol silyl ether, R^2, OSiMe$_3$) →(Lewis acid) R–CH(OH)–CH(R^1)–C(O)–R^2

R–CH=O⁺–LA⁻ + enol silyl ether (R^1, R^2, O–SiMe$_3$), X^- → R–CH(OH)–CH(R^1)–C(O)–R^2 + X–SiMe$_3$

Mukaiyama Michael addition

R–CH=CH–C(O)–CH$_3$ + R^1 (enol silyl ether, R^2, OSiMe$_3$) →(Lewis acid) R^2–C(O)–CH(R^1)–CH(R)–CH$_2$–C(O)–CH$_3$

References

1. Mukaiyama, T.; Narasaka, K.; Banno, K. *Chem. Lett.* **1973**, 1011.
2. Mukaiyama, T.; Narasaka, K.; Banno, K. *J. Am. Chem. Soc.* **1974**, *96*, 7503.
3. Langer, P.; Koehler, V. *Org. Lett.* **2000**, *2*, 1597.
4. Matsukawa, S.; Okano, N.; Imamoto, T. *Tetrahedron Lett.* **2000**, *41*, 103.
5. Delas, C.; Blacque, O.; Moise, C. *ibid.* **2000**, *41*, 8269.
6. Ishihara, K.; Kondo, S.; Yamamoto, H. *J. Org. Chem.* **2000**, *65*, 9125.
7. Kumareswaran, R.; Reddy, B. G.; Vankar, Y. D. *Tetrahedron Lett.* **2001**, *42*, 7493.

Mukaiyama esterification

General scheme:

$R_1CO_2H + R_2OH$ — base (Mukaiyama reagent) →

X = F, Cl, Br

e.g.

F_4B^-, Bu_3N, CH_2Cl_2

S_NAr, $-H^+$

$-Br^-$

Amide formation using the Mukaiyama reagent follows a similar mechanistic pathway [4].

References

1. Mukaiyama, T.; Usui, M.; Shimada, E.; Saigo, K. *Chem. Lett.* **1975**, 1045.
2. Hojo, K.; Kobayashi, S.; Soai, K.; Ikeda, S.; Mukaiyama, T. *ibid.* **1977**, 635.
3. Mukaiyama, T. *Angew. Chem., Int. Ed. Engl.* **1979**, *18*, 707.
4. For amide formation, see: Huang, H.; Iwasawa, N.; Mukaiyama, T. *Chem. Lett.* **1984**, 1465.
5. Nicolaou, K. C.; Bunnage, M. E.; Koide, K. *J. Am. Chem. Soc.* **1994**, *116*, 8402.
6. Yong, Y. F.; Kowalski, J. A.; Lipton, M. A. *J. Org. Chem.* **1997**, *62*, 1540.
7. Folmer, J. J.; Acero, C.; Thai, D. L.; Rapoport, H. *ibid.* **1998**, *63*, 8170.

Myers–Saito cyclization

Sometimes known as "Schmittel" cyclization, *Cf.* Bergman cyclization.

Δ or

hv

hydrogen
atom donor

reversible

allenyl enyne

diradical

References

1. Myers, A. G.; Proteau, P. J.; Handel, T. M. *J. Am. Chem. Soc.* **1988**, *110,* 7212.
2. Myers, A. G.; Dragovich, P. S.; Kuo, E. Y. *ibid.* **1992**, *114,* 9369.
3. Saito, K.; Watanabe, T.; Takahashi, K. *Chem. Lett.* **1989**, 2099.
4. Saito, I.; Nagata, R.; Yamanaka, H.; Murahashi, E. *Tetrahedron Lett.* **1990**, *31* 2907.
5. Schmittel, M.; Strittmatter, M.; Kiau, S. *Tetrahedron Lett.* **1995**, *36*, 4975.
6. Engels, B.; Lennartz, C.; Hanrath, M.; Schmittel, M.; Strittmatter, M. *Angew. Chem., Int. Ed.* **1998**, *37*, 1960.
7. Ferri, F.; Bruckner, R.; Herges, R. *New J. Chem.* **1998**, *22*, 531.
8. Wu, M.-J.; Lin, C.-F.; Chen, S.-H.; Lee, F.-C. *J. Chem. Soc., Perkin Trans. 1* **1999**, 2875.
9. Kim, C.-S.; Diez, C.; Russell, K. C. *Chem.--Eur. J.* **2000**, *6*, 1555.
10. Cramer, C. J.; Kormos, B. L.; Seierstad, M.; Sherer, E. C.; Winget, P. *Org. Lett.* **2001**, *3*, 1881.

Nametkin rearrangement (Retropinacol rearrangement)

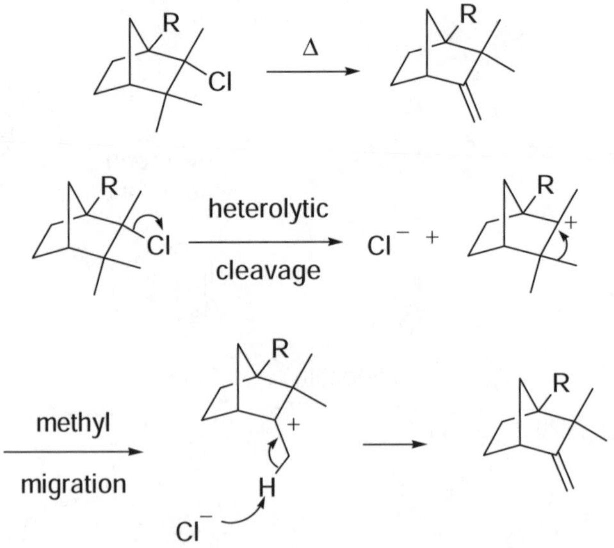

Δ

heterolytic

cleavage

$Cl^- +$

methyl

migration

References

1. Nametkin, S. S. *Liebigs Ann. Chem.* **1923**, *432*, 207.
2. Bernstein, D. *Tetrahedron Lett.* **1967**, 2281.
3. Kossanyi, J.; Furth, B.; Morizur, J. P. *Tetrahedron* **1970**, *26*, 395.
4. Moews, P. C.; Knox, J. R.; Vaughan, W. R. *J. Am. Chem. Soc.* **1978**, *100*, 260.
5. Starling, S. M.; Vonwiller, S. C.; Reek, J. N. H. *J. Org. Chem.* **1998**, *63*, 2262.
6. Martinez, A. G.; Vilar, E. T.; Fraile, A. G.; Fernandez, A. H.; De La Moya, C. S. *Tetrahedron* **1998**, *54*, 4607.

Nazarov cyclization

O

H^+

O

H^+

O

protonation

OH

+

4π *e* conrotory

electrocyclic cyclization

HO

+

HO

+

H

OH

tautomerization

O

References

1. Nazarov, I. N. Torgov, I. B.; Tcrckhova, L. N. *Bull. Acad. Sci. (USSR)* **1942**, 2000.
2. Habermas, K. L.; Denmark, S. E.; Jones, T. K. *Org. React.* **1994**, *45,* 1.
3. Kuroda, C.; Koshio, H.; Koito, A.; Sumiya, H.; Murase, A.; Hitono, Y. *Tetrahedron* **2000**, *56,* 6441.
4. Giese, S.; Kastrup, L.; Stiens, D.; West, F. G. *Angew. Chem., Int. Ed.* **2000**, *39,* 1970.
5. Kim, S.-H.; Cha, J. K. *Synthesis* **2000**, 2113.
6. Giese, S.; West, F. G. *Tetrahedron* **2000**, *56,* 10221.
7. Fernández M., A.; Martin de la Nava, E. M.; González, R. R. *ibid.* **2001**, *57,* 1049.

Neber rearrangement

ketoxime $\xrightarrow[\text{2. } H_2O]{\text{1. KOEt}}$ α-aminoketone + $C_7H_7SO_3H$

deprotonation

cyclization

$C_7H_7SO_3H$ + azirine intermediate

hydrolysis

References

1. Neber, P. W.; v. Friedolsheim, A. *Liebigs Ann. Chem.* **1926**, *449*, 109.
2. O'Brien, C. *Chem. Rev.* **1964**, *64*, 81.
3. Kakehi, A.; Ito, S.; Manabe, T.; Maeda, T.; Imai, K. *J. Org. Chem.* **1977**, *42*, 2514.
4. Friis, P.; Larsen, P. O.; Olsen, C. E. *J. Chem. Soc., Perkin Trans. 1* **1977**, 661.
5. Corkins, H. G.; Storace, L.; Osgood, E. *J. Org. Chem.* **1980**, *45*, 3156.
6. Parcell, R. F.; Sanchez, J. P. *ibid.* **1981**, *46*, 5229.
7. Verstappen, M. M. H.; Ariaans, G. J. A.; Zwanenburg, B. *J. Am. Chem. Soc.* **1996**, *118*, 8491.
8. Mphahlele, M. J. *Phosphorus, Sulfur Silicon Relat. Elem.* **1999**, *144–146*, 351.
9. Banert, K.; Hagedorn, M.; Liedtke, C.; Melzer, A.; Schoffler, C. *Eur. J. Org. Chem.* **2000**, 257.

Nef reaction

Treatment of a primary or secondary nitroalkane with an acid, yielding the corresponding carbonyl compound.

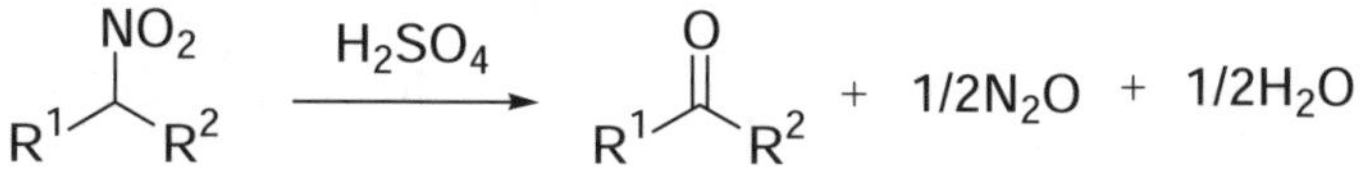

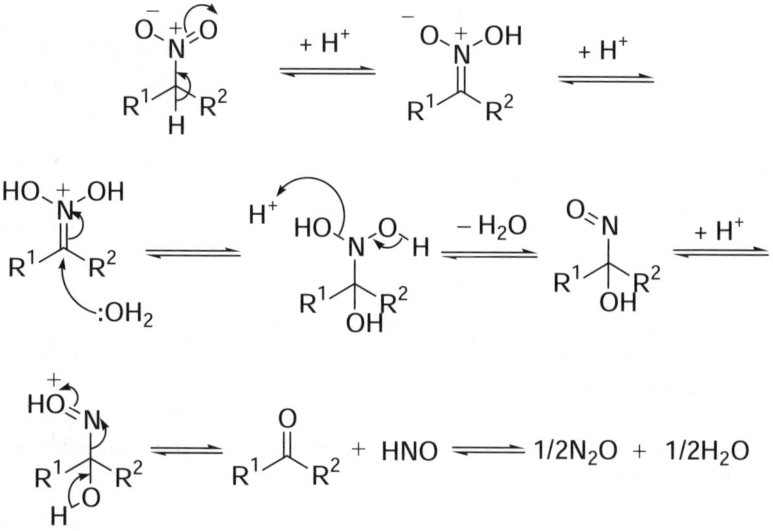

References

1. Nef, J. U. *Liebigs Ann. Chem.* **1894**, *280*, 263.
2. Pinnick, H. W. *Org. React.* **1990**, *38*, 655.
3. Adam, W.; Makosza, M.; Saha-Moeller, C. R.; Zhao, C.-G. *Synlett* **1998**, 1335.
4. Shahi, S. P.; Vankar, Y. D. *Synth. Commun.* **1999**, *29*, 4321.
5. Capecchi, T.; de Koning, C. B.; Michael, J. P. *Perkin 1* **2000**, 2681.

Negishi cross-coupling reaction

Palladium-catalyzed cross-coupling reaction of organozinc reagents with organic halides, triflates, *etc.* For the catalytic cycle, see the Kumada coupling on page 208

$$R{-}X \;+\; R^1{-}ZnX \xrightarrow{Pd(0)} R{-}R^1 \;+\; ZnX_2$$

$$R{-}X \;+\; L_2Pd(0) \xrightarrow[\text{addition}]{\text{oxidative}} \text{R(L)(L)Pd{-}X} \xrightarrow[\text{transmetallation, isomerization}]{R^1{-}ZnX}$$

$$ZnX_2 \;+\; \text{L}_2\text{Pd(R)(R}^1) \xrightarrow[\text{elimination}]{\text{reductive}} R{-}R^1 \;+\; L_2Pd(0)$$

References

1. Negishi, E.-I.; Baba, S. *J. Chem. Soc., Chem. Commun.* **1976**, 596.
2. Negishi, E.-I. *Acc. Chem. Res.* **1982**, *15,* 340.
3. Erdik, E. *Tetrahedron* **1992**, *48,* 9577.
4. Negishi, E.-I.; Liu, F. In *Metal-Catalyzed Cross-Coupling Reactions* **1998**, Diederich, F.; Stang, P. J. eds.;Wiley–VCH Verlag GmbH: Weinheim, Germany, pp 0–47.
5. Yus, M.; Gomis, J. *Tetrahedron Lett.* **2001**, *42,* 5721.

Nenitzescu indole synthesis

O R1 O + H CO2R3 HN R2 Δ HO CO2R3 R1 N R2

O R1 O H CO2R3 HN R2 conjugate addition HO CO2R3 H R1 O N R2 H+

HO H CO2R3 R1 N OH R2 H+ HO CO2R3 R1 N R2

Alternatively:

O R1 O H CO2R3 HN R2 conjugate addition HO H CO2R3 H R1 O N R2 H+

isomerization HO CO2R3 R1 OH HN R2 oxidation (see below) O CO2R3 R1 O :NH R2

The internal oxidation-reduction process might involve a bimolecular face-to-face electronic transfer complex (in nitromethane) [3]:

References

1. Nenitzescu, C. D. *Bull. Soc. Chim. Romania* **1929**, *11*, 37.
2. Allen, Jr. G. R. *Org. React.* **1973**, *20*, 337.
3. Bernier, J. L.; Henichart, J. P. *J. Org. Chem.* **1981**, *46*, 4197.
4. Kinugawa, M.; Arai, H.; Nishikawa, H.; Sakaguchi, A.; Ogasa, T.; Tomioka, S.; Kasai, M. *J. Chem. Soc., Perkin Trans. 1* **1995**, 2677.
5. Mukhanova, T. I.; Panisheva, E. K.; Lyubchanskaya, V. M.; Alekseeva, L. M.; Sheinker, Y. N.; Granik, V. G. *Tetrahedron* **1997**, *53*, 177.
6. Ketcha, D. M.; Wilson, L. J.; Portlock, D. E. *Tetrahedron Lett.* **2000**, *41*, 6253.

Nicholas reaction

$$R^1-C\equiv C-C(OR^2)(R^3)(R^4) \xrightarrow[\text{2. H}^+ \text{ or Lewis acid}]{\text{1. } Co_2(CO)_8} \xrightarrow[\text{2. [O]}]{\text{1. NuH}} R^1-C\equiv C-C(Nu)(R^3)(R^4)$$

$$(CO)_4Co-Co(CO)_3 + R^1-C\equiv C-C(OR^2)(R^3)(R^4) \xrightarrow{-CO} (CO)_3Co-Co(CO)_3(CO)\text{-alkyne complex} \xrightarrow{-CO}$$

$$(CO)_3Co-Co(CO)_3\text{-}[R^1C\equiv C-C(R^4)(R^3)OR^2] \xrightarrow{+H^+} (CO)_3Co-Co(CO)_3\text{-}[R^1C\equiv C-C(R^4)(R^3)-O^+(H)R^2] \xrightarrow{E1}$$

$$(CO)_3Co-Co(CO)_3\text{-}[R^1C\equiv C-C^+(R^4)(R^3)] \xrightarrow{NuH} (CO)_3Co-Co(CO)_3\text{-}[R^1C\equiv C-C(R^4)(R^3)Nu] \xrightarrow[\text{demetallation}]{[O]}$$

propargyl cation intermediate (stabilized by the hexacarbonyldicobalt complex).

$$O{=}C{=}O\uparrow + (CO)_3Co-Co(CO)_2\text{-}[R^1C\equiv C-C(R^4)(R^3)Nu] \rightleftharpoons R^1-C\equiv C-C(Nu)(R^3)(R^4)$$

References

1. Lockwood, R. F.; Nicholas, K. M. *Tetrahedron Lett.* **1977**, 4163.
2. Nicholas, K. M. *Acc. Che. Res.* **1992**, 435.
3. Roth, K. D. *Synlett* **1992**, 435.
4. Iqbal, J.; Bhatia, B.; Khanna, V. *J. Indian Inst. Sci.* **1994**, *74*, 411.
5. Jacobi, P. A.; Zheng, W. In *Enantiosel. Synth. β-Amino Acids* Juaristi, E., ed.; Wiley-VCH: New York, N. Y., **1997**, 359.
6. Diaz, D.; Martin, V. S. *Tetrahedron Lett.* **2000**, *41*, 743.
7. Guo, R.; Green, J. R. *Synlett* **2000**, 746.
8. Green, J. R. *Curr. Org. Chem.* **2001**, *5*, 809.

Noyori asymmetric hydrogenation

$RCOCH_2CO_2R^1 \xrightarrow[(R)\text{-BINAP-Ru}]{H_2} RCH(OH)CH_2CO_2R^1$

(*R*)-BINAP-Ru =

PPh_2 $RuCl_2L_2$ PPh_2

$$[RuCl_2(binap)(solv)_2] \xrightarrow[-\ HCl]{H_2} [RuHCl(binap)(solv)_2]$$

The catalytic cycle:

OR^1 (binap)ClHRu R H^+ solv R OR^1 [RuHCl(binap)(solv)$_2$] OR^1 (binap)ClHRu O H H R H$^+$, solv solv H_2 [RuCl(binap)(solv)$_2$]$^+$ OH O R OR^1

References

1. Noyori, R.; Ohkuma, T.; Kitamura, H.; Takaya, H.; Sayo, H.; Kumobayashi, S.; Akutagawa, S. *J. Am. Chem. Soc.* **1987**, *109*, 5856.
2. Case-Green, S. C.; Davies, S. G.; Hedgecock, C. J. R. *Synlett* **1991**, 781.
3. King, S. A.; Thompson, A. S.; King, A. O.; Verhoeven, T. R. *J. Org. Chem.* **1992**, *57*, 6689.
4. Noyori, R. In *Asymmetric Catalysis in Organic Synthesis* Ojima, I., ed.; Wiley: New York, **1994**, chapter 2.
5. Chung, J. Y. L.; Zhao, D.; Hughes, D. L.; Mcnamara, J. M.; Grabowski, E. J. J.; Reider, P. J. *Tetrahedron Lett.* **1995**, *36*, 7379.
6. Bayston, D. J.; Travers, C. B.; Polywka, M. E. C. *Tetrahedron: Asymmetry* **1998**, *9*, 2015.
7. Noyori, R.; Ohkuma, T. *Angew. Chem., Int. Ed.* **2001**, *40*, 40.

Nozaki–Hiyama–Kishi reaction

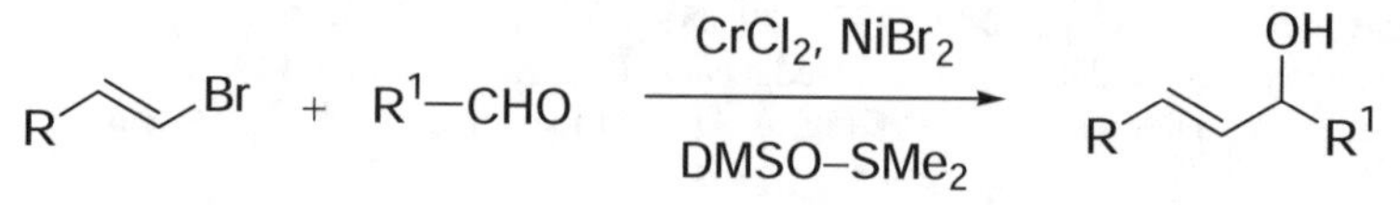

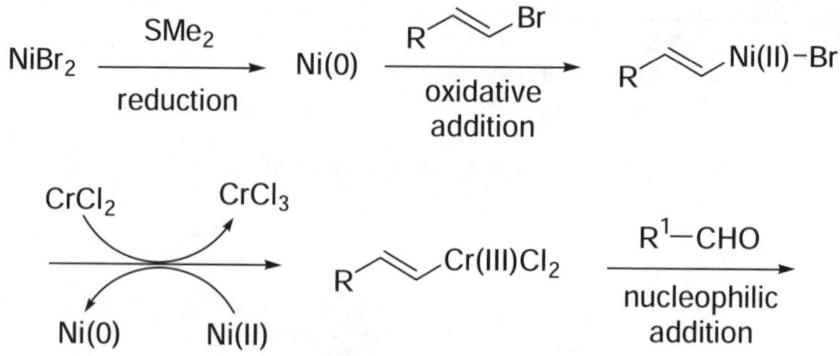

Transmetallation and then reduction by Me_2S

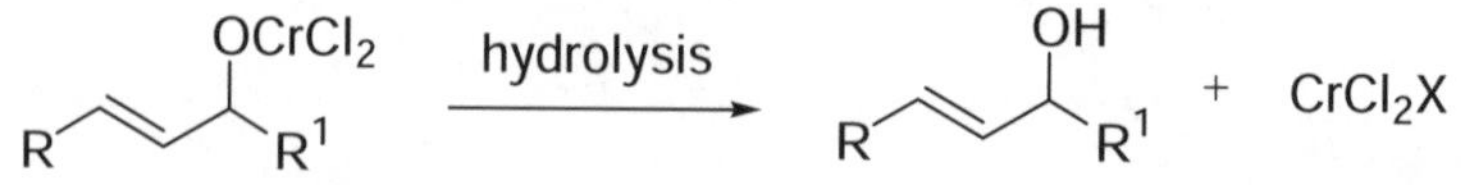

References

1. Takai, K.; Tagahira, M.; Kuroda, T.; Oshima, K.; Utimoto, K.; Nozaki, H. *J. Am. Chem. Soc.* **1986**, *108*, 6048.
2. Cintas, P. *Synthesis* **1992**, 248.
3. Kress, M. H.; Ruel, R.; Miller, L. W. H.; Kishi, Y. *Tetrahedron Lett.* **1993**, *34*, 5999.
4. Boeckman, R. K., Jr.; Hudack, R. A., Jr. *J. Org. Chem.* **1998**, *63*, 3524.
5. Kuroboshi, M.; Tanaka, M.; Kishimoto, S. Goto, K.; Mochizuki, M.; Tanaka, H. *Tetrahedron Lett.* **2000**, *41*, 81.
6. Dai, W.-M.; Wu, A.; Hamaguchi, W. *Tetrahedron Lett.* **2001**, *42*, 4211.

Oppenauer oxidation

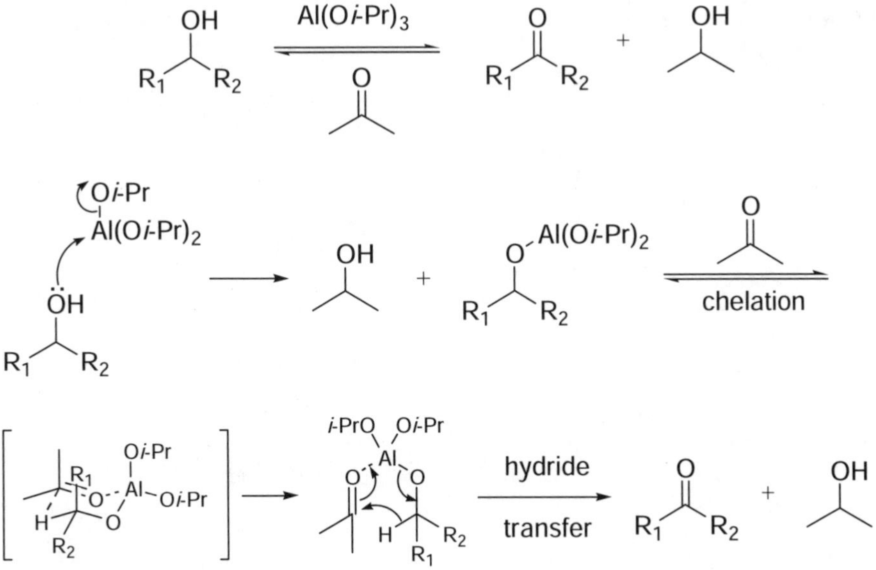

References

1. Oppenauer, R. V. *Rec. Trav. Chim.* **1937**, *56*, 137.
2. de Graauw, C. F.; Peters, J. A.; van Bekkum, H.; Huskens, J. *Synthesis* **1994**, 1007.
3. Almeida, M. L. S.; Kocovsky, P.; Baeckvall, J.-E. *J. Org. Chem.* **1996**, *61*, 6587.
4. Akamanchi, K. G.; Chaudhari, B. A. *Tetrahedron Lett.* **1997**, *38*, 6925.
5. Raja, T.; Jyothi, T. M.; Sreekumar, K.; Talawar, M. B.; Santhanalakshmi, J.; Rao, B. S. *Bull. Chem. Soc. Jpn.* **1999**, *72*, 2117.
6. Nait Ajjou, A. *Tetrahedron Lett.* **2001**, *42*, 13.

Orton rearrangement

Transformation of *N*-chloroanilides to the corresponding *para*-chloroanilides. *Cf.* Fischer–Hepp rearrangement.

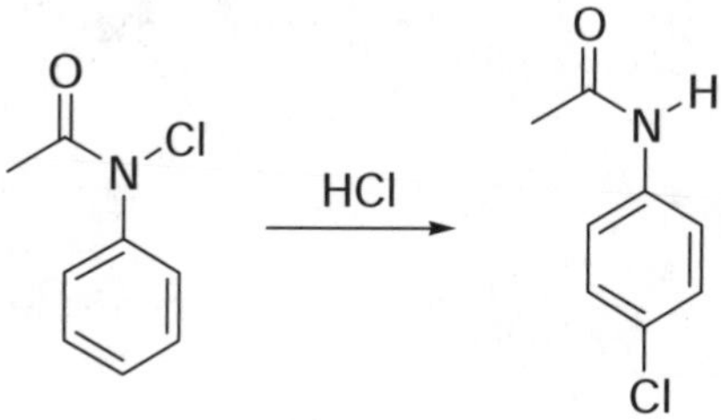

HCl

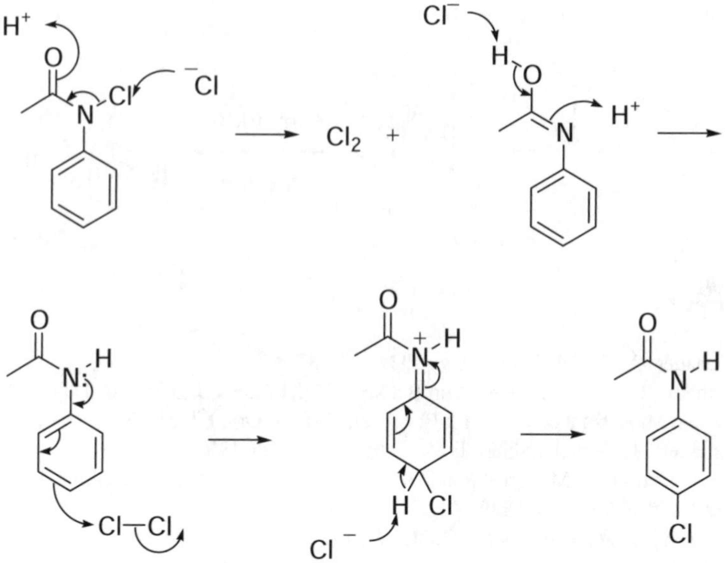

Cl_2 +

H^+

Cl^-

Alternatively:

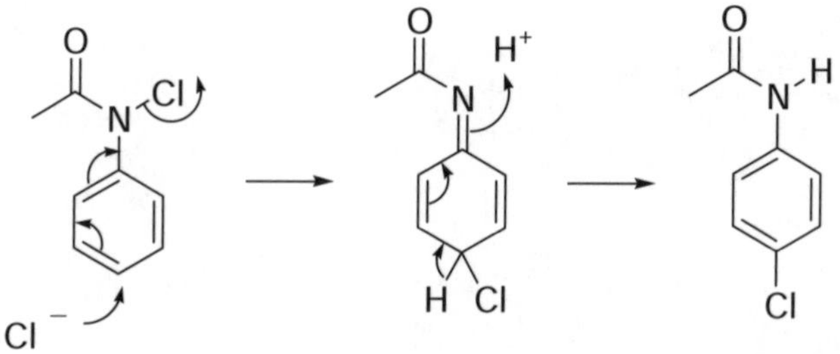

References

1. Shine, H. J. *Aromatic Rearrangement* Elsevier: New York, **1967**, *221*, 362.
2. Scott, J. M. W.; Martin, J. G. *Can. J. Chem.* **1966**, *44*, 2901.
3. Golding, P. D.; Reddy, S.; Scott, J. M. W.; White, V. A.; Winter, J. G. *Can. J. Chem.* **1981**, *59*, 839.
4. Yamamoto, J.; Matsumoto, H. *Chem. Express* **1988**, *3*, 419.
5. Kannan, P.; Venkatachalaphathy, C.; Pitchumani, K. *Indian J. Chem., Sect. B* **1999**, *38B*, 384.
6. Ghosh, S.; Baul, S. *Synth. Commun.* **2001**, *31*, 2783.

Overman rearrangement

Stereoselective transformation of allylic alcohol to allylic trichloroacetimide *via* trichloroacetimidate intermediate.

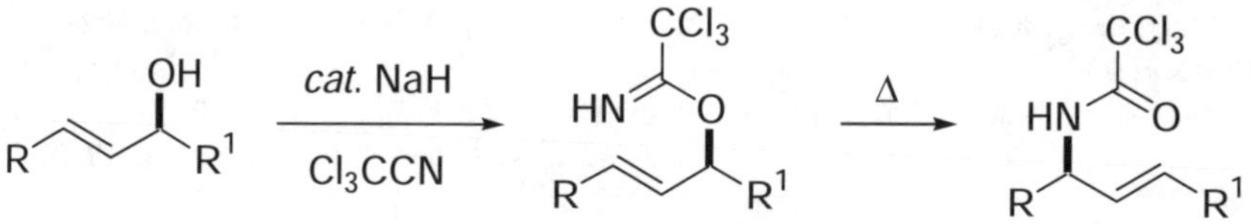

cat. NaH, Cl_3CCN → trichloroacetimidate → Δ

$H_2\uparrow$ +

Δ, [3,3]-sigmatropic rearrangement

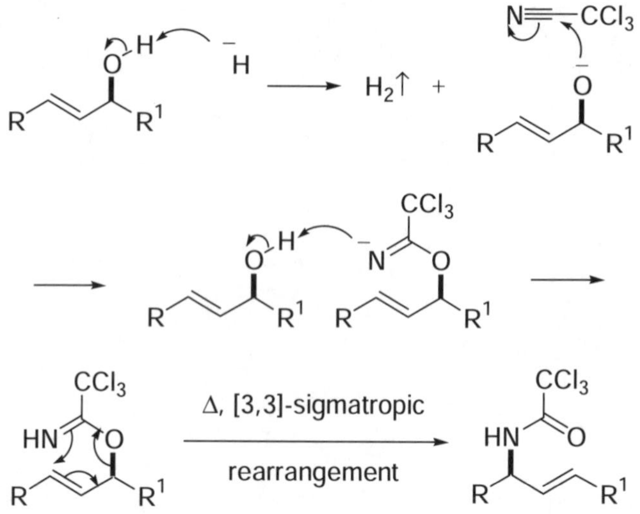

References

1. Overman, L. E. *J. Am. Chem. Soc.* **1974**, *96*, 597.
2. Overman, L. E. *Acc. Chem. Res.* **1971**, *4*, 49.
3. Isobe, M.; Fukuda, Y.; Nishikawa, T.; Chabert, P.; Kawai, T.; Goto, T. *Tetrahedron Lett.* **1990**, *31*, 3327.
4. Eguchi, T.; Koudate, T.; Kakinuma, K. *Tetrahedron* **1993**, *49*, 4527.
5. Toshio, N.; Masanori, A.; Norio, O.; Minoru, I. *J. Org. Chem.* **1998**, *63*, 188.
6. Cho, C.-G.; Lim, Y.-K.; Lee, K.-S.; Jung, I.-H.; Yoon, M.-Y. *Synth. Commun.* **2000**, *30*, 1643.
7. Martin, C.; Prunck, W.; Bortolussi, M.; Bloch, R. *Tetrahedron: Asymmetry* **2000**, *11*, 1585.
8. Demay, S.; Kotschy, A.; Knochel, P. *Synthesis* **2001**, 863.

Paal–Knorr furan synthesis

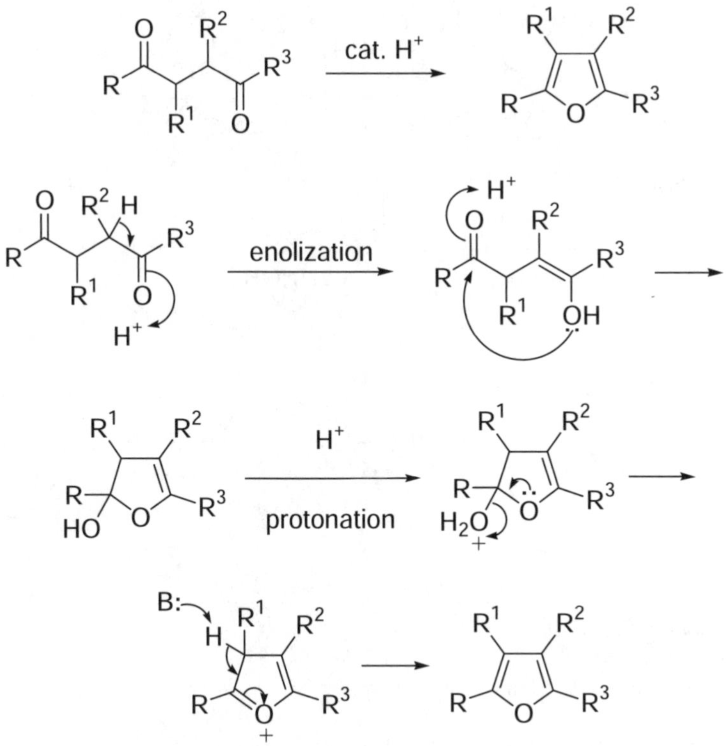

References

1. Haley, J. F., Jr.; Keehn, P. M. *Tetrahedron Lett.* **1973**, 4017.
2. Amarnath, V.; Amarnath, K. *J. Org. Chem.* **1995**, *60*, 301.
3. Truel, I.; Mohamed-Hachi, A.; About-Jaudet, E.; Collignon, N. *Synth. Commun.* **1997**, *27*, 1165.
4. Friedrichsen, W. In *Comprehensive Heterocyclic Chemistry II*; Katritzky, A. R.; Rees, C. W.; Scriven, E. F. V. eds; Pergamon: Oxford, **1996**, *Vol. 2*, p352.
5. Stauffer, F.; Neier, R. *Org. Lett.* **2000**, *2*, 3535.

Paal–Knorr pyrrole synthesis

R^2–NH_2

imine formation

References

1. Paal, C. *Ber.* **1885**, *18*, 367.
2. Hori, I.; Igarashi, M. *Bull. Chem. Soc. Jpn.* **1971**, *44*, 2856.
3. Chiu, P. K.; Lui, K. H.; Maini, P. N.; Sammes, M. P. *J. Chem. Soc., Chem. Commun.* **1987**, 109.
4. Chiu, P. K.; Sammes, M. P. *Tetrahedron* **1988**, *44*, 3531.
5. Chiu, P. K.; Sammes, M. P. *ibid.* **1990**, *46*, 3439.
6. Yu, S.-X.; Le Quesne, P. W. *Tetrahedron Lett.* **1995**, *36*, 6205.
7. Robertson, J.; Hatley, R. J. D.; Watkin, D. J. *Perkin 1* **2000**, 3389.
8. Braun, R. U.; Zeitler, K.; Mueller, T. J. J. *Org. Lett.* **2001**, *3*, 3297.

Parham cyclization

CH_3O CH_3O I N NEt_2 O

2.2 eq. *t*-BuLi

THF, –78 °C→rt

CH_3O CH_3O O N

CH_3O CH_3O I N NEt_2 O

halogen-metal

exchange

I +

Et_2N O Li CH_3O CH_3O N

cyclization

Et_2N O⁻ CH_3O CH_3O N

CH_3O CH_3O O N

The fate of the second equivalent of *t*-BuLi:

H I

+ + I^-

References

1. Parham, W. E.; Jones, L. D. *J. Org. Chem.* **1976**, *41*, 1184.
2. Bradsher, C. K.; Hunt, D. A. *Org. Prep. Proced. Int.* **1978**, *10*, 267.
3. Bradsher, C. K.; Hunt, D. A. *J. Org. Chem.* **1981**, *46*, 4608.
4. Parham, W. E.; Bradsher, C. K. *Acc. Chem. Res.* **1982**, *15*, 305.
5. Quallich, G. J.; Fox, D. E.; Friedmann, R. C.; Murtiashaw, C. W. *J. Org. Chem.* **1992**, *57*, 761.
6. Couture, A.; Deniau, E.; Grandclaudon, P. *J. Chem. Soc., Chem. Commun.* **1994**, 1329.

7. Collado, M. I.; Manteca, I.; Sotomayor, N.; Villa, M.-J.; Lete, E. *J. Org. Chem.* **1997**, *62*, 2080.
8. Osante, I.; Collado, M. I.; Lete, E.; Sotomayor, N. *Synlett* **2000**, 101.
9. Ardeo, A.; Lete, E.; Sotomayor, N. *Tetrahedron Lett.* **2000**, *41*, 5211.
10. Osante, I.; Collado, M. I.; Lete, E.; Sotomayor, N. *Eur. J. Org. Chem.* **2001**, 1267.

Passerini reaction

Three-component condensation (3CC) of carboxylic acids, *C*-isocyanides, and oxo compounds to afford α-acyloxycarboxamides. *Cf.* Ugi reaction.

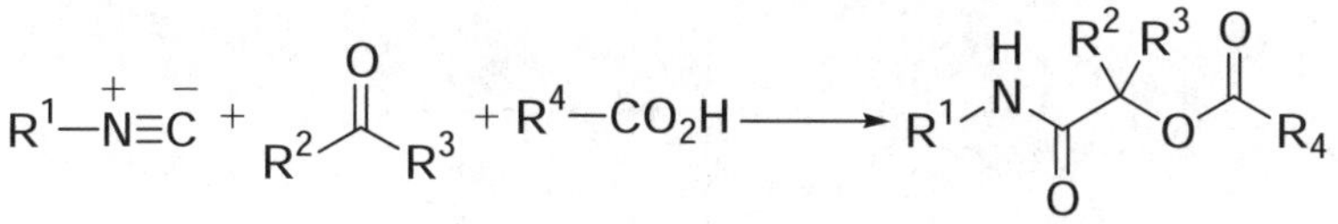

isocyanide

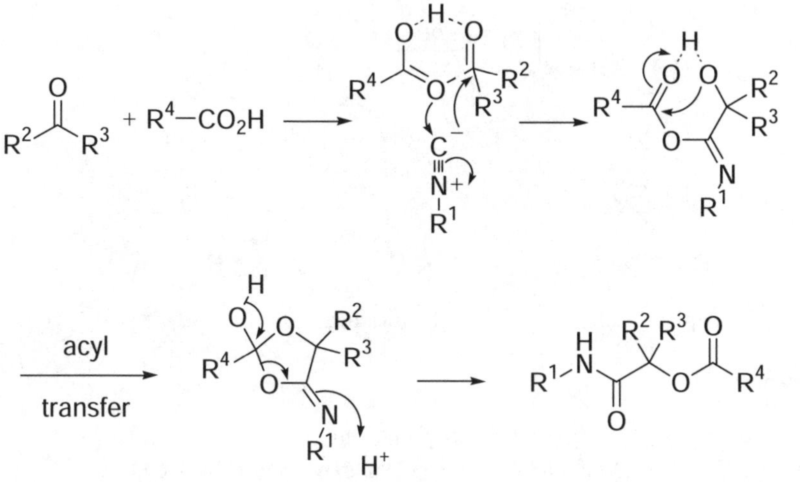

acyl

transfer

H^+

References

1. Passerini, M. *Gazz. Chim. Ital.* **1921**, *51*, 126, 181.
2. Ferosie, I. *Aldrichimica Acta* **1971**, *4*, 21.
3. Ugi, I.; Lohberger, S.; Karl, R. In *Comprehensive Organic Synthesis*, Trost, B. M.; Fleming, I., Eds, Pergamon: Oxford, **1991**, Vol. 2, p.1083.
4. Ziegler, T.; Kaisers, H.-J.; Schlomer, R.; Koch, C. *Tetrahedron* **1999**, *55*, 8397.
5. Banfi, L.; Guanti, G.; Riva, R. *Chem. Commun.* **2000**, 985.
6. Semple, J. E.; Owens, T. D.; Nguyen, K.; Levy, O. E. *Org. Lett.* **2000**, *2*, 2769.
7. Owens, T. D.; Semple, J. E. *Org. Lett.* **2001**, *3*, 3301.

Paterno–Büchi reaction

Photo-induced oxetane formation from a ketone and an olefin.

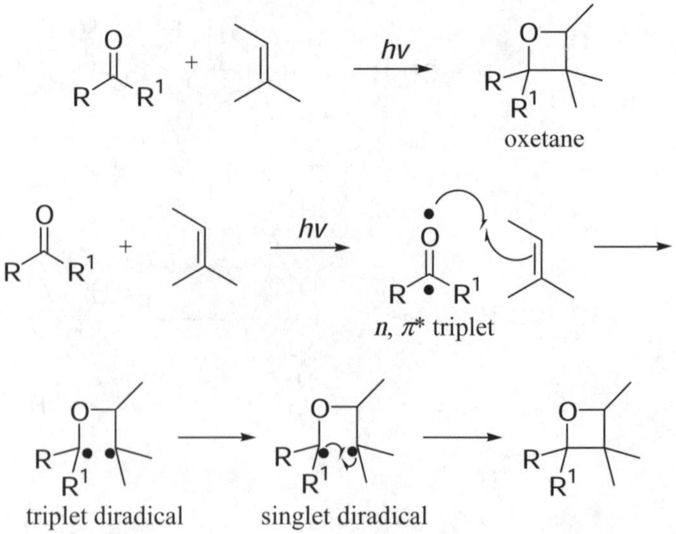

References

1. Paterno, E.; Chieffi, G. *Gazz. Chim. Ital.* **1909**, *39*, 341.
2. Büchi, G.; Inman, C. G.; Lipinsky, E. S. *J. Am. Chem. Soc.* **1954**, *76*, 4327.
3. Porco, J. A., Jr.; Schreiber, S. L. In *Comprehensive Organic Synthesis* Trost, B. M.; Fleming, I., Eds.; Pergamon: Oxford, **1991**, *Vol. 5*, 151–192.
4. Fleming, S. A.; Gao, J. J. *Tetrahedron Lett.* **1997**, *38*, 5407.
5. Hubig, S. M.; Sun, D.; Kochi, J. K. *J. Chem. Soc., Perkin Trans. 2* **1999**, 781.
6. D'Auria, M.; Racioppi, R.; Romaniello, G. *Eur. J. Org. Chem.* **2000**, 3265.
7. Bach, T.; Brummerhop, H.; Harms, K. *Chem.--Eur. J.* **2000**, *6*, 3838.
8. Bach, T. *Synlett* **2000**, 1699.
9. Abe, M.; Tachibana, K.; Fujimoto, K.; Nojima, M. *Synthesis* **2001**, 1243.

Pauson–Khand cyclopentenone synthesis

$Co_2(CO)_8$

toluene, 60–80 °C
4–6 d, 60%

hexacarbonyldicobalt complex

exo complex

sterically-favored isomer

reductive elimination
– $Co_2(CO)_6$

References

1. Bladon, P.; Khand, M. J.; Pauson, P. L. *J. Chem. Res. (M)*, **1977**, 153.
2. Pauson, P. L. *Tetrahedron* **1985**, *41*, 5855.
3. Schore, N. E. *Chem. Rev.* **1988**, *88*, 1081.
4. Schore, N. E. In *Comprehensive Organic Synthesis*, Paquette, L. A.; Fleming, I.; Trost, B. M. Eds, Pergamon: Oxford, **1991**, Vol. 5, p.1037.

5. Schore, N. E. *Org. React.* **1991**, *Vol. 40*, pp 1–90.
6. Brummond, K. M.; Kent, J. L. *Tetrahedron* **2000**, *56*, 3263.
7. Son, S. U.; Lee, S. I.; Chung, Y. K. *Angew. Chem., Int. Ed.* **2000**, *39*, 4158.
8. Kraft, M. E.; Fu, Z.; Boñaga, L. V. R. *Tetrahedron Lett.* **2001**, *42*, 1427.
9. Muto, R.; Ogasawara, K. *Tetrahedron Lett.* **2001**, *42*, 4143.

Payne rearrangement

References

1. Payne, G. B. *J. Org. Chem.* **1962**, *27*, 3819.
2. Page, P. C. B.; Rayner, C. M.; Sutherland, I. O. *J. Chem. Soc., Perkin Trans. 1*, **1990**, 1375.
3. Konosu, T.; Miyaoka, T.; Tajima, Y.; Oida, S. *Chem. Pharm. Bull.* **1992**, *40*, 562.
4. Dols, P. P. M. A.; Arnouts, E. G.; Rohaan, J.; Klunder, A. J. H.; Zwanenburg, B. *Tetrahedron* **1994**, *50*, 3473.
5. Ibuka, T. *Chem. Soc. Rev.* **1998**, *27*, 145.
6. Bickley, J. F.; Gillmore, A. T.; Roberts, S. M.; Skidmore, J.; Steiner, A. *J. Chem. Soc., Perkin Trans. 1* **2001**, 1109.

Pechmann condensation (coumarin synthesis)

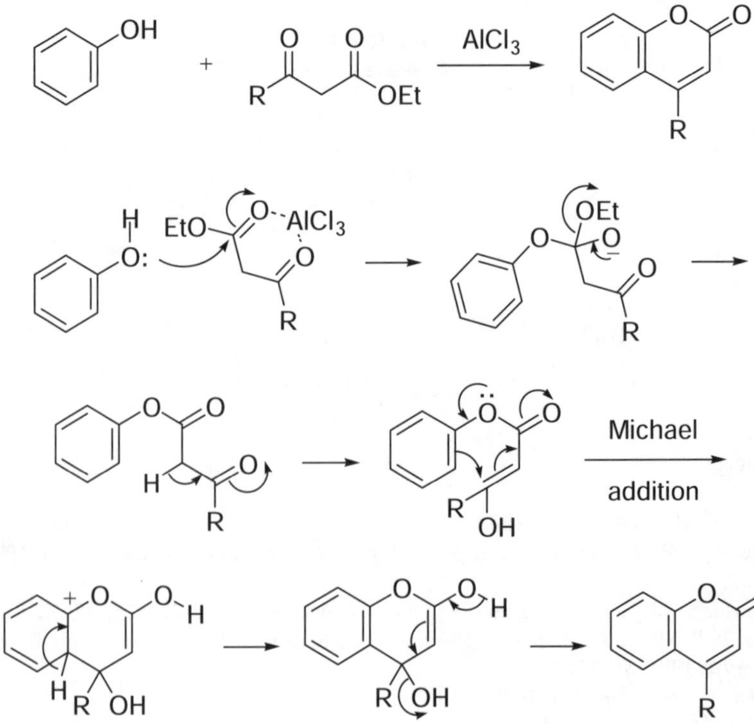

OH + R–CO–CH$_2$–CO–OEt $\xrightarrow{AlCl_3}$

Michael addition

References

1. v. Pechmann, H.; Duisberg, C. *Ber.* **1883**, *16*, 2119.
2. Hirata, T.; Suga, T. *Bull. Chem. Soc. Jpn.* **1974**, *47*, 244.
3. Chaudhari, D. D. *Chem. Ind.* **1983**, 568.
4. Holden, M. S.; Crouch, R. D. *J. Chem. Educ.* **1998**, *75*, 1631.
5. Corrie, J. E. T. *J. Chem. Soc., Perkin Trans. 1* **1990**, 2151.
6. Hua, D. H.; Saha, S.; Roche, D.; Maeng, J. C.; Iguchi, S.; Baldwin, C. *J. Org. Chem.* **1992**, *57*, 399.
7. Biswas, G. K.; Basu, K.; Barua, A. K.; Bhattacharyya, P. *Indian J. Chem., Sect. B* **1992**, *31B*, 628.
8. Li, T.-S.; Zhang, Z.-H.; Yang, F.; Fu, C.-G. *J. Chem. Res., (S)* **1998**, 38.
9. Sugino, T.; Tanaka, K. *Chem. Lett.* **2001**, 110.

Pechmann pyrazole synthesis

$H{-}C{\equiv}C{-}H \; + \; H_2C{=}\overset{+}{N}{=}\overset{-}{N} \longrightarrow$ NH N

acetylene diazomethane

H CH$_2$ N+ N− H

[3 + 2]

cycloaddition

H N N

NH N

Reference

v. Pechmann, H.; Duisberg, C. *Ber.* **1898**, *31*, 2950.

Perkin reaction (cinnamic acid synthesis)

Ar—CHO + Ac_2O AcONa → $^-$OH, H_2O → cinnamic acid

enolate formation

aldol condensation

intermolecular acyl transfer

acyl transfer

E2 elimination

hydrolysis

HOAc

References

1. Perkin, W. H. *J. Chem. Soc.* **1868**, *21,* 53.
2. Pohjala, E. *Heterocycles* **1975**, *3,* 615.
3. Poonia, N. S.; Sen, S.; Porwal, P. K.; Jayakumar, A. *Bull. Chem. Soc. Jpn.* **1980**, *53,* 3338.
4. Gaset, A.; Gorrichon, J. P. *Synth. Commun.* **1982**, *12,* 71.

5. Kinastowski, S.; Nowacki, A. *Tetrahedron Lett.* **1980**, *23,* 3723.
6. Koepp, E.; Voegtle, F. *Synthesis* **1987**, 177.
7. Brady, W. T.; Gu, Y.-Q. *J. Heterocycl. Chem.* **1988**, *25,* 969.
8. Palinko, I.; Kukovecz, A.; Torok, B.; Kortvelyesi, T. *Monatsh. Chem.* **2001**, *131,* 1097.

Perkow reaction

Enol phosphate synthesis from α-halocarbonyls and trialkylphosphites.

General scheme:

$$P(OR^1)(OR^2)(OR^3) + R-C(=O)-C(CH_3)_2X \longrightarrow (R^1O)(R^2O)P(=O)-O-C(R)=C(CH_3)_2 + R^3-X$$

X = Cl, Br, I, secondary or tertiary halides are required to prevent the Michaelis–Arbuzov reaction.

e.g.

$$(EtO)_3P + Br-C(CH_3)_2-C(=O)Ph \longrightarrow (EtO)_2P(=O)-O-C(Ph)=C(CH_3)_2 + EtBr\uparrow$$

$(EtO)_3P$: + Br–C(CH$_3$)$_2$–C(=O)Ph $\xrightarrow{S_N2'}$ Br$^-$ + $(EtO)_2P^+(OEt)-O-C(Ph)=C(CH_3)_2$

$\xrightarrow{S_N2}$ $(EtO)_2P(=O)-O-C(Ph)=C(CH_3)_2$ + EtBr↑

References

1. Perkow, W.; Ullrich, K.; Meyer, F. *Nasturwiss.* **1952**, *39*, 353.
2. Perkow, W. *Ber.* **1954**, *87*, 755.
3. Borowitz, G. B.; Borowitz, I. J. *Handb. Organophosphorus Chem.* **1992**, 115.
4. Hudson, H. R.; Matthews, R. W.; McPartlin, M.; Pryce, M. A.; Shode, O. O. *J. Chem. Soc., Perkin Trans. 2* **1993**, 1433.
5. Janecki, T.; Bodalski, R. *Heteroat. Chem.* **2000**, *11*, 115.

Peterson olefination

O R1 R2 + M+ – H SiR3 R3 → HO SiR3 R2 H R1 R3 acid or base → R1 H R2 R3

Basic conditions:

O R1 R2 M+ – H SiR3 R3 → O– SiR3 R1 H R2 R3 ⇌

O–SiR3 R1 H R2 R3 *cis*- elimination → R1 H R2 R3

Acidic conditions:

HO SiR3 R1 H R2 R3 ← rotation → HO R3 R1 H R2 SiR3 H+ ⇌

H2O+ R3 R1 H R2 SiR3 :OH2 E2 *anti*- elimination → R1 R3 R2 H

References

1. Peterson, D. J. *J. Org. Chem.* **1968**, *33*, 780.
2. Ager, D. J. *Org. React.* **1990**, *38*, 1.
3. Waschbusch, R.; Carran, J.; Savignac, P. *Tetrahedron* **1996**, *52*, 14199.
4. Barrett, A. G. M.; Hill, J. M.; Wallace, E. M.; Flygare, J. A. *Synlett* **1991**, 764.
5. Fassler, J.; Linden, A.; Bienz, S. *Tetrahedron* **1999**, *55*, 1717.
6. Chiang, C.-C.; Chen, Y.-H.; Hsieh, Y.-T.; Luh, T.-Y. *J. Org. Chem.* **2000**, *65*, 4694.
7. Galano, J.-M.; Audran, G.; Monti, H. *Tetrahedron Lett.* **2001**, *42*6125.

Pfau–Plattner azulene synthesis

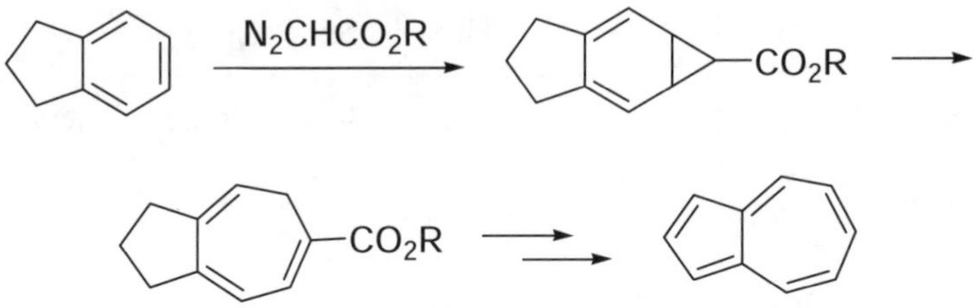

N_2CHCO_2R

CO_2R

CO_2R

N_2 CO_2CH_3 ≡ $\pm$ $CHCO_2CH_3$ ≡ $:CHCO_2CH_3$

The diazo compound is depicted as a carbene equivalent in the mechanism

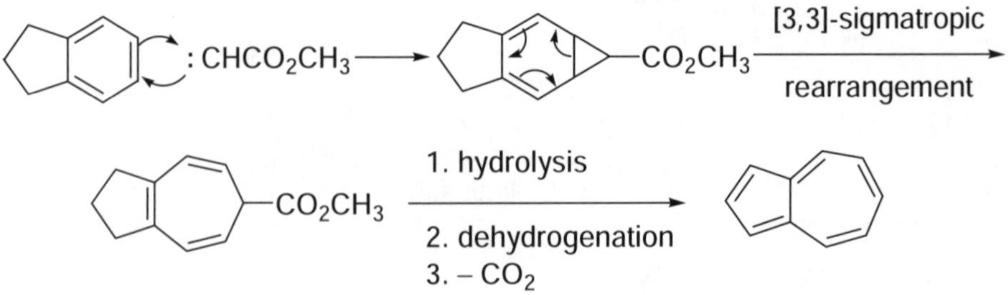

$: CHCO_2CH_3$

CO_2CH_3

[3,3]-sigmatropic rearrangement

CO_2CH_3

1. hydrolysis
2. dehydrogenation
3. – CO_2

References

1. St. Pfau, A.; Plattner, P. A. *Helv. Chim. Acta* **1939**, *22*, 202.
2. Hansen, H. J. *Chimia* **1996**, *50*, 489.
3. Hansen, H. J. *ibid.* **1997**, *51*, 147.

Pfitzinger quinoline synthesis

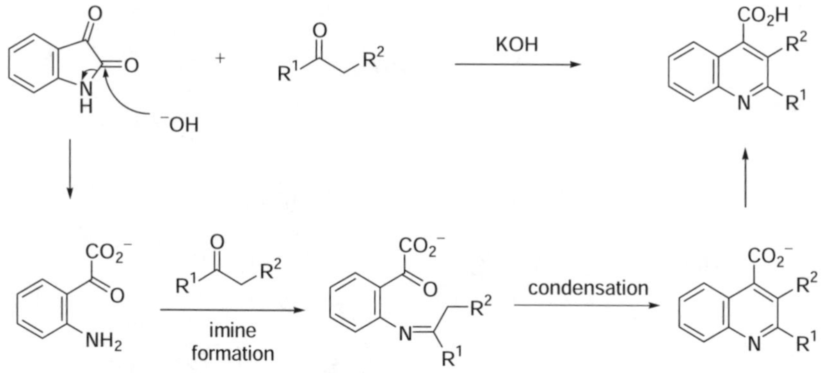

References

1. Buu-Hoi, N. P.; Royer, R.; Nuong, N. D.; Jacquhnos, P. *J. Org. Chem.* **1953**, *18*, 1209.
2. Cragoe, E. J., Jr.; Robb,C. M. *Org. Synth.* **1973**, *Coll. Vol. 5*, 635.
3. Cragoe, E. J., Jr.; Robb,C. M.; Bealor, M. D. *J. Am. Chem. Soc.* **1982**, *53*, 552.
4. Gainer, J. A.; Weinreb, S. M. *J. Org. Chem.* **1982**, *47*, 2833.
5. Lasikova, A.; Vegh, D. *Chem. Pap.* **1997**, *51*, 408.

Pictet–Gams isoquinoline synthesis

P_2O_5 actually exists as P_4O_{10}, an adamantane-like structure.

Reference

1. Pictet, A.; Gams, A. *Ber.* **1910**, *43*, 2384.
2. Ardabilchi, N.; Fitton, A. O.; Frost, J. R.; Oppong-Boachie, F. K.; Hadi, A. Hamid, A.; Sharif, A. .M. *J. Chem. Soc., Perkin Trans. 1* **1979**, 539.
3. Poszavacz, L.; Simig, G. *J. Heterocycl. Chem.* **2000**, *37*, 343.
4. Poszavacz, L.; Simig, G. *Tetrahedron* **2001**, *57*, 8573.

Pictet–Spengler isoquinoline synthesis

References

1. Pictet, A.; Spengler, T. *Ber.* **1911**, *44*, 2030.
2. Hudlicky, T.; Kutchan, T. M.; Shen, G.; Sutliff, V. E.; Coscia, C. J. *J. Org. Chem.* **1981**, *46*, 1738.
3. Miller, R. B.; Tsang, T. *Tetrahedron Lett.* **1988**, *29*, 6715.
4. Rozwadowska, M. D. *Heterocycles* **1994**, *39*, 903.
5. Cox, E. D.; Cook, J. M. *Chem. Rev.* **1995**, *95*, 1797.
6. Yokoyama, A.; Ohwada, T.; Shudo, K. *J. Org. Chem.* **1999**, *64*, 611.
7. Singh, K.; Deb, P. K.; Venugopalan, P. *Tetrahedron* **2001**, *57*, 7939.

Pinacol rearrangement

HO OH R R^1 R^2 R^3 $\xrightleftharpoons{H^+}$ O R R^1 R^2 R^3

HO OH R R^1 R^2 R^3 $\xrightleftharpoons{H^+}$ HO $\overset{+}{O}H_2$ R R^1 R^2 R^3 $\xrightarrow[\text{migration}]{-H_2O\text{, alkyl}}$

H O R + R^1 R^2 R^3 $\rightleftharpoons$ H^+ + O R R^1 R^2 R^3

References

1. Fittig, R. *Liebigs Ann. Chem.* **1860**, *114*, 54.
2. Toda, F.; Shigemasa, T. *J. Chem. Soc., Perkin Trans. 1* **1989**, 209.
3. Nakamura, K.; Osamura, Y. *J. Am. Chem. Soc.* **1993**, *115*, 9112.
4. Paquette, L. A.; Lord, M. D.; Negri, J. T. *Tetrahedron Lett.* **1993**, *34*, 5693.
5. Jabur, F. A.; Penchev, V. J.; Bezoukhanova, C. P. *J. Chem. Soc., Chem. Commun.* **1994**, 1591.
6. Patra, D.; Ghosh, S. *J. Org. Chem.* **1995**, *60*, 2526.
7. Magnus, P.; Diorazio, L.; Donohoe, T. J.; Giles, M.; Pye, P.; Tarrant, J.; Thom, S. *Tetrahedron* **1996**, *52*, 14147.
8. Bach, T.; Eilers, F. *J. Org. Chem.* **1999**, *64*, 8041.
9. Razavi, H.; Polt, R. *ibid.* **2000**, *65*, 5693.
10. Rashidi-Ranjbar, P.; Kianmehr, E. *Molecules* **2001**, *6*, 442.

Pinner synthesis

Transformation of a nitrile into an imino ether, which can be converted to either an ester or an amidine.

$$R{-}CN + R^1{-}OH \xrightarrow{HCl\,(g)} R{-}C(=NH_2^+\,Cl^-){-}OR^1 \begin{cases} \xrightarrow{H^+,\ H_2O} R{-}C(=O){-}OR^1 \\ \xrightarrow{NH_3,\ EtOH} R{-}C(=NH){-}NH_2{\cdot}HCl \end{cases}$$

protonation

nucleophilic addition

common intermediate

hydrolysis

nucleophilic addition

References

1. Pinner, A.; Klein, F. *Ber.* **1877**, *10*, 1889.
2. Poupaert, J.; Bruylants, A.; Crooy, P. *Synthesis* **1972**, 622.
3. Lee, Y. B.; Goo, Y. M.; Lee, Y. Y.; Lee, J. K. *Tetrahedron Lett.* **1990**, *31*, 1169.
4. Cheng, C. C. *Org. Prep. Proced. Int.* **1990**, *22*, 643.
5. Neugebauer, W.; Pinet, E.; Kim, M.; Carey, P. R. *Can. J. Chem.* **1996**, *74*, 341.
6. Spychala, J. *Synth. Commun.* **2000**, *30*, 1083.
7. Kigoshi, H.; Hayashi, N.; Uemura, D. *Tetrahedron Lett.* **2001**, *42*, 7469.

Polonovski reaction

Treatment of a tertiary *N*-oxide with an activating agent such as acetic anhydride, resulting in rearrangement where an *N,N*-disubstituted acetamide and an aldehyde are generated.

$(CH_3CO)_2O$

pyr. CH_2Cl_2

acylation

$CH_3CO_2^-$

elimination

of HOAc

$CH_3CO_2^-$

iminium ion

OAc

The intramolecular pathway is also possible:

References

1. Polonovski, M.; Polonovski, M. *Bull. Soc. Chim. Fr.* **1927**, *41*, 1190.
2. Michelot, R. *ibid.* **1969**, 4377.
3. Volz, H.; Ruchti, L. *Ann.* **1972**, *763*, 184.
4. Hayashi, Y.; Nagano, Y.; Hongyo, S.; Teramura, K. *Tetrahedron Lett.* **1974**, 1299.
5. M'Pati, J.; Mangeney, P.; Langlois, Y. *ibid.* **1981**, *22*, 4405.
6. Lounasmaa, M.; Koskinen, A. *ibid.* **1982**, *23*, 349.
7. Manninen, K.; Hakala, E. *Acta Chem. Scand.* **1986**, *B40*, 598.
8. Grierson, D. *Org. React.* **1990**, *39*, 85.
9. Lounasmaa, M.; Jokela, R.; Halonen, M.; Miettinen, J. *Heterocycles* **1993**, *36*, 2523.
10. Thomas, O. P.; Zaparucha, A.; Husson, H.-P. *Tetrahedron Lett.* **2001**, *42*, 3291.

Polonovski–Potier reaction

A modification of the Polonovski reaction where trifluoroacetic anhydride is used in place of acetic anhydride.

$(CF_3CO)_2O$

pyr. CH_2Cl_2

tertiary *N*-oxide

acylation

$CF_3CO_2^-$

iminium ion

enamine

References

1. Lewin, G.; Poisson, J.; Schaeffer, C.; Volland, J. P. *Tetrahedron* **1990**, *46*, 7775.
2. Kende, A. S.; Liu, K.; Jos Brands, K. M. *J. Am. Chem. Soc.* **1995**, *117*, 10597.
3. Sundberg, R. J.; Gadamasetti, K. G.; Hunt, P. J. *Tetrahedron* **1992**, *48*, 277.
4. Lewin, G.; Schaeffer, C.; Morgant, G.; Nguyen-Huy, D. *J. Org. Chem.* **1996**, *61*, 9614.
5. Renko, D.; Mary, A.; Guillou, C.; Potier, P.; Thal, C. *Tetrahedron Lett.* **1998**, *39*, 4251.
6. Suau, R.; Najera, F.; Rico, R. *Tetrahedron* **2000**, *56*, 9713.

Pomeranz–Fritsch reaction

Isoquinoline synthesis from benzaldehyde and aminoacetal.

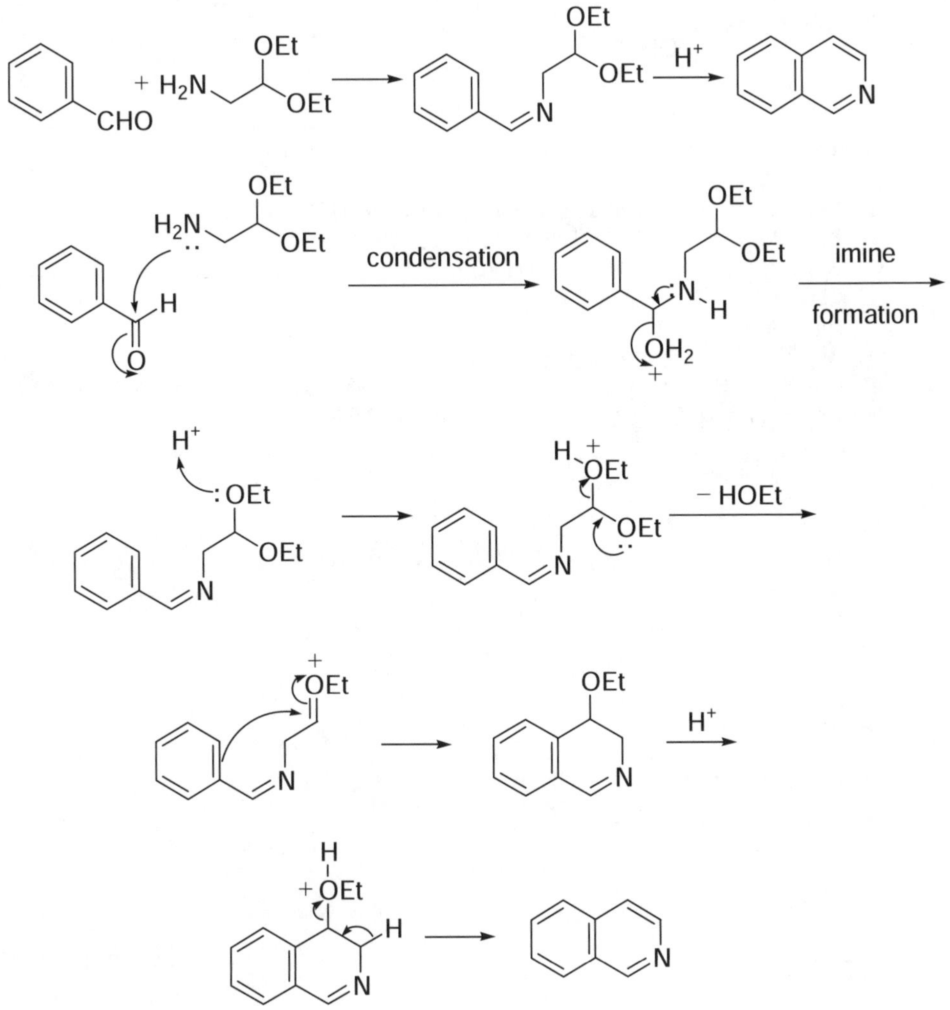

Schilittle–Müller modification

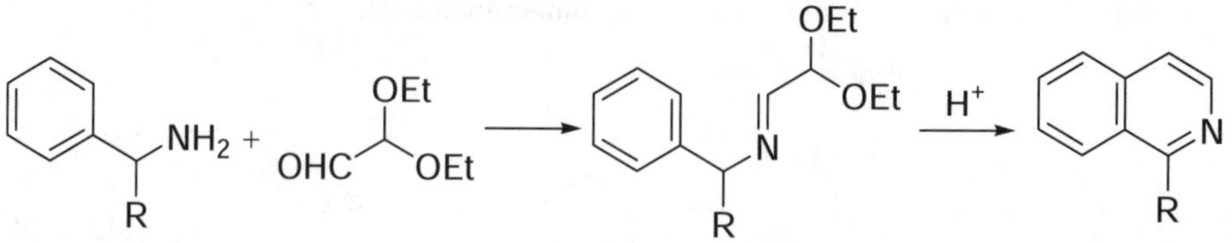

References

1. Bevis, M. J.; Forbes, Eric J.; Uff, B. C. *Tetrahedron* **1969**, *25*, 1585.
2. Bevis, M. J.; Forbes, E. J.; Naik, N. N.; Uff, B. C. *ibid.* **1971**, *27*, 1253.
3. Birch, A. J.; Jackson, A. H.; Shannon, P. V. R. *J. Chem. Soc., Perkin Trans. 1* **1974**, 2185.
4. Birch, A. J.; Jackson, A. H.; Shannon, P. V. R. *ibid.* **1974**, 2190.
5. Brown, E. V. *J. Org. Chem.* **1977**, *42*, 3208.
6. Gill, E. W.; Bracher, A. W. *J. Heterocycl. Chem.* **1983**, *20*, 1107.
7. Ishii, H.; Ishida, T. *Chem. Pharm. Bull.* **1984**, *32*, 3248.

Prévost *trans*-dihydroxylation

Cf. Woodward *cis*-dihydroxylation

I_2, AgO_2CPh → O_2CPh / O_2CPh → hydrolysis → OH / OH

I—I → S_N2 ← ^-O_2CPh

cyclic iodonium ion intermediate

I, O, O, Ph → S_N2 → ^-O_2CPh, +O, O, Ph → S_N2 →

neighboring group assistance

O_2CPh / O_2CPh → hydrolysis → OH / OH

References

1. Prévost, C. *Compt. Rend.* **1933**, *196* 1129.
2. Brimble, M. A.; Nairn, M. R. *J. Org. Chem.* **1996**, *61*, 4801.
3. Hamm, S.; Hennig, L.; Findeisen, M.; Muller, D.; Welzel, P. *Tetrahedron* **2000**, *56*, 1345.

Prilezhaev reaction

Epoxidation of olefins using peracids.

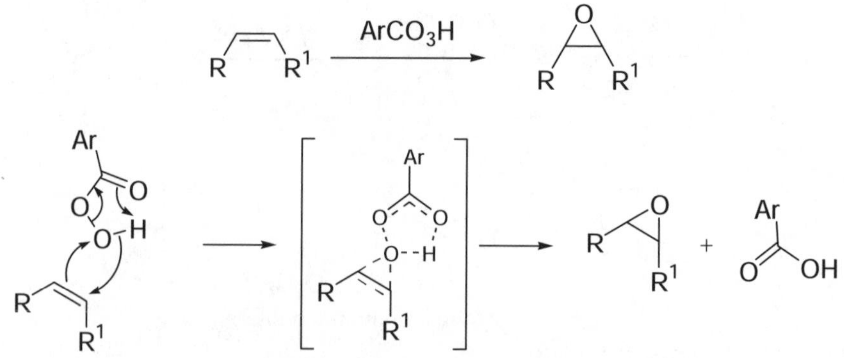

The "butterfly" transition state

References

1. Prilezhaev, N. *Ber.* **1909**, *64*, 8041.
2. Rebek, J., Jr.; Marshall, L.; McManis, J.; Wolak, R. *J. Org. Chem.* **1986**, *51*, 1649.
3. Kaneti, I. *Tetrahedron* **1986**, *42*, 4017.
4. De Cock, C. J. C.; De Keyser, J. L.; Poupaert, J. H.; Dumont, P. *Bull. Soc. Chim. Belg.* **1987**, *96*, 783.
5. Hilker, I.; Bothe, D.; Pruss, J.; Warnecke, H.-J. *Chem. Eng. Sci.* **2001**, *56*, 427.

Prins reaction

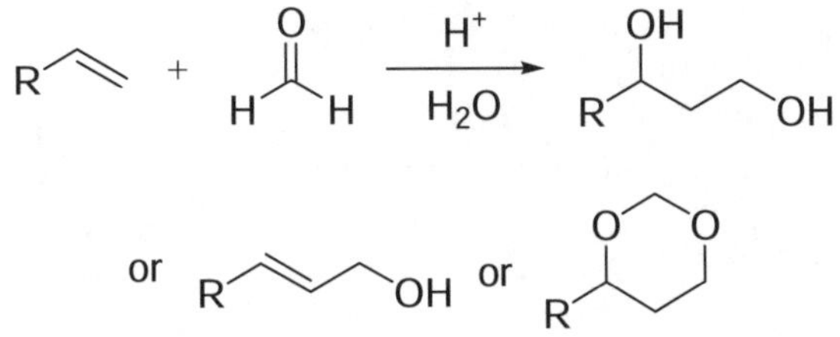

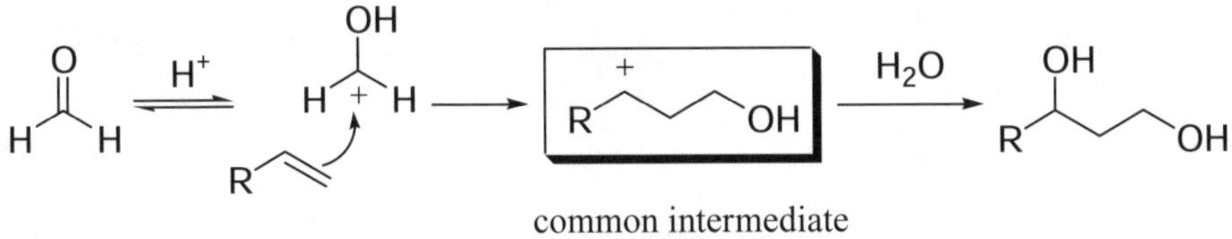

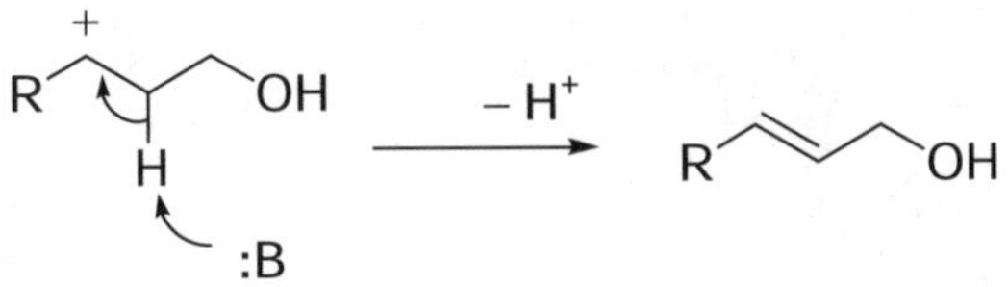

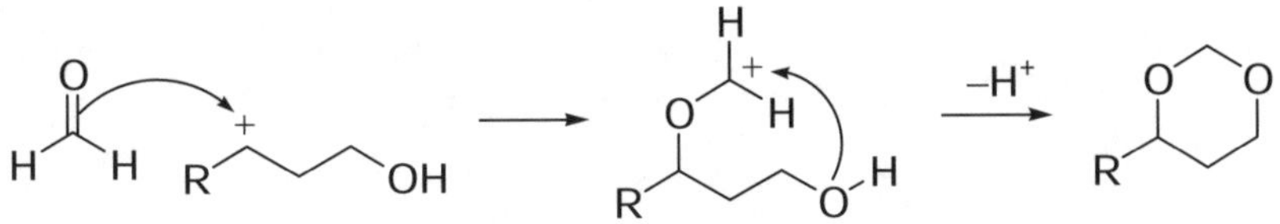

common intermediate

References

1. Prins, H. J. *Chem. Weekblad* **1919**, *16*, 64, 1072.
2. Adam, D. R.; Bhtnagar, S. P. *Synthesis* **1977**, 661.
3. El Gharbi, R. *Synthesis* **1981**, 361.
4. Hanaki, N.; Link, J. T.; MacMillan, D. W. C.; Overman, L. E.; Trankle, W. G.; Wurster, J. A. *Org. Lett.* **2000**, *2*, 223.
5. Yadav, J. S.; Reddy, B. V. S.; Kumar, G. M.; Murthy, Ch. V. S. R. *Tetrahedron Lett.* **2001**, *42*, 89.

Pschorr ring closure

The intramolecular version of the Gomberg–Bachmann reaction.

CO_2H

$NaNO_2$, HCl

Cu

NH_2

CO_2H

H^+

$-H_2O$

$-NO_2^-$

H^+

H^+

$-H_2O$

Cu(0)

Cu(I) +

+ $N_2\uparrow$

6-*exo-trig*

radical cyclization

Cu(I)

$-H^+$

Cu(0) +

CO_2H

References

1. Pschorr, R. *Ber.* **1896**, *29*, 496.
2. Kametani, T.; Fukumoto, K. *J. Heterocycl. Chem.* **1971**, *8*, 341.
3. Kupchan, S. M.; Kameswaran, V.; Findlay, J. W. A. *J. Org. Chem.* **1973**, *38*, 405.
4. Daidone, G. *J. Heterocycl. Chem.* **1980**, *17*, 1409.
5. Buck, K. T.; Edgren, D. L.; Blake, G. W.; Menachery, M. D. *Heterocycles* **1993**, *36*, 2489.
6. Wassmundt, F. W.; Kiesman, W. F. *J. Org. Chem.* **1995**, *60*, 196.

Pummerer rearrangement

The transformation of sulfoxides into α-acyloxythioethers using acetic anhydride.

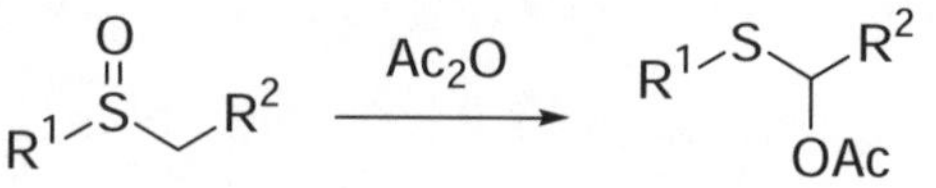

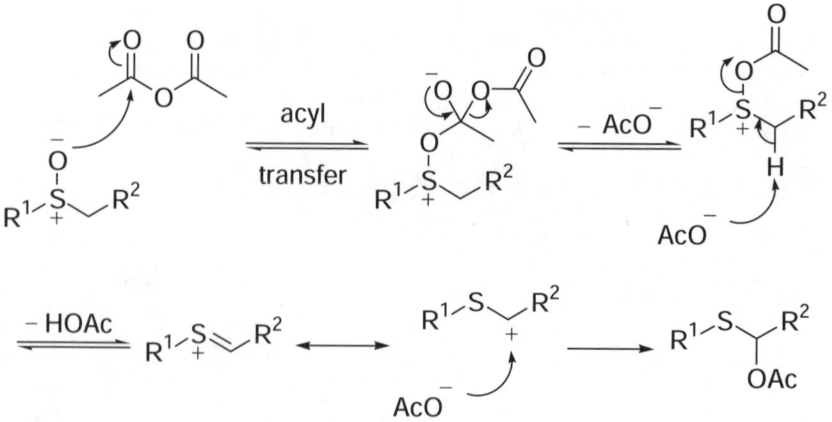

References

1. Pummerer, R. *Ber.* **1910**, *43*, 1401.
2. De Lucchi, O.; Miotti, U.; Modena, G. *Org. React.* **1991**, *40*, 157.
3. Kita, Y. *Phosphorus, Sulfur Silicon Relat. Elem.* **1991**, *120 & 121*, 145.
4. Padwa, A.; Gunn, D. E., Jr.; Osterhout, M. H. *Synthesis* **1997**, 1353.
5. Padwa, A.; Waterson, A. G. *Curr. Org. Chem.* **2000**, *4*, 175.
6. Marchand, P.; Gulea, M.; Masson, S.; Averbuch-Pouchot, M.-T. *Synthesis* **2001**, 1623.

Ramberg–Bäcklund olefin synthesis

Olefin synthesis by treatment of an α-halosulfone with base.

episulfone intermediate

References

1. Ramberg, L.; Bäcklund, B. *Arkiv. Kemi, Mineral Geol.* **1940**, *13A*, 50.
2. Paquette, L. A. *Acc. Chem. Res.* **1968**, *1*, 209.
3. Braveman, S.; Zafrani, Y. *Tetrahedron* **1998**, *54*, 1901.
4. Taylor, R. J. K. *Chem. Commun.* **1999**, 217.
5. McGee, D. I.; Beck, E. J. *Can. J. Chem.* **2000**, *78*, 1060.
6. McAllister, G. D.; Taylor, R. J. K. *Tetrahedron Lett.* **2001**, *42*, 1197.

Reformatsky reaction

Nucleophilic addition of organozinc reagents (generated from α-haloesters) to carbonyls.

R(C=O)R[1] + Br–CH2–C(=O)OR[2] —Zn→ HO–C(R)(R[1])–CH2–C(=O)OR[2]

Br–CH2–C(=O)OR[2] —Zn(0), oxidative addition or electron transfer→ R(C=O)R[1] + CH2=C(OR[2])–O–ZnBr

—nucleophilic addition→ BrZnO–C(R)(R[1])–CH2–C(=O)OR[2]

—H_2O, hydrolysis during workup→ HO–C(R)(R[1])–CH2–C(=O)OR[2]

References

1. Reformatsky, S. *Ber.* **1887**, *20,* 1210.
2. Gaudemar, M. *Organometal. Chem. Rev., Sect. A* **1972**, 8, 183.
3. Fürstner, A. *Synthesis* **1989**, 571.
4. Fürstner, A. In *Organozinc Reagents* Knochel, P.; Jones, P. eds; Oxford University Press: New York, **1999**, pp 287–305.
5. Hirashita, T.; Kinoshita, K.; Yamamura, H.; Kawai, M.; Araki, S. *J. Chem. Soc., Perkin Trans. 1* **2000**, 825.
6. Kurosawa, T.; Fujiwara, M.; Nakano, H.; Sato, M.; Yoshimura, T.; Murai, T. *Steroids* **2001**, *66,* 499.

Regitz diazo synthesis

The synthesis of 2-diazo-1,3-dicarbonyl or 2-diazo-3-ketoesters using tosyl azide or mesyl azide.

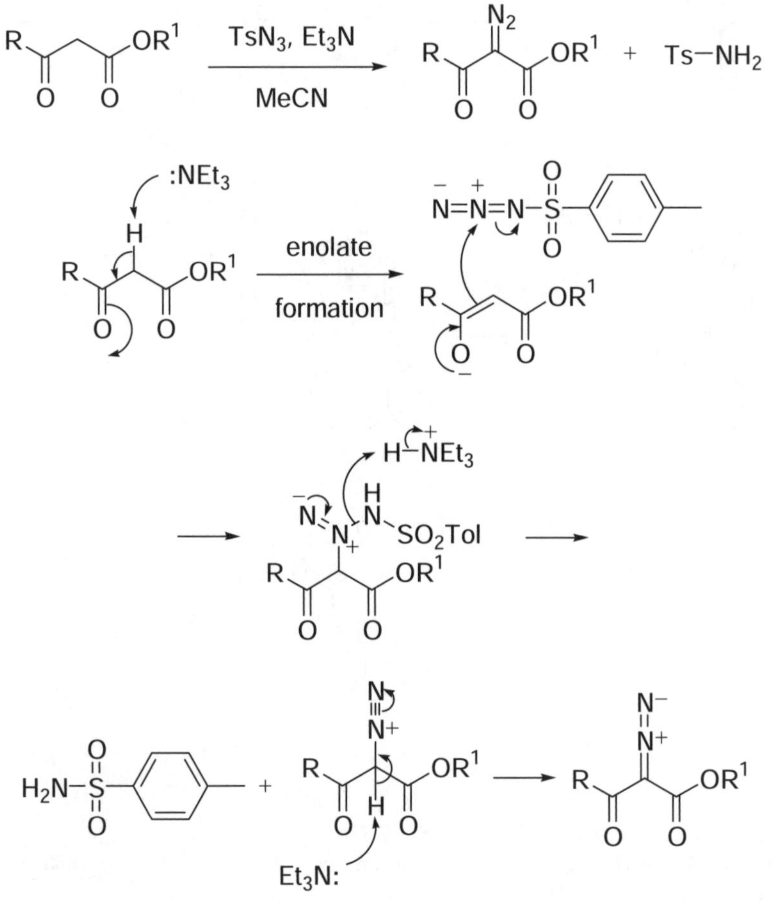

When only one carbonyl is present, ethylformate can be used as an activating auxiliary [6–9]:

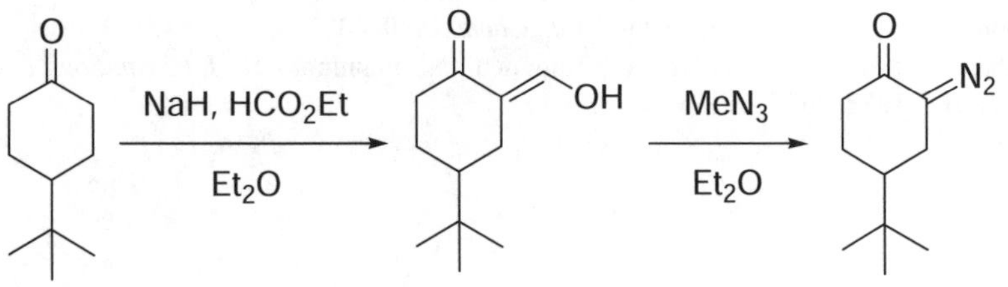

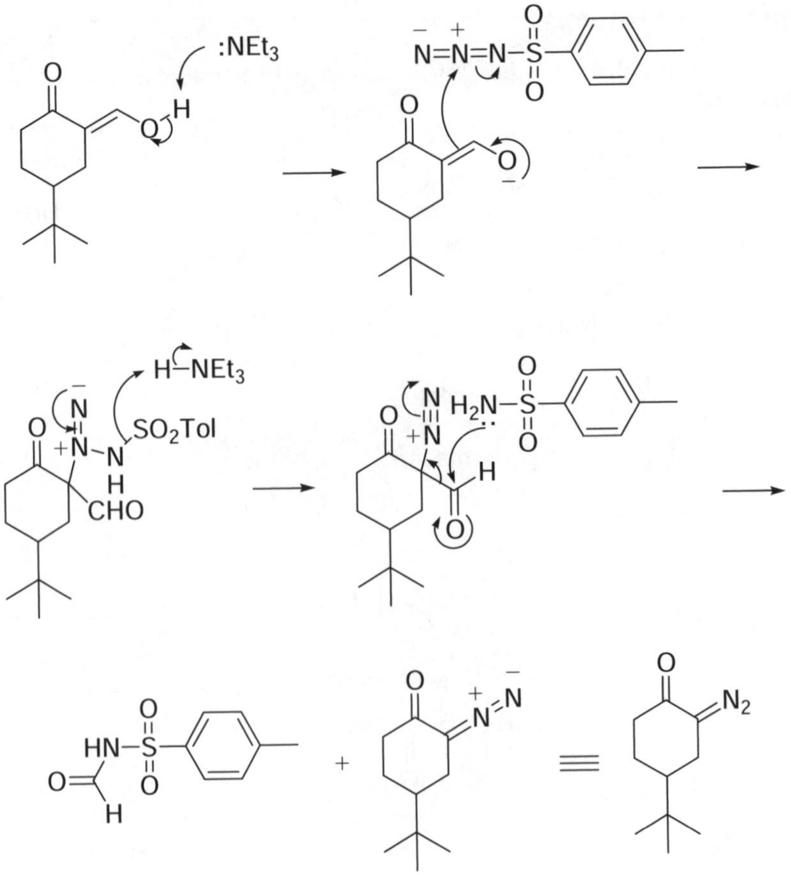

References

1. Regitz, M. *Angew. Chem., Int. Ed.* **1967**, *6*, 733.
2. Regitz, M.; Anschuetz, W.; Bartz, W.; Liedhegener, A. *Tetrahedron Lett.* **1968**, 3171.
3. Regitz, M. *Synthesis* **1972**, 351.
4. Hoffmann, R. W.; Gerlach, R.; Goldmann, S. *Ber.* **1980**, *113*, 856.
5. Charette, A. B.; Wurz, R. P.; Ollevier, T. *J. Org. Chem.* **2000**, *65*, 9252.
6. Taber, D. F.; Ruckle, R. E., Jr.; Hennessy, M. J. *J. Org. Chem.* **1986**, *51*, 4077.
7. Taber, D. F.; Schuchardt, J. L. *Tetrahedron* **1987**, *43*, 5677.
8. Pudleiner, H.; Laatsch, H. *Liebigs Ann. Chem.* **1990**, 423.
9. Ihara, M.; Suzuki, T.; Katogi, M.; Taniguchi, N.; Fukumoto, K. *J. Chem. Soc., Perkin Trans. 1* **1992**, 865.

Reimer–Tiemann reaction

Synthesis of *o*-formylphenol from phenols and chloroform in alkaline medium.

OH + $CHCl_3$ + 3 KOH ⟶ (OH, CHO) + 3 KCl + 2 H_2O

1. Carbene generation:

$Cl_3C{-}H$ + ^{-}OH $\xrightleftharpoons{\text{fast}}$ H_2O + $^{-}CCl_2{-}Cl$ $\xrightarrow[\alpha\text{-elimination}]{-\ Cl^{-},\ \text{slow}}$ $:CCl_2 \equiv$ "$\pm CCl_2$"

2. Addition of dichlorocarbene and hydrolysis:

OH ⟶(KOH) H_2O + O^{-} + $\pm CCl_2$ ⟶ ... CCl_2 ⟶

O^{-} ... Cl, CHCl ⟶ O ... Cl, ^{-}OH, H ⟶ O^{-} ... Cl, OH, H

⟶ O ... H ... O, H ⟶ OH, CHO

References

3. Reimer, K.; Tiemann, F. *Ber.* **1876**, *9*, 824.
4. Wynberg, H.; Meijer, E. W. *Org. React.* **1982**, *28*, 1.
5. Smith, K. M.; Bobe, F. W.; Minnetian, O. M.; Hope, H.; Yanuck, M. D. *J. Org. Chem.* **1985**, *50*, 790.
6. Bird, C. W.; Brown, A. L.; Chan, C. C. *Tetrahedron* **1985**, *41*, 4685.
7. Cochran, J. C.; Melville, M. G. *Synth. Commun.* **1990**, *20*, 609.
8. Langlois, B. R. *Tetrahedron Lett.* **1991**, *32*, 3691.

9. Jimenez, M. Co.; Miranda, M. A.; Tormos, R. *Tetrahedron* **1995**, *51*, 5825.
10. Jung, M. E.; Lazarova, T. I. *J. Org. Chem.* **1995**, *62*, 1553.
11. Castillo, R.; Moliner, V.; Andres, J. *Chem. Phys. Lett.* **2000**, *318*, 270.

Reissert reaction (aldehyde synthesis)

Aldehyde synthesis from the corresponding acid chloride, isoquinoline, and KCN.

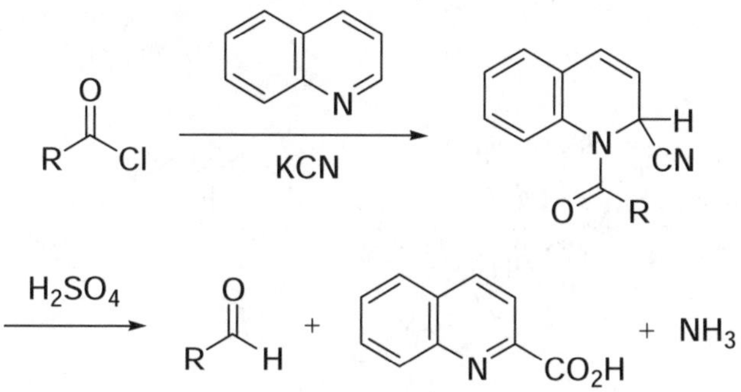

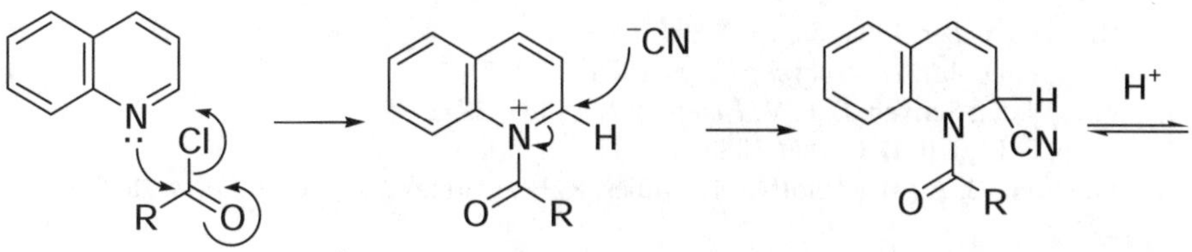

R–C(=O)–Cl + quinoline, KCN → Reissert compound

$\xrightarrow{H_2SO_4}$ R–CHO + quinoline-2-carboxylic acid (CO_2H) + NH_3

Reissert compound

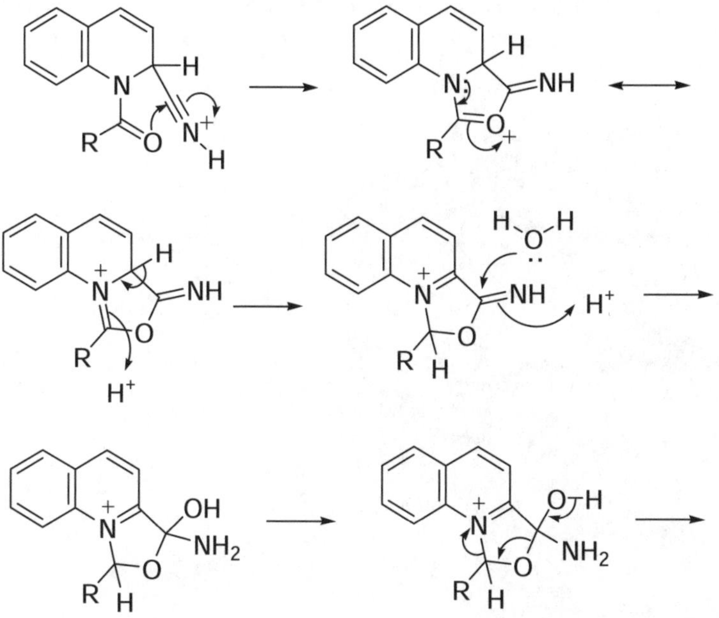

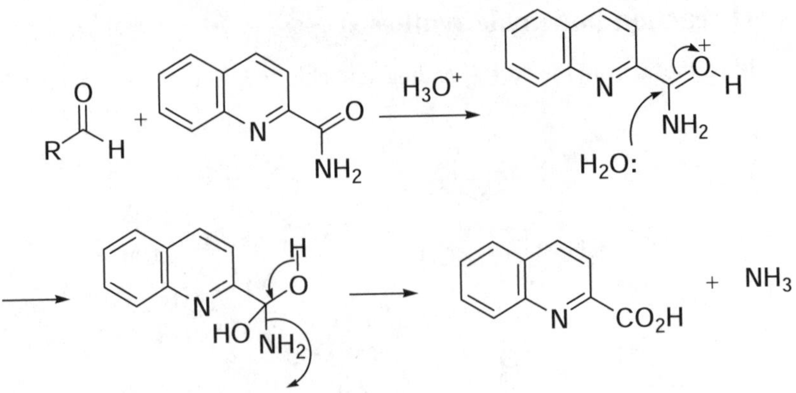

References

1. Reissert, A. *Ber.* **1905**, *38*, 1603, 3415.
2. Popp, F. D. *Adv. Heterocyclic Chem.* **1979**, *24*, 187.
3. Fife, W. K.; Scriven, E. F. V. *Heterocycles* **1984**, *22*, 2375.
4. Popp, F. D.; Uff, B. C. *ibid.* **1985**, *23*, 731.
5. Lorsbach, B. A.; Bagdanoff, J. T.; Miller, R. B.; Kurth, M. J. *J. Org. Chem.* **1998**, *63*, 2244.
6. Perrin, S.; Monnier, K.; Laude, B.; Kubicki, M.; Blacque, O. *Eur. J. Org. Chem.* **1999**, 297.
7. Takamura, M.; Funabashi, K.; Kanai, M.; Shibasaki, M. *J. Am. Chem. Soc.* **2001**, *123*, 6801.

Riley oxidation (Selenium dioxide oxidation)

A selenium dioxide oxidation of activated methylenes into ketones.

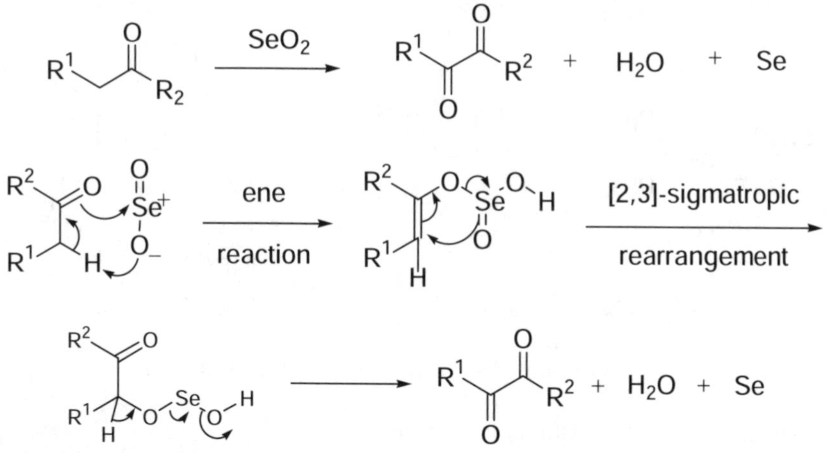

SeO_2 + H_2O + Se

ene reaction

[2,3]-sigmatropic rearrangement

+ H_2O + Se

References

1. Riley, H. L.; Morley, J. F.; Friend, N. A. C. *J. Chem. Soc.* **1932**, 1875.
2. Rabjohn, N. *Org. React.* **1976**, *24*, 261.
3. Goudgaon, N. M.; Nayak, U. R. *Indian J. Chem., Sect. B* **1985**, *24B*, 589.
4. Dalavoy, V. S.; Deodhar, V. B.; Nayak, U. R. *ibid.* **1987**, *26B*, 1.

Ring-closing metathesis (RCM) using Grubbs and Schrock catalysts

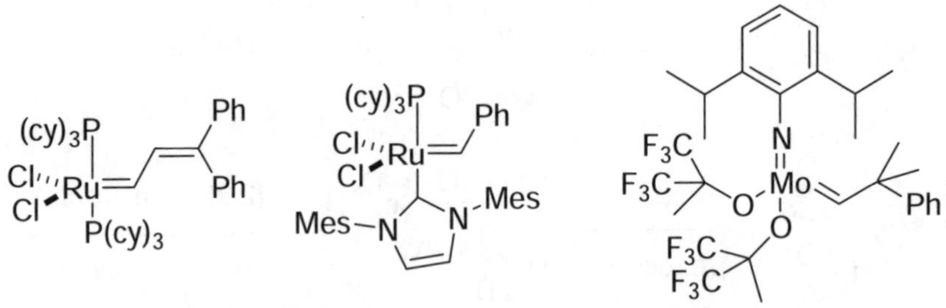

(cy)3P Ph Cl Ru Cl P(cy)3 Ph

(cy)3P Ph Cl Ru Cl Mes N N Mes

F3C F3C N Mo O O Ph F3C F3C

Grubbs' reagents
Mes = mesityl

Schrock's reagent

All three catalysts are illustrated as "L_nM=CHR" in the mechanism below.

Generation of the catalyst from the precatalysts:

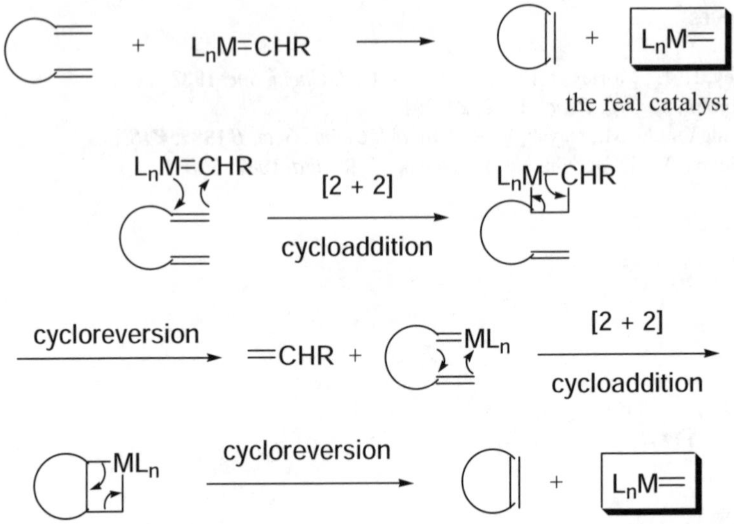

+ L_nM=CHR → + L_nM=
the real catalyst

L_nM=CHR [2 + 2] cycloaddition L_nM–CHR

cycloreversion =CHR + =ML_n [2 + 2] cycloaddition

ML_n cycloreversion + L_nM=

Catalytic cycle:

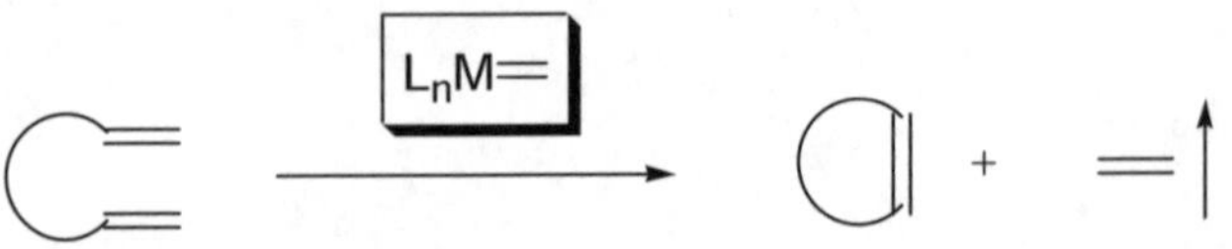

L_nM=
+ =↑

L_nM= [2 + 2] cycloaddition L_nM cycloreversion

+ =ML_n [2 + 2] cycloaddition

ML_n cycloreversion + L_nM=

References

1. Schrock R. R.; Murdzek, JS.; Bazan, G. C.; Robbins, J.; DiMare, M.; O'Reagan, M. *J. Am. Chem. Soc.* **1990**, *112*, 3875.
2. Grubbs, R. H.; Miller, S. J.; Fu, G. C. *Acc. Chem. Res.* **1995**, *28,* 446.
3. Armstrong, S. K. *J. Chem. Soc., Perkin Trans. 1* **1998**, 371.
4. Morgan, J. P.; Grubbs, R. H. *Org. Lett.* **2000**, *2*, 3153.
5. Renaud, J.; Graf, C.-D.; Oberer, L. *Angew. Chem., Int. Ed.* **2000**, *39*, 3101.
6. Lane, C.; Snieckus, V. *Synlett* **2000**, 1294.
7. Fellows, I. M.; Kaelin, D. E., Jr.; Martin, S. F. *J. Am. Chem. Soc.* **2000**, *122*, 10781.
8. Timmer, M. S. M.; Ovaa, H.; Filippov, D. V.; Van der Marel, G. A.; Van Boom, J. H. *Tetrahedron Lett.* **2000**, *41*, 8635.
9. Lee, C. W.; Grubbs, R. H. *J. Org. Chem.* **2001**, *66*, 7155.

Ritter reaction

General scheme:

$$R^1{-}OH + R^2{-}CN \xrightarrow{H^+} R^1{-}NH{-}C(=O){-}R^2$$

e.g.:

$$(CH_3)_3C{-}OH + H_3C{-}CN \xrightarrow[H_2O]{H_2SO_4} (CH_3)_3C{-}NH{-}C(=O){-}CH_3$$

Similarly:

$$(CH_3)_2C{=}CH_2 + H_3C{-}CN \xrightarrow[H_2O]{H_2SO_4} (CH_3)_3C{-}NH{-}C(=O){-}CH_3$$

$$(CH_3)_3C{-}OH \xrightarrow{H_3O^+} (CH_3)_3C{-}\overset{+}{O}H_2 \xrightarrow{E1} H_2O + (CH_3)_3C^+ \quad (:N{\equiv}C{-}CH_3)$$

$$\longrightarrow (CH_3)_3C{-}\overset{+}{N}{\equiv}C{-}CH_3 \longleftrightarrow (CH_3)_3C{-}N{=}\overset{+}{C}{-}CH_3 \quad (:OH_2) \longrightarrow$$

$$(CH_3)_3C{-}N{=}C(\overset{+}{O}H_2){-}CH_3 \xrightarrow{H^+} (CH_3)_3C{-}N{=}C(OH){-}CH_3 \longrightarrow (CH_3)_3C{-}NH{-}C(=O){-}CH_3$$

References

1. Ritter, J. J.; Minieri, P. P. *J. Am. Chem. Soc.* **1948**, *70*, 4045.
2. Ritter, J. J.; Kalish, J. *ibid.* **1948**, *70*, 4048.
3. Krimen, L. I.; Cota, D. J. *Org. React.* **1969**, *17*, 2123.
4. Djaidi, D.; Leung, I. S. H.; Bishop, R.; Craig, D. C.; Scudder, M. L. *Perkin 1* **2000**, 2037.
5. Jirgensons, A.; Kauss, V.; Kalvinsh, I.; Gold, M. R. *Synthesis* **2001**, 1709.
6. Le Goanic, D.; Lallemand, M.-C.; Tillequin, F.; Martens, T. *Tetrahedron Lett.* **2001**, *42*, 5175.

Robinson annulation

base

methyl vinyl ketone (MVK)

enolate formation

Michael addition

isomerization

aldol condensation

$- H_2O$

dehydration

References

1. Rapson, W. S.; Robinson, R. *J. Chem. Soc.* **1935**, 1285.
2. Gawley, R. E. *Synthesis* **1996**, 777.
3. Bui, T.; Barbas, C. F., III *Tetrahedron Lett.* **2000**, *41*, 6951.
4. Jansen, B. J. M.; Hendrix, C. C. J.; Masalov, N.; Stork, G. A.; Meulemans, T. M.; Macaev, F. Z.; De Groot, A. *Tetrahedron* **2000**, *56*, 2075.
5. Guarna, A.; Lombardi, E.; Machetti, F.; Occhiato, E. G.; Scarpi, D. *J. Org. Chem.* **2000**, *65*, 8093.
6. Tai, C.-L.; Ly, T. W.; Wu, J.-D.; Shia, K.-S,; Liu, H.-J. *Synlett* **2001**, 214.

Robinson–Schöpf reaction

Tropinone synthesis.

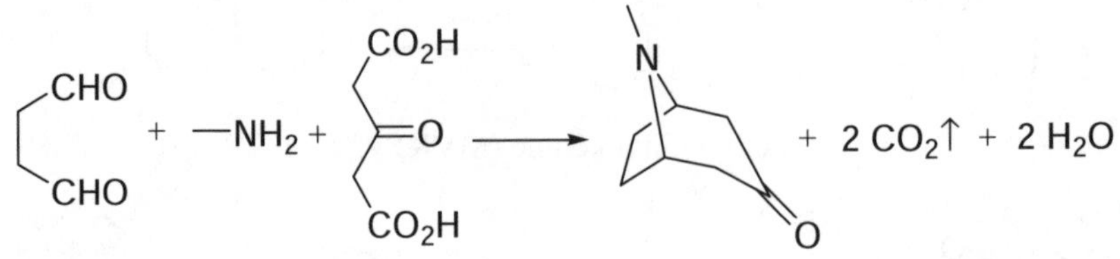

CHO CHO + —NH_2 + CO_2H =O CO_2H → N O + 2 CO_2↑ + 2 H_2O

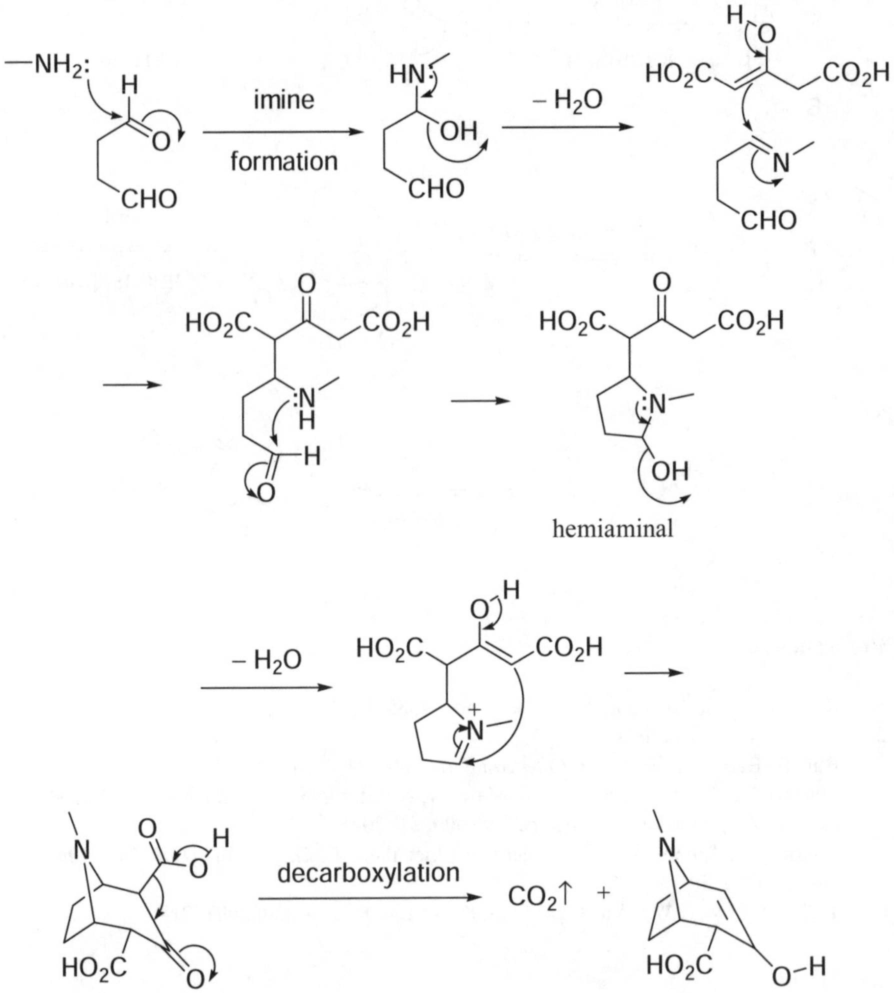

—NH_2: H O CHO —imine formation→ HN OH CHO —– H_2O→ H O HO_2C CO_2H N CHO

→ O HO_2C CO_2H N H O H → O HO_2C CO_2H N OH

hemiaminal

—– H_2O→ O H HO_2C CO_2H N+ →

N O O H HO_2C O —decarboxylation→ CO_2↑ + N HO_2C O–H

References

1. Robinson, R. *J. Chem. Soc.* **1917**, *111*, 762.
2. Büchi, G.; Fliri, H.; Shapiro, R. *J. Org. Chem.* **1978**, *43*, 4765.
3. Guerrier, L.; Royer, J.; Grierson, D. S.; Husson, H. P. *J. Am. Chem. Soc.* **1983**, *105*, 7754.
4. Royer, J.; Husson, H. P. *Tetrahedron Lett.* **1987**, *28*, 6175.
5. Langlois, M.; Yang, D.; Soulier, J. L.; Florac, C. *Synth. Commun.* **1992**, *22*, 3115.
6. Jarevang, T.; Anke, H.; Anke, T.; Erkel, G.; Sterner, O. *Acta Chem. Scand.* **1998**, *52*, 1350.

Roush allylboronate reagent

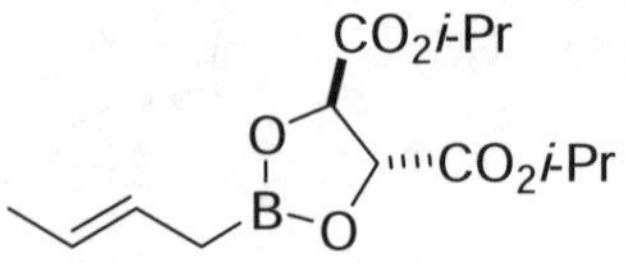

Tartrate allyl boronate, asymmetric allylation agent

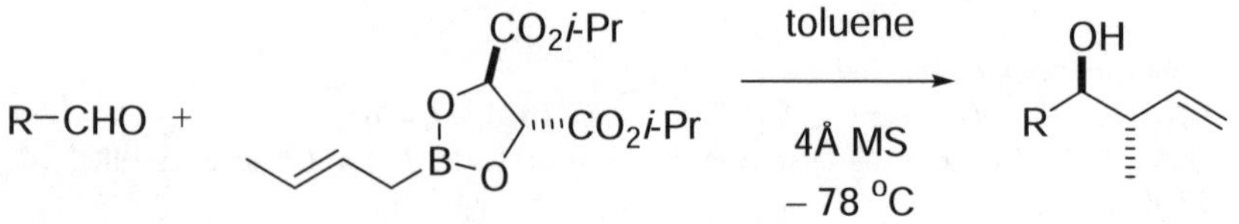

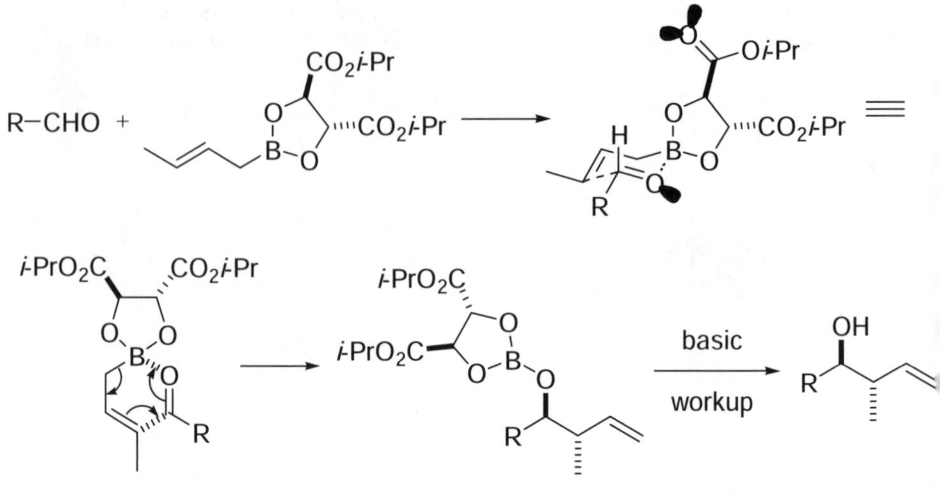

References

1. Roush, W. R.; Walts, A. E.; Hoong, L. K. *J. Am. Chem. Soc.* **1985**, *107*, 8186.
2. Roush, W. R.; Adam, M. A.; Walts, A. E.; Harris, D, J. *ibid.* **1986**, *108*, 3422.
3. Roush, W. R.; Ando, K.; Powers, D. B.; Halterman, R. L.; Palkowitz, A. D. *Tetrahedron Lett.* **1988**, *29*, 5579.
4. Brown, H. C.; Racherla, U. S.; Pellechia, P. J. *J. Org. Chem.* **1990**, *55*, 1868.
5. Kadota, I.; Yamamoto, Y. *Chemtracts: Org. Chem.* **1992**, *5*, 242.

Rubottom oxidation

α-Hydroxylation of enolsilanes.

1. *m*-CPBA

2. K_2CO_3, H_2O

Prilezhaev

epoxidation

The "butterfly" transition state

H^+

acyl

transfer

K_2CO_3, H_2O

hydrolysis

References

1. Gleiter, R.; Kraemer, R.; Irngartinger, H.; Bissinger, C. *J. Org. Chem.* **1992**, *57*, 252.
2. Johnson, C. R.; Golebiowski, A.; Steensma, D. H. *J. Am. Chem. Soc.* **1992**, *114*, 9414.
3. Jauch, J. *Tetrahedron* **1994**, *50*, 1203.
4. Gleiter, R.; Staib, M.; Ackermann, U. *Liebigs Ann.* **1995**, 1655.
5. Xu, Y.; Johnson, C. R. *Tetrahedron Lett.* **1997**, *38*, 1117.

Rupe rearrangement

The acid-catalyzed rearrangement of tertiary α-acetylenic (terminal) alcohols, leading to the formation of α,β-unsaturated ketones rather than the corresponding α,β-unsaturated aldehydes. *Cf.* Meyer–Schuster rearrangement.

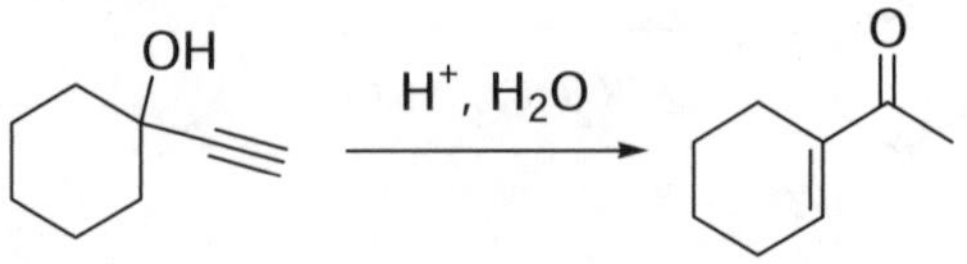

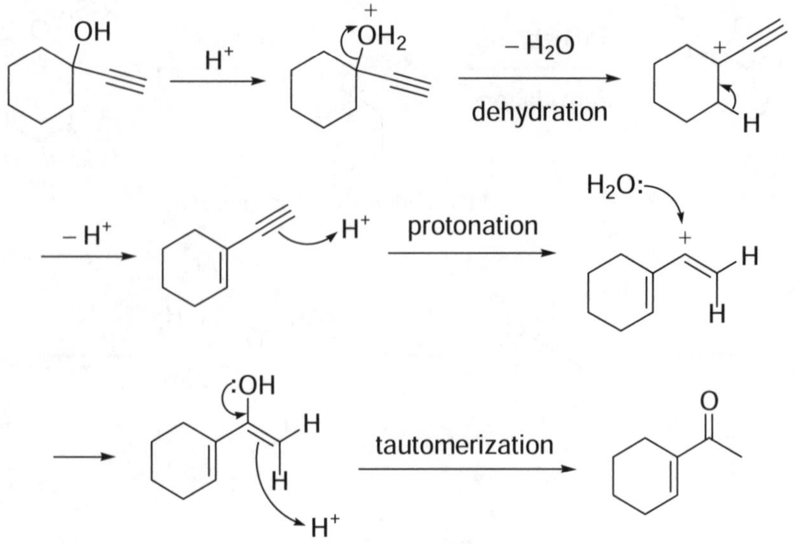

References

1. Schmidt, C.; Thazhuthaveetil, J. *Tetrahedron Lett.* **1970**, 2653.
2. Swaminathan, S.; Narayanan, K. V. *Chem. Rev.* **1971**, *71*, 429.
3. Hasbrouck, R. W.; Anderson, A. D. *J. Org. Chem.* **1973**, *38*, 2103.
4. Barre, V.; Massias, F.; Uguen, D. *Tetrahedron Lett.* **1989**, *30*, 7389.
5. An, J.; Bagnell, L.; Cablewski, T.; Strauss, C. R.; Trainor, R. W. *J. Org. Chem.* **1997**, *62*, 2505.
6. Strauss, C. R. *Aust. J. Chem.* **1999**, *52*, 83.

Rychnovsky polyol synthesis

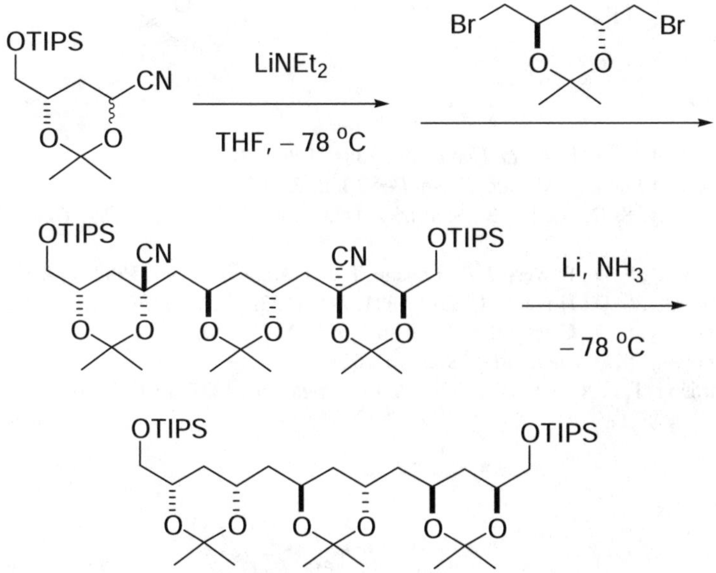

OTIPS CN $LiNEt_2$ THF, – 78 °C Br Br O O

OTIPS CN CN OTIPS Li, NH_3 – 78 °C

OTIPS OTIPS

The stereochemical outcome of the reductive decyanation:

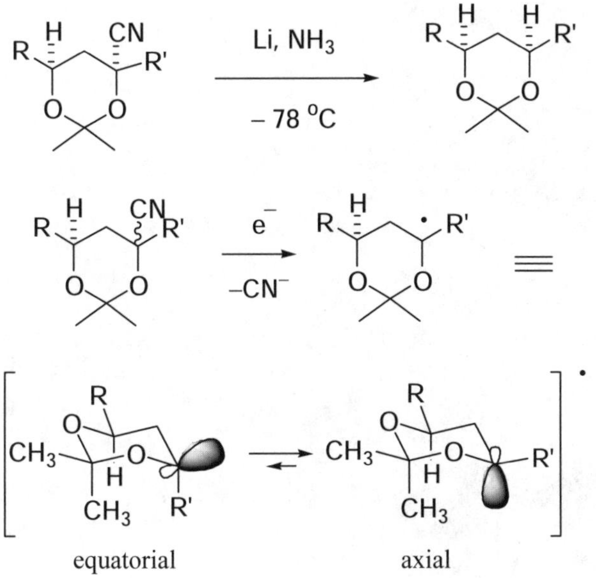

R H CN R' Li, NH_3 – 78 °C H H R R'

R H CN R' e⁻ –CN⁻ R H R' ≡

R O CH₃ O H CH₃ R' ⇌ R O CH₃ O H CH₃ R'

equatorial axial

References

1. Cohen, T.; Lin, M. T. *J. Am. Chem. Soc.* **1984**, *106*, 1130.
2. Cohen, T.; Bhupathy, M. *Acc. Chem. Res.* **1989**, *22*, 152.
3. Rychnovsky, S. D.; Zeller, S.; Skalitzky, D. J.; Griesgraber, G. *J. Org. Chem.* **1990**, *55*, 5550.
4. Rychnovsky, S. D.; Powers, J. P.; Lepage, T. J. *J. Am. Chem. Soc.* **1992**, *114*, 8375.
5. Rychnovsky, S. D.; Hoye, R. C. *ibid.* **1994**, *116*, 1753.
6. Rychnovsky, S. D.; Griesgraber, G.; Kim, J. *ibid.* **1994**, *116*, 2621.
7. Rychnovsky, S. D. *Chem. Rev.* **1995**, *95*, 2021.
8. Richardson, T. I.; Rychnovsky, S. D. *J. Am. Chem. Soc.* **1997**, *119*, 12360.

Sakurai allylation reaction (Hosomi–Sakurai reaction)

The β-carbocation is stabilized by the silicon group

References

1. Hisomi, A.; Sakurai, H. *Tetrahedron Lett.* **1976**, 1295.
2. Marko, I. E.; Mekhalfia, A.; Murphy, F.; Bayston, D. J.; Bailey, M.; Janoouusek, Z.; Dolan, S. *Pure Appl. Chem.* **1997**, *69*, 565.
3. Bonini, B. F.; Comes-Franchini, M.; Fochi, M.; Mazzanti, G.; Ricci, A.; Varchi, G. *Tetrahedron: Asymmetry* **1998**, *9*, 2979.
4. Wang, D.-K.; Zhou, Y.-G.; Tang, Y.; Hou, X.-L.; Dai, L.-X. *J. Org. Chem.* **1999**, *64*, 4233.
5. Sugita, Y.; Kimura, Y.; Yokoe, I. *Tetrahedron Lett.* **1999**, *40*, 5877.

6. Wang, M. W.; Chen, Y. J.; Wang, D. *Synlett* **2000**, 385.
7. Organ, M. G.; Dragan, V.; Miller, M.; Froese, R. D. J.; Goddard, J. D. *J. Org. Chem.* **2000**, *65*, 3666.
8. Tori, M.; Makino, C.; Hisazumi, K.; Sono, M.; Nakashima, K. *Tetrahedron: Asymmetry* **2001**, *12*, 301.

Sandmeyer reaction

Haloarenes from the reaction of diazonium salt and CuX.

$$ArN_2^+\ Y^- \xrightarrow{CuX} Ar{-}X \qquad X = Cl, Br, CN$$

e.g.:

$$ArN_2^+\ Cl^- \xrightarrow{CuCl} Ar{-}Cl$$

$$ArN_2^+\ Cl^- \xrightarrow{CuCl} N_2\uparrow + Ar\bullet + CuCl_2 \longrightarrow Ar{-}Cl + CuCl$$

References

1. Sandmeyer, T. *Ber.* **1884**, *17*, 1633.
2. Galli, C. *J. Chem. Soc., Perkin Trans. 2* **1984**, 897.
3. Suzuki, N.; Azuma, T.; Kaneko, Y.; Izawa, Y.; Tomioka, H.; Nomoto, T. *J. Chem. Soc., Perkin Trans. 1* **1987**, 645.
4. Merkushev, E. B. *Synthesis* **1988**, 923.
5. Obushak, M. D.; Lyakhovych, M. B.; Ganushchak, M. I. *Tetrahedron Lett.* **1998**, *39*, 9567.
6. Hanson, P.; Lovenich, P. W.; Rowell, S. C.; Walton, P. H.; Timms, A. W. *J. Chem. Soc., Perkin Trans. 2* **1999**, 49.
7. Chandler, St. A.; Hanson, P.; Taylor, A. B.; Walton, P. H.; Timms, A. W. *J. Chem. Soc., Perkin Trans. 2* **2001**, 214.

Sarett oxidation

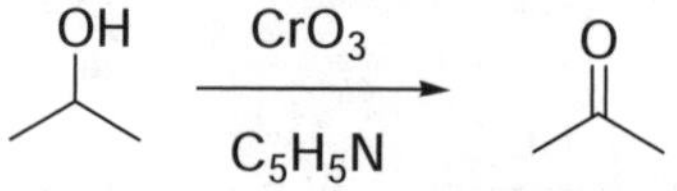

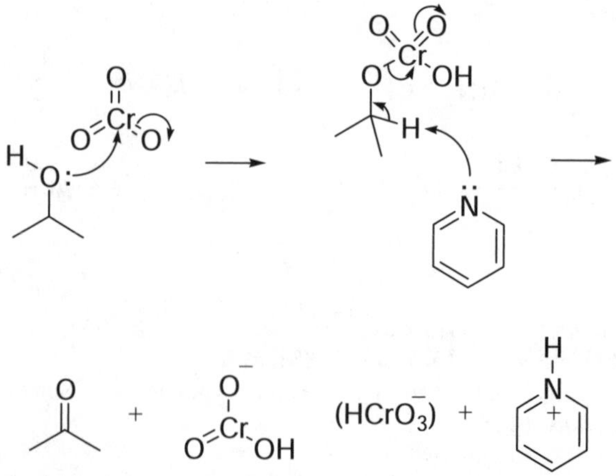

$$\text{(CH}_3)_2\text{CHOH} \xrightarrow[C_5H_5N]{CrO_3} (CH_3)_2C{=}O$$

(HCrO$_3^-$)

The intramolecular mechanism is also operative:

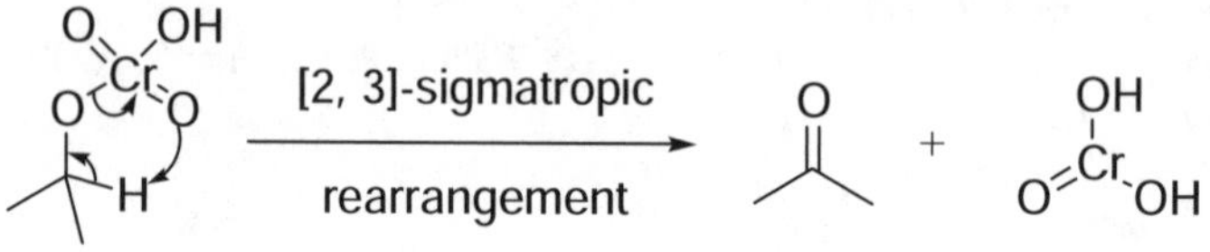

[2, 3]-sigmatropic rearrangement

The Collins oxidation, Jones oxidation, and Corey´s PCC (pyridinium chlorochromate) and PDC (pyridinium dichromate) oxidations follow a similar pathway.

References

1. Poos, G. I.; Arth, G. E.; Beyler, R. E.; Sarett, L. H. *J. Am. Chem. Soc.* **1953**, *75*, 422.
2. Ratcliffe, R. W. *Org. Syn.* **1973**, *53*, 1852.
3. Andrieux, J.; Bodo, B.; Cunha, H.; Deschamps-Vallet, C.; Meyer-Dayan, M.; Molho, D. *Bull. Soc. Chim. Fr.* **1976**, 1975.
4. Glinski, J. A.; Joshi, B. S.; Jiang, Q. P.; Pelletier, S. W. *Heterocycles* **1988**, *27*, 185.
5. Caamano, O.; Fernandez, F.; Garcia-Mera, X.; Rodriguez-Borges, J. E. *Tetrahedron Lett.* **2000**, *41*, 4123.

Schiemann reaction (Balz–Schiemann reaction)

Fluoroarene formation from arylamines.

$$\mathrm{Ar{-}NH_2 + HNO_2 + HBF_4 \longrightarrow ArN_2^{+}\ BF_4^{-} \xrightarrow{\Delta} Ar{-}F + N_2\uparrow + BF_3}$$

$$\mathrm{HO{-}N{=}O \xrightarrow{HBF_4} H_2\overset{+}{O}{-}N{=}O \longrightarrow H_2O + N{\equiv}O^{+}}$$

$$\mathrm{Ar{-}NH_2 + N{\equiv}O^{+} \xrightarrow{-H^{+}} Ar{-}NH{-}N{=}O \longrightarrow Ar{-}N{=}N{-}OH}$$

$$\mathrm{\xrightarrow{H^{+}} Ar{-}N{=}N{-}\overset{+}{O}H_2 \longrightarrow H_2O + Ar{-}\overset{+}{N}{\equiv}N \longleftrightarrow}$$

$$\mathrm{ArN_2^{+}\ BF_4^{-} \xrightarrow{\Delta} N_2\uparrow + Ar^{+} + \overset{-}{F}{-}BF_3 \longrightarrow Ar{-}F + BF_3}$$

References

1. Balz, G.; Schiemann. G. *Ber.* **1927**, *60*, 1186.
2. Sharts, C. M. *J. Chem. Educ.* **1968**, *45*, 185.
3. Matsumoto, J.; Miyamoto, T.; Minamida, A.; Nishimura, Y.; Egawa, H.; Nishimura, H. *J. Heterocycl. Chem.* **1984**, *21*, 673.
4. Corral, C.; Lasso, A.; Lissavetzky, J.; Sanchez Alvarez-Insua, A.; Valdeolmillos, A. M. *Heterocycles* **1985**, *23*, 1431.
5. Tsuge, A.; Moriguchi, T.; Mataka, S.; Tashiro, M. *J. Chem. Res., (S)* **1995**, 460.
6. Saeki, K.-i.; Tomomitsu, M.; Kawazoe, Y.; Momota, K.; Kimoto, H. *Chem. Pharm. Bull.* **1996**, *44*, 2254.
7. Laali, K. K.; Gettwert, V. J. *J. Fluorine Chem.* **2001**, *107*, 31.

Schlosser modification of the Wittig reaction

The normal Wittig reaction of nonstabilized ylides with aldehydes gives *Z*-olefins. The Schlosser modification of the Wittig reaction of nonstabilized ylides furnishes *E*-olefins instead.

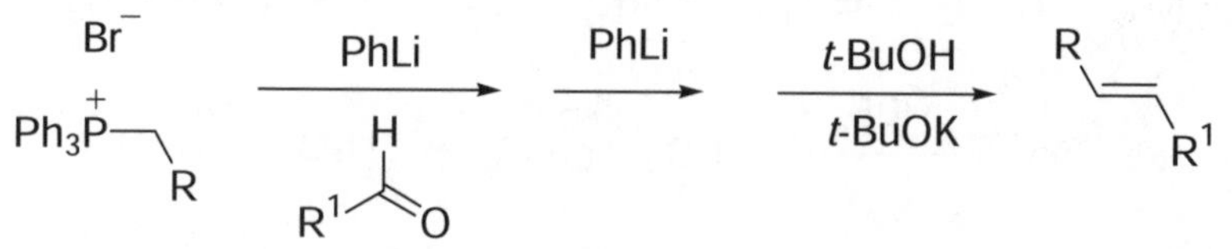

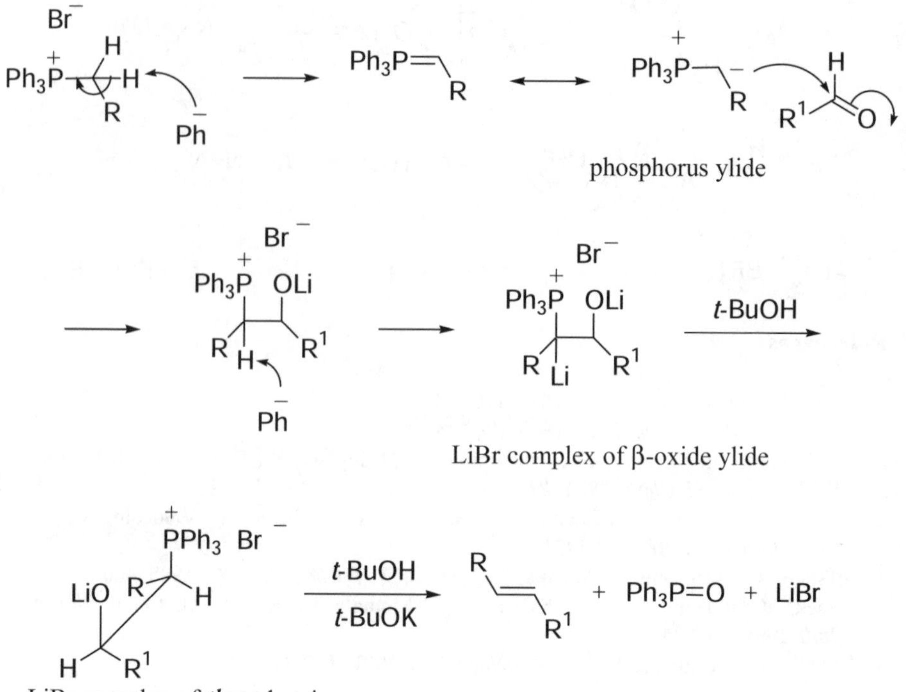

phosphorus ylide

LiBr complex of β-oxide ylide

LiBr complex of *threo*-betaine

References

1. Schlosser, M.; Christmann, K. F. *Angew. Chem., Int. Ed. Engl.* **1966**, *5*, 126.
2. Schlosser, M.; Christmann, K. F. *Liebigs Ann. Chem.* **1967**, *708*, 35.
3. Schlosser, M.; Christmann, K. F.; Piskala, A.; Coffinet, D. *Synthesis* **1971**, 29.
4. Deagostino, A.; Prandi, C.; Tonachini, G.; Venturello, P. *Trends Org. Chem.* **1995**, *5*, 103.
5. Celatka, C. A.; Liu, P.; Panek, J. S. *Tetrahedron Lett.* **1997**, *38*, 5449.

Schmidt reaction

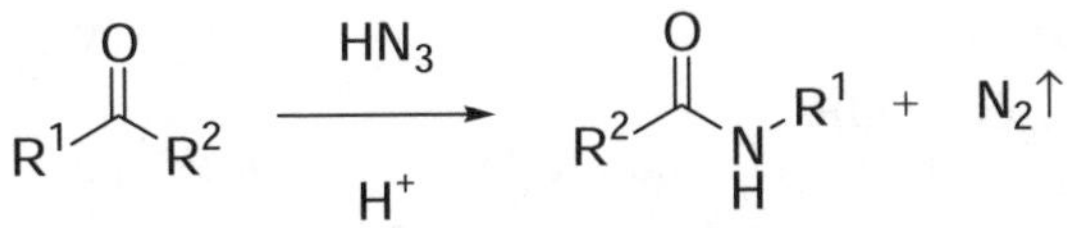

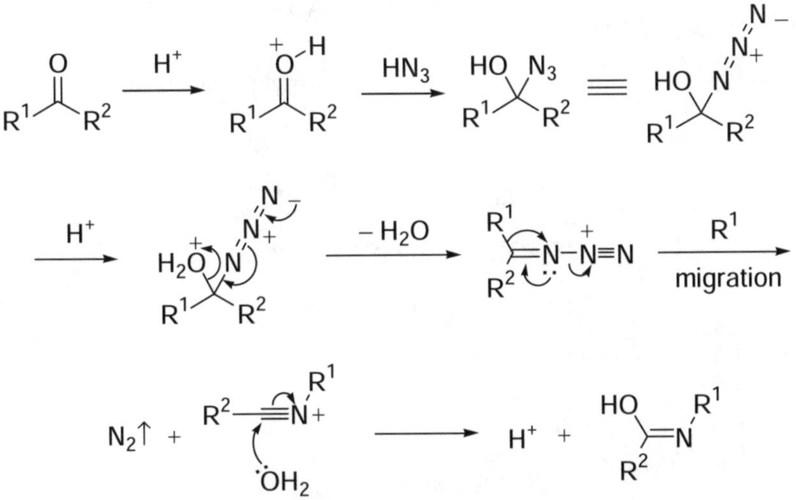

nitrilium ion intermediate (*Cf.* Ritter intermediate)

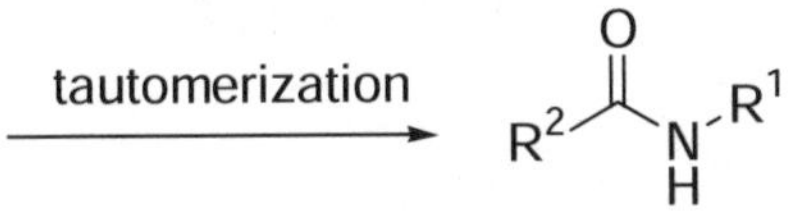

References

1. Schmidt, R. F. *Ber.* **1924**, *57*, 704.
2. Richard, J. P.; Amyes, T. L.; Lee, Y.-G.; Jagannadham, V. *J. Am. Chem. Soc.* **1994**, *116*, 10833.
3. Kaye, P. T.; Mphahlele, M. J. *Synth. Commun.* **1995**, *25*, 1495.
4. Krow, G. R.; Szczepanski, S W.; Kim, J. Y.; Liu, N.; Sheikh, A.; Xiao, Y.; Yuan, J. *J. Org. Chem.* **1999**, *64*, 1254.
5. Mphahlele, M. J. *Phosphorus, Sulfur Silicon Relat. Elem.* **1999**, *144-146*, 351.
6. Mphahlele, M. J. *J. Chem. Soc., Perkin Trans. 1* **1999**, 3477.
7. Iyengar, R.; Schildknegt, K.; Aubé, J. *Org. Lett.* **2000**, *2*, 1625.
8. Pearson, W. H.; Hutta, D. A.; Fang, W.-k. *J. Org. Chem.* **2000**, *65*, 8326.
9. Pearson, W. H.; Walavalkar, R. *Tetrahedron* **2001**, *57*, 5081.

Schmidt's trichloroacetimidate glycosidation reaction

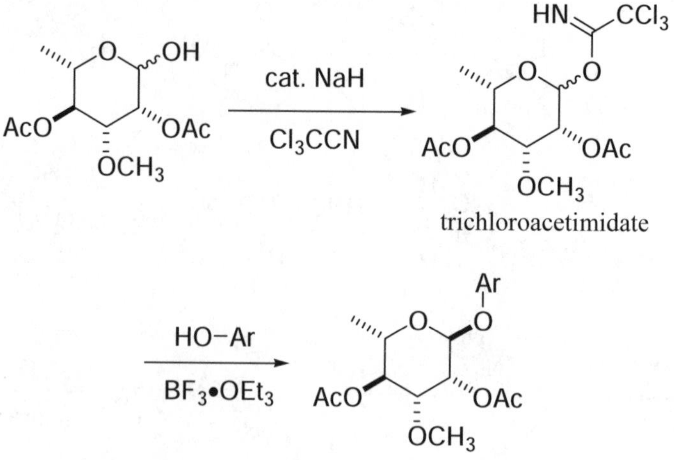

cat. NaH

Cl_3CCN

trichloroacetimidate

HO–Ar

$BF_3 \bullet OEt_3$

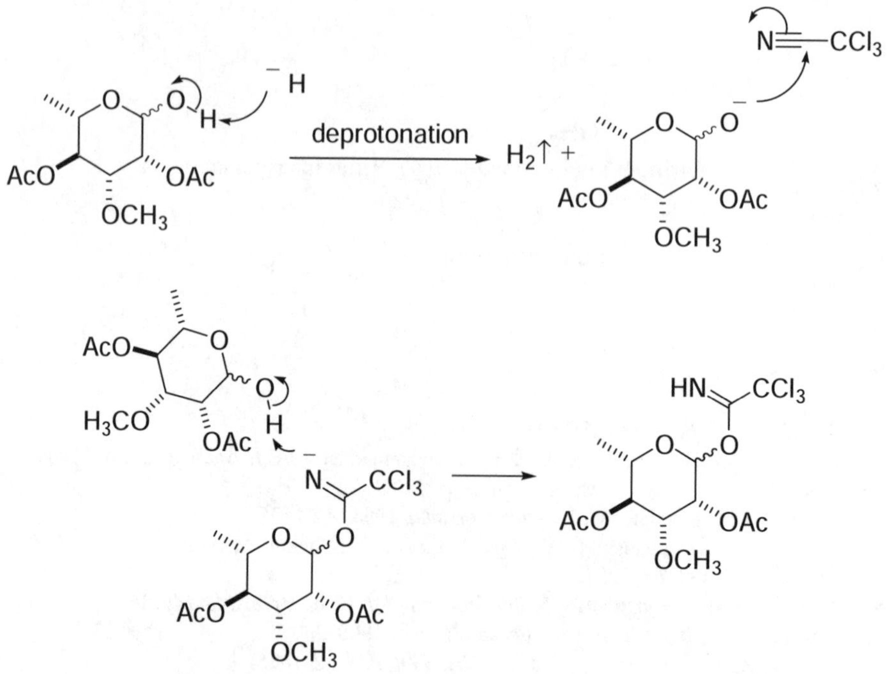

deprotonation

$H_2\uparrow$ +

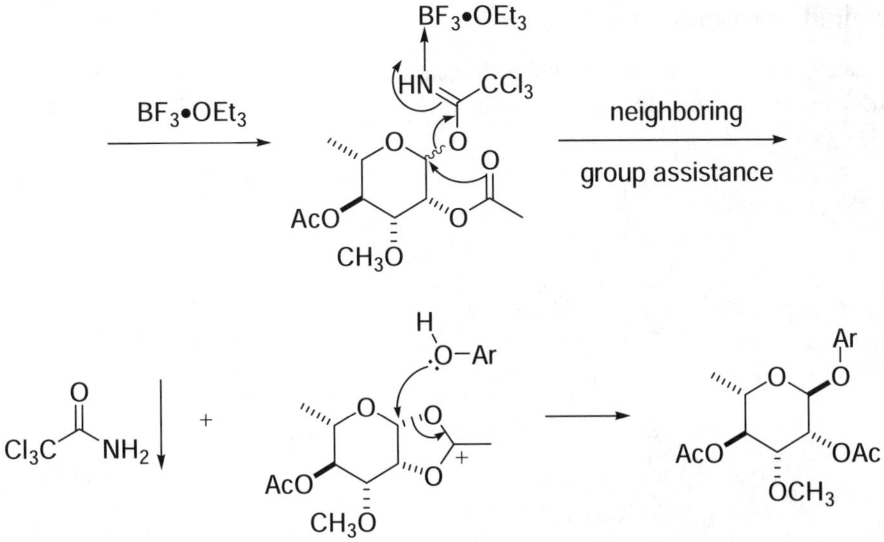
BF3•OEt3
HN
CCl3
BF3•OEt3
neighboring
group assistance
O
O
O
AcO
O
CH3O
H
O−Ar
O
Cl3C
NH2
+
O
O
AcO
O
CH3O
Ar
O
O
AcO
OAc
OCH3

References

1. Grundler, G.; Schmidt, R. R. *Carbohydr. Res.* **1985**, *135*, 203.
2. Schmidt, R. R. *Angew. Chem., Int. Ed. Engl.* **1986**, *25*, 212.
3. Toshima, K.; Tatsuta, K. *Chem. Rev.* **1993**, *93*, 1503.
4. Nicolaou, K. C. *Angew. Chem., Int. Ed. Engl.* **1993**, *32*, 1377.
5. Weingart, R.; Schmidt, R. R. *Tetrahedron Lett.* **2000**, *41*, 8753.

Scholl reaction

The elimination of two aryl-bound hydrogens accompanied by the formation of an aryl-aryl bond under the influence of Friedel–Crafts catalysts. *Cf.* Friedel–Crafts reaction.

$AlCl_3$

H^+

$AlCl_3$

$\bar{A}lCl_3$

$-\,H^+$

H_2O

workup

$-\,H_2$

autooxidation or DDQ

References

1. Scholl, R.; Seer, C. *Ann,* **1912**, *394*, 111.
2. Clowes, G. A. *J. Chem. Soc., C* **1968**, 2519.
3. Olah, G. A.; Schilling, P.; Gross, I. M. *J. Am. Chem. Soc.* **1974**, *96*, 876.
4. Dopper, J. H.; Oudman, D.; Wynberg, H. *J. Org. Chem.* **1975**, *40*, 3398.

5. Poutsma, M. L.; Dworkin, A. S.; Brynestad, J.; Brown, L. L.; Benjamin, B. M.; Smith, G. P. *Tetrahedron Lett.* **1978**, 873.
6. Youssef, A. K.; Vingiello, F. A.; Ogliaruso, M. A. *Org. Prep. Proced. Int.* **1979**, *11*, 17.
7. Pritchard, R. G.; Steele, M.; Watkinson, M.; Whiting, A. *Tetrahedron Lett.* **2000**, *41*, 6915.
8. Ma, C.; Liu, X.; Li, X.; Flippen-Anderson, J.; Yu, Sh.; Cook, J. M. *J. Org. Chem.* **2001**, *66*, 4525.

Schöpf reaction

CO_2Na

enolate formation

CO_2H

nucleophilic addition

decarboxylation

$CO_2\uparrow$ +

acidic workup

References

1. Schöpf, C.; Braun, F.; Burkhardt, K.; Dummer, G.; Müller, H. *Ann,* **1959**, *626*, 123.
2. Guerrier, L.; Royer, J.; Grierson, D. S.; Husson, H. P. *J. Am. Chem. Soc.* **1983**, *105*, 7754.
3. Bermudez, J.; Gregory, J. A.; King, F. D.; Starr, S.; Summersell, R. J. *Bioorg. Med. Chem. Lett.* **1992**, *2*, 519.
4. Jarevang, T.; Anke, H.; Anke, T.; Erkel, G.; Sterner, O. *Acta Chem. Scand.* **1998**, *52*, 1350.
5. Bender, D. R.; Bjelfdanes, L. F.; Knapp, D. R.; Rapopport, H. *J. Org. Chem.* **1975**, *40*, 1264.

Schotten–Baumann reaction

Esterification or amidation of acid chloride with alcohol or amine under basic conditions.

$$RCOX + R^1{-}OH \xrightarrow{NaOH} RCOOR^1 \quad , X = Cl, Br$$

$$RCOOR^1 + HX \xrightarrow{NaOH} NaX + H_2O$$

References

1. Schotten, C. *Ber.* **1884**, *17*, 2544.
2. Altman, J.; Ben-Ishai, D. *J. Heterocycl. Chem.* **1968**, *5*, 679.
3. Babad, E.; Ben-Ishai, D. *ibid.* **1969**, *6*, 235.
4. Tsuchiya, M.; Yoshida, H.; Ogata, T.; Inokawa, S. *Bull. Chem. Soc. Jpn.* **1969**, *42*, 1756.
5. Gutteridge, N. J. A.; Dales, J. R. M. *J. Chem. Soc., C* **1971**, 122.
6. Low, C. M. R.; Broughton, H. B.; Kalindjian, S. B.; McDonald, I. M. *Bioorg. Med. Chem. Lett.* **1992**, *2*, 325.
7. Sano, T.; Sugaya, T.; Inoue, K.; Mizutaki, S.-i.; Ono, Y.; Kasai, M. *Org. Process Res. Dev.* **2000**, *4*, 147.

Shapiro reaction

The Shapiro reaction is a variant of the Bamford–Stevens reaction. The former uses bases such as alkyllithiums and Grignard reagents whereas the latter employs bases such as Na, NaOMe, LiH, NaH, $NaNH_2$, *etc.* As a result, the Shapiro reaction generally affords the less-substituted olefins as the kinetic products, while the Bamford–Stevens reaction delivers the more-substituted olefins as the thermodynamic products.

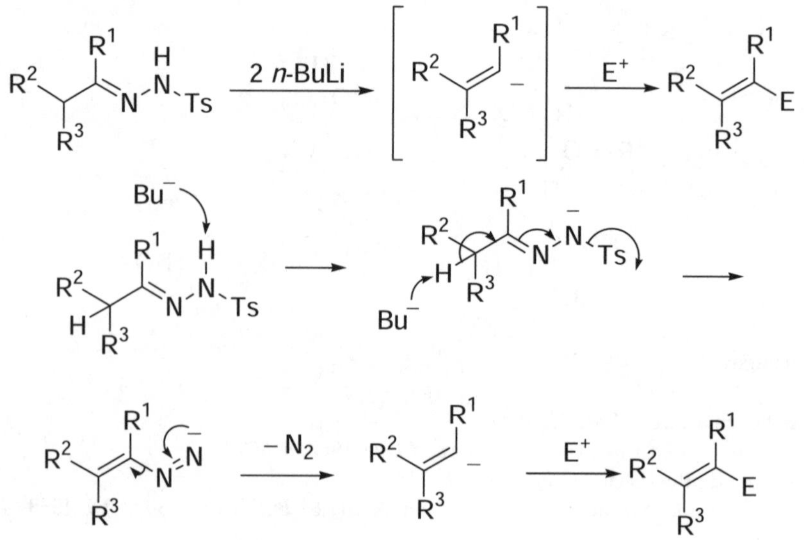

References

1. Bamford, W. R.; Stevens, T. S. M. *J. Chem. Soc.* **1952**, 4735.
2. Casanova, J.; Waegell, B. *Bull. Soc. Chim. Fr.* **1975**, 922.
3. Shapiro, R. H. *Org. React.* **1976**, *23*, 405.
4. Adlington, R. M.; Barrett, A. G. M. *Acc. Chem. Res.* **1983**, *16*, 55.

Sharpless asymmetric aminohydroxylation

Osmium-mediated *cis*-addition of nitrogen and oxygen to olefins. Nitrogen sources (X–NClNa) include:

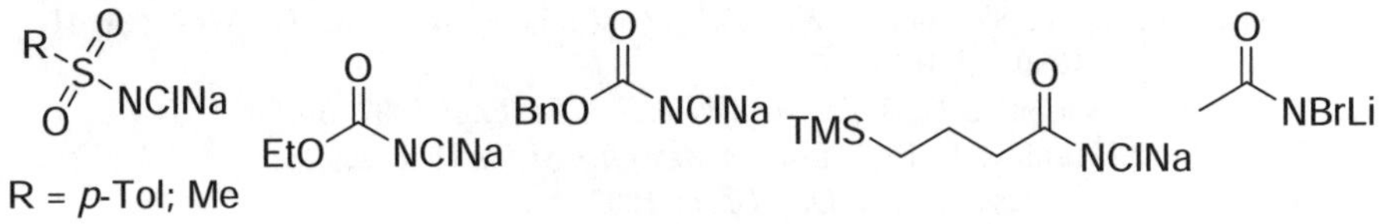

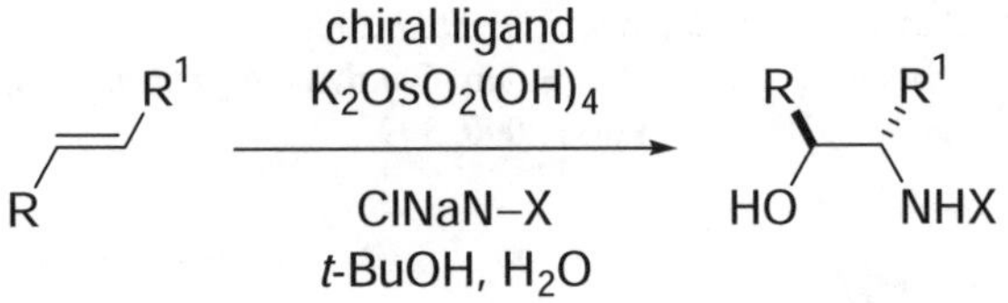

$R-SO_2-NClNa$ (R = *p*-Tol; Me)

EtO–C(=O)–NClNa

BnO–C(=O)–NClNa

TMS–$CH_2CH_2CH_2$–C(=O)–NClNa

CH_3–C(=O)–NBrLi

R = *p*-Tol; Me

$$\text{R-CH=CH-R}^1 \xrightarrow[\text{ClNaN–X, } t\text{-BuOH, } H_2O]{\text{chiral ligand, } K_2OsO_2(OH)_4} \text{R(HO)CH–CH(NHX)R}^1$$

chiral ligand

$K_2OsO_2(OH)_4$

ClNaN–X

t-BuOH, H_2O

R, R^1, HO, NHX

The catalytic cycle:

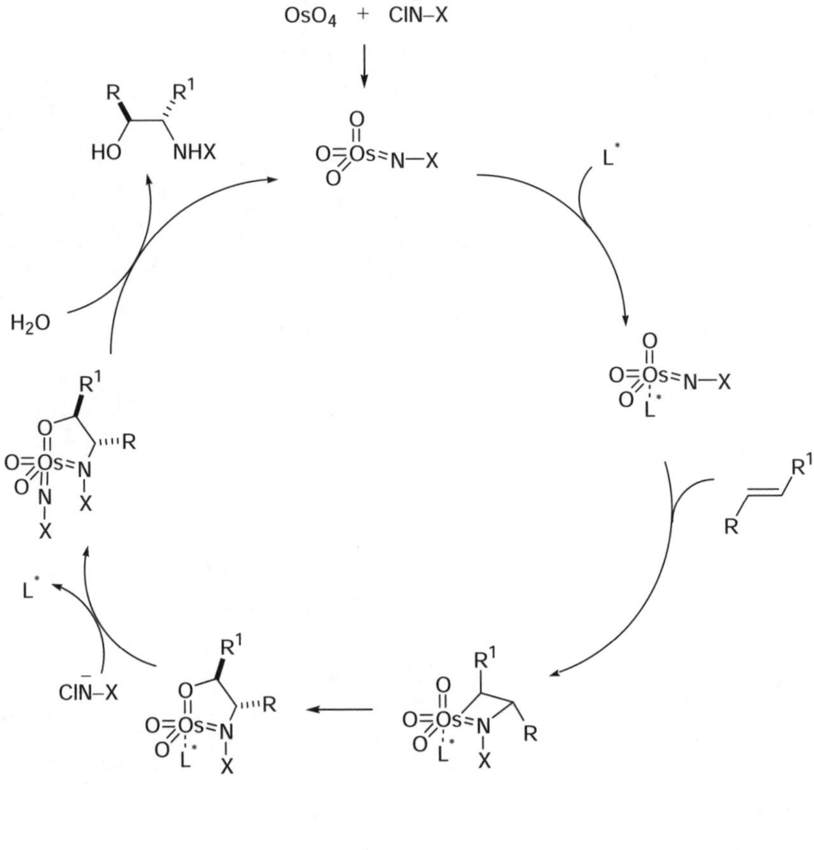

OsO_4 + ClN–X

O=Os(=O)(=O)=N–X

L*

O=Os(=O)(=O)=N–X · L*

R–CH=CH–R^1

O=Os(=O)(L*)=N(X) [2+2] adduct with R and R^1

O=Os(=O)(L*)(O)=N(X) azaosmaglycolate with R and R^1

ClN–X

L*

O=Os(=O)(=N–X)(O)=N(X) with R and R^1

H_2O

R, R^1, HO, NHX

References

1. Herranz, E.; Sharpless, K. B. *J. Org. Chem.* **1978**, *43*, 2544.
2. Mangatal, L.; Adeline, M. T.; Guenard, D.; Gueritte-Voegelein, F.; Potier, P. *Tetrahedron* **1989**, *45*, 4177.
3. Engelhardt, L. M.; Skelton, B. W.; Stick, R. V.; Tilbrook, D. M. G.; White, A. H. *Aust. J. Chem.* **1990**, *43*, 1657.
4. Rubin, A. E.; Sharpless, K. B. *Angew. Chem., Int. Ed. Engl.* **1997**, *36*, 2637.
5. Kolb, H. C.; Sharpless, K. B. *Transition Met. Org. Synth.* **1998**, *2*, 243.
6. Thomas, A.; Sharpless, K. B. *J. Org. Chem.* **1999**, *64*, 8279.
7. Gontcharov, A. V.; Liu, H.; Sharpless, K. B. *Org. Lett.* **1999**, *1*, 783.
8. Demko, Z. P.; Bartsch, M.; Sharpless, K. B. *ibid.* **2000**, *2*, 2221.
9. Bolm, C.; Hildebrand, J. P.; Muñiz, K. In *Catalytic Asymmetric Synthesis* 2nd ed., Ojima, I., ed.; Wiley-VCH: New York, **2000**, 399.

Sharpless asymmetric epoxidation

Enantioselective epoxidation of allylic alcohols using *t*-butyl peroxide, titanium tetra-*iso*-propoxide, and optically pure diethyl tartrate.

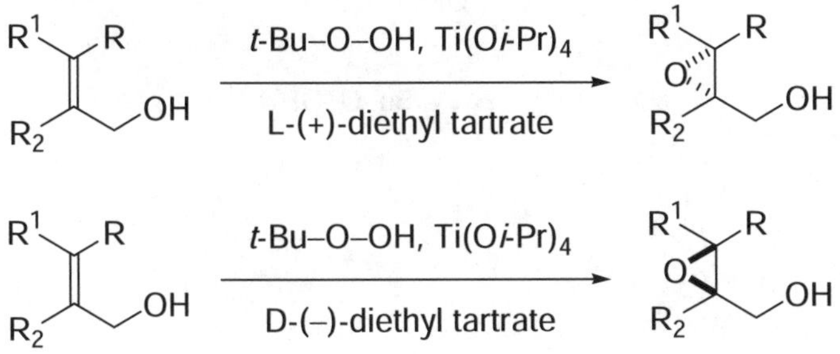

t-Bu–O–OH, $Ti(Oi\text{-}Pr)_4$

L-(+)-diethyl tartrate

t-Bu–O–OH, $Ti(Oi\text{-}Pr)_4$

D-(–)-diethyl tartrate

The putative active catalyst [2]:

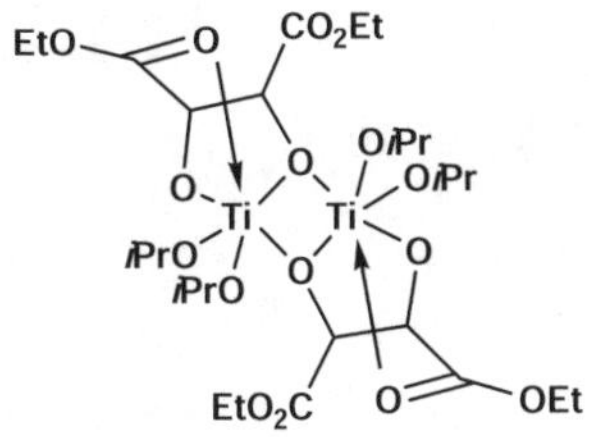

The transition state:

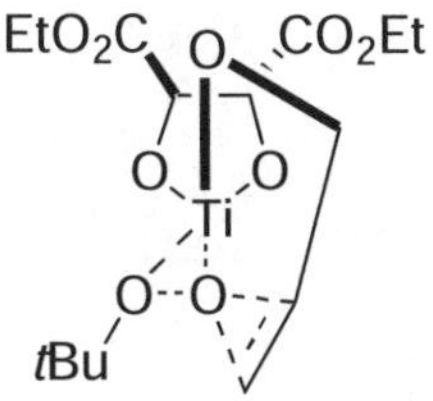

The catalytic cycle:

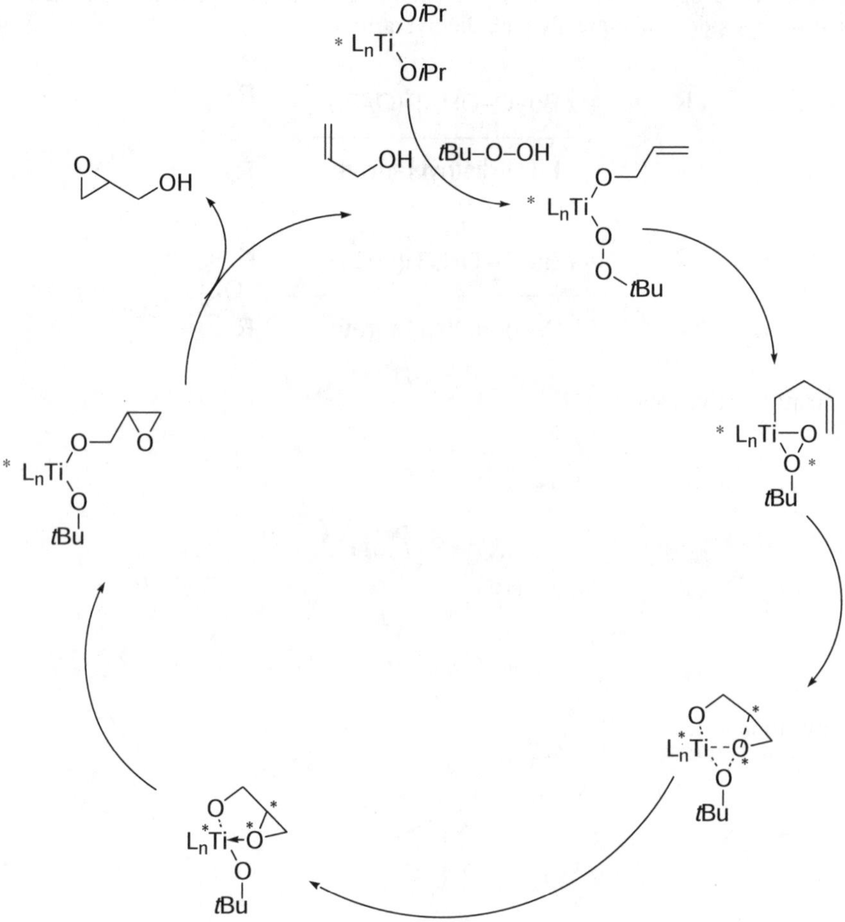

References

1. Katsuki, T.; Sharpless, K. B. *J. Am. Chem. Soc.* **1980**, *102*, 5974.
2. Williams, I. D.; Pedersen, S. F.; Sharpless, K. B.; Lippard, S. J. *ibid.* **1984**, *106*, 6430.
3. Rossiter, B. E. *Chem. Ind.* **1985**, *22(Catal. Org. React.),* 295.
4. Pfenninger, A. *Synthesis* **1986**, 89.
5. Corey, E. J. *J. Org. Chem.* **1990**, *55*, 1693.
6. Woodard, S. S.; Finn, M. G.; Sharpless, K. B. *J. Am. Chem. Soc.* **1991**, *113*, 106.
7. Schinzer, D. *Org. Synth. Highlights II* **1995**, 3.
8. Katsuki, T.; Martin, V. S. *Org. React.* **1996**, *48*, 1–299.
9. Johnson, R. A.; Sharpless, K. B. In *Catalytic Asymmetric Synthesis* 2nd ed., Ojima, I., ed. Wiley-VCH: New York, **2000**, 231.

Sharpless dihydroxylation

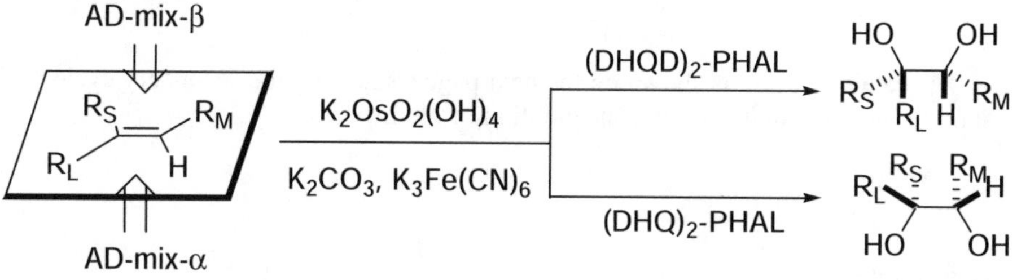

AD-mix-β
R_S R_M R_L H
AD-mix-α
$K_2OsO_2(OH)_4$
K_2CO_3, $K_3Fe(CN)_6$
$(DHQD)_2$-PHAL
$(DHQ)_2$-PHAL

$(DHQD)_2$-PHAL = 1,4-bis(9-*O*-dihydroquinidine)phthalazine:

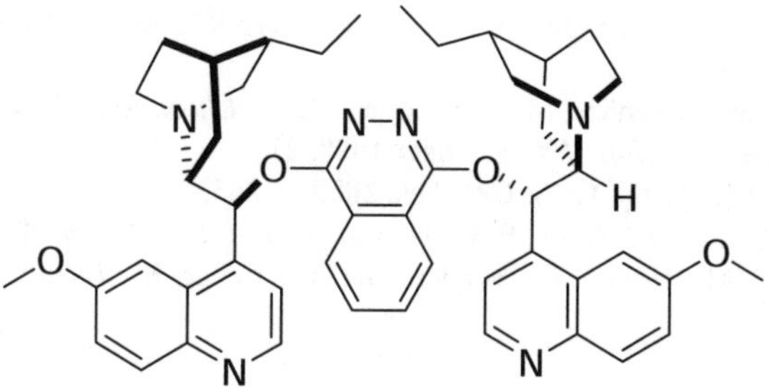

$(DHQ)_2$-PHAL = 1,4-bis(9-*O*-dihydroquinine)phthalazine:

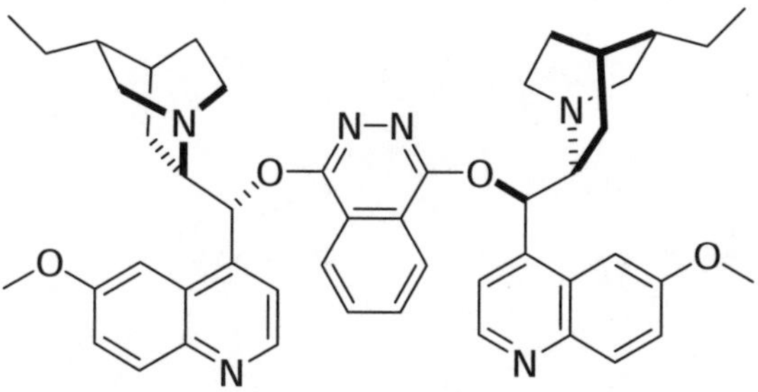

A stepwise mechanism involving osmaoxetane seems to be more consistent with the experimental data than the corresponding concerted [3 + 2] mechanism:

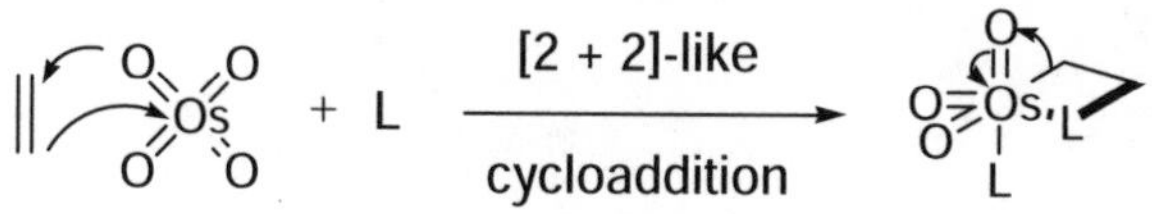

[2 + 2]-like
cycloaddition

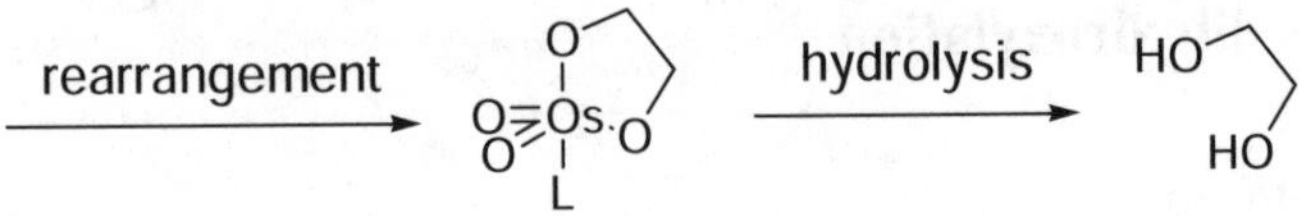

rearrangement → [osmate ester with L] → hydrolysis → HO / HO (diol)

The catalytic cycle is shown on the next page (page 337, the secondary cycle is shut off by maintaining a low concentration of olefin):

References

1. Jacobsen, E. N.; Markó, I.; Mungall, W. S.; Schröder, G.; Sharpless, K. B. *J. Am. Chem. Soc.* **1988**, *110*, 1968.
2. Wai, J. S. M.; Markó, I.; Svenden, J. S.; Finn, M. G.; Jacobsen, E. N.; Sharpless, K. B. *ibid.* **1989**, *111*, 1123.
3. Kolb, H. C.; VanNiewenhze, M. S.; Sharpless, K. B. *Chem. Rev.* **1994**, *94*, 2483.
4. Bolm, C.; Gerlach, A. *Eur. J. Org. Chem.* **1998**, 21.
5. Balachari, D.; O'Doherty, G. A. *Org. Lett.* **2000**, *2*, 863.
6. Liang, J.; Moher, E. D.; Moore, R. E.; Hoard, D. W. *J. Org. Chem.* **2000**, *65*, 3143.
7. Mehltretter, G. M.; Dobler, C.; Sundermeier, U.; Beller, M. *Tetrahedron Lett.* **2000**, *41*, 8083.

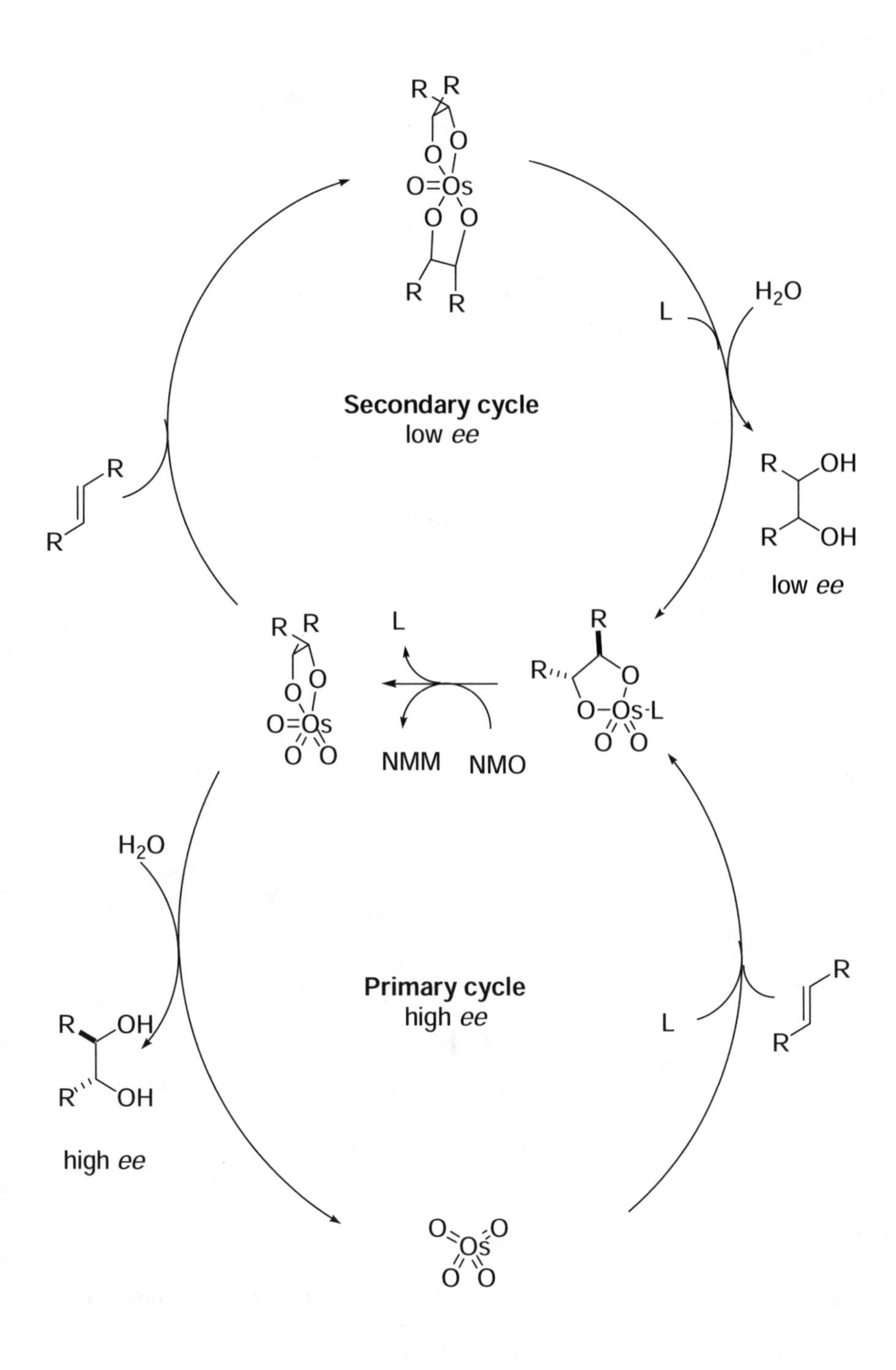

Secondary cycle
low *ee*
L
H_2O
low *ee*
L
NMM
NMO
H_2O
Primary cycle
high *ee*
high *ee*
L

Shi asymmetric epoxidation

An asymmetric epoxidation using fructose-derived chiral ketone.

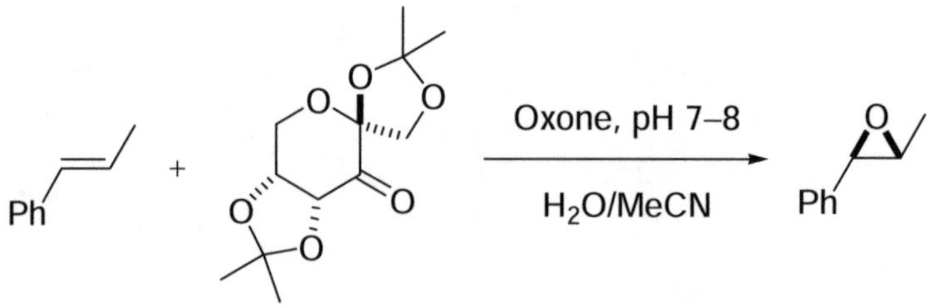

Oxone, pH 7–8

H_2O/MeCN

Ph

The catalytic cycle:

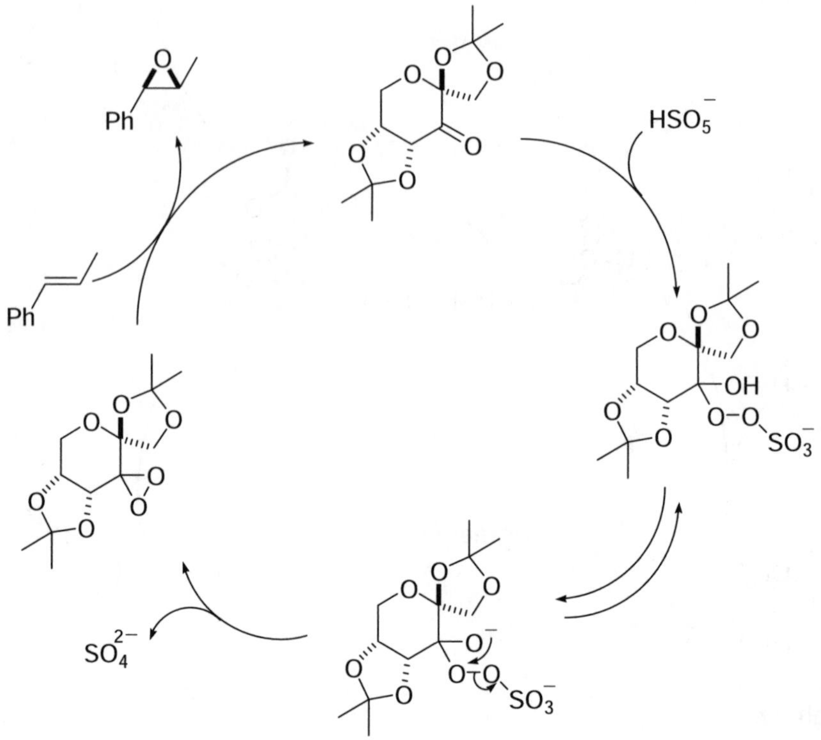

HSO_5^-

SO_4^{2-}

References

1. Wang, Z.-X.; Tu, Y.; Frohn, M.; Zhang, J.-R.; Shi, Y. *J. Am. Chem. Soc.* **1997**, *119*, 11224.
2. Wang, Z.-X.; Shi, Y. *J. Org. Chem.* **1997**, *62*, 8622.

3. Tu, Y.; Wang, Z.-X.; Frohn, M.; He, M.; Yu, H.; Tang, Y.; Shi, Y. *ibid.* **1998**, *63*, 8475.
4. Tian, H.; She, X.; Shu, L.; Yu, H.; Shi, Y. *J. Am. Chem. Soc.* **2000**, *1229*, 11551.
5. Katsuki, T. . In *Catalytic Asymmetric Synthesis* 2nd ed., Ojima, I., ed.; Wiley-VCH: New York, **2000**, 287.

Simmons–Smith reaction

Cyclopropanation of olefins using CH_2I_2 and Zn(Cu).

$$CH_2I_2 + Zn(Cu) \longrightarrow ICH_2ZnI \longrightarrow$$

$$I{-}CH_2{-}I \xrightarrow[\text{addition}]{\text{Zn, Oxidative}} ICH_2ZnI$$

Simmons–Smith reagent

$$2\,ICH_2ZnI \rightleftharpoons (ICH_2)_2Zn + ZnI_2$$

$$\longrightarrow \longrightarrow + ZnI_2$$

References

1. Simmons, H. E.; Smith, R. D. *J. Am. Chem. Soc.* **1958**, *80*, 5323.
2. Kaltenberg, O. P. *Wiad. Chem.* **1972**, *26*, 285.
3. Takai, K.; Kakiuchi, T.; Utimoto, K. *J. Org. Chem.* **1994**, *59*, 2671.
4. Takahashi, H.; Yoshioka, M.; Shibasaki, M.; Ohno, M.; Imai, N.; Kobayashi, S. *Tetrahedron* **1995**, *51*, 12013.
5. Nakamura, E.; Hirai, A.; Nakamura, M. *J. Am. Chem. Soc.* **1998**, *120*, 5844.
6. Kaye, P. T.; Molema, W. E. *Chem. Commun.* **1998**, 2479.
7. Kaye, P. T.; Molema, W. E. *Synth. Commun.* **1999**, *29*, 1889.
8. Baba, Y.; Saha, G.; Nakao, S.; Iwata, C.; Tanaka, T.; Ibuka, T.; Ohishi, H.; Takemoto, Y. *J. Org. Chem.* **2001**, *66*, 81.

Simonini reaction

Ester formation when silver carboxylate is treated with iodine. On the other hand, when silver carboxylate is treated with bromine, the product is alkyl bromide, R–Br (**Hunsdiecker reaction**, page 178).

$$2\ RCO_2Ag \xrightarrow{I_2} RCO_2R + CO_2\uparrow + AgI$$

$$RCO_2Ag + I{-}I \xrightarrow{S_N2} AgI + RCO_2I \xrightarrow[\text{cleavage}]{\text{homolytic}}$$

$$I\bullet + RCO_2\bullet \xrightarrow{\text{decarboxylation}} CO_2\uparrow + R\bullet \xrightarrow{RCO_2I} R{-}I + RCO_2\bullet$$

$$RCO_2Ag + R{-}I \xrightarrow{S_N2} RCO_2R$$

References

1. Simonini, A. *Monatsch.* **1892**, *13*, 320.
2. Wasserman, H. H.; Precopio, F. M. *J. Am. Chem. Soc.* **1954**, *76*, 1242.
3. Chalmers, D. J.; Thomson, R. H. *J. Chem. Soc. (C)* **1968**, 848.
4. Bunce, N. J.; Murray, N. G. *Tetrahedron* **1971**, *27*, 5323.

Simonis chromone cyclization

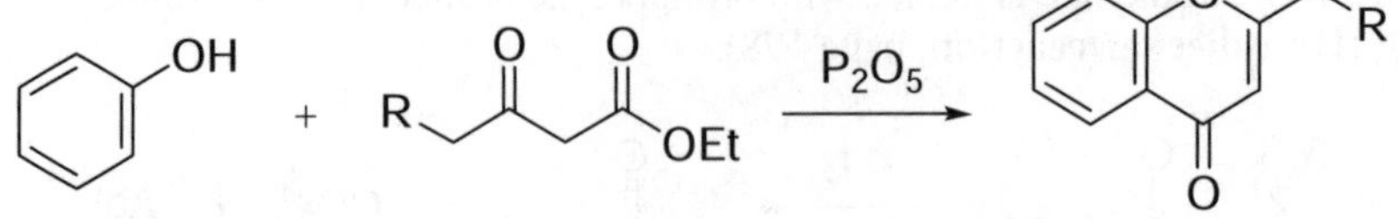

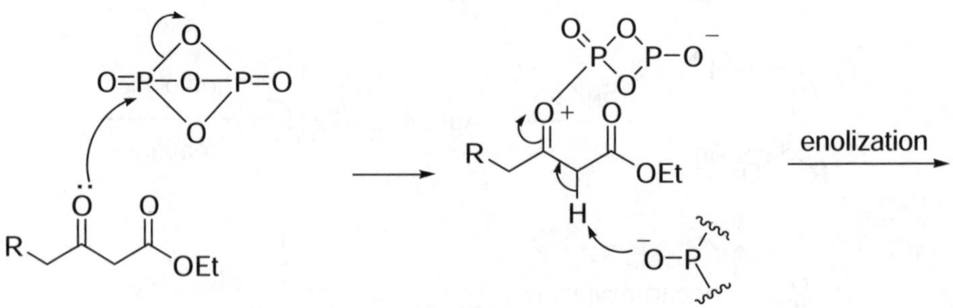

P_2O_5 actually exists as P_4O_{10}, an adamantane-like structure.

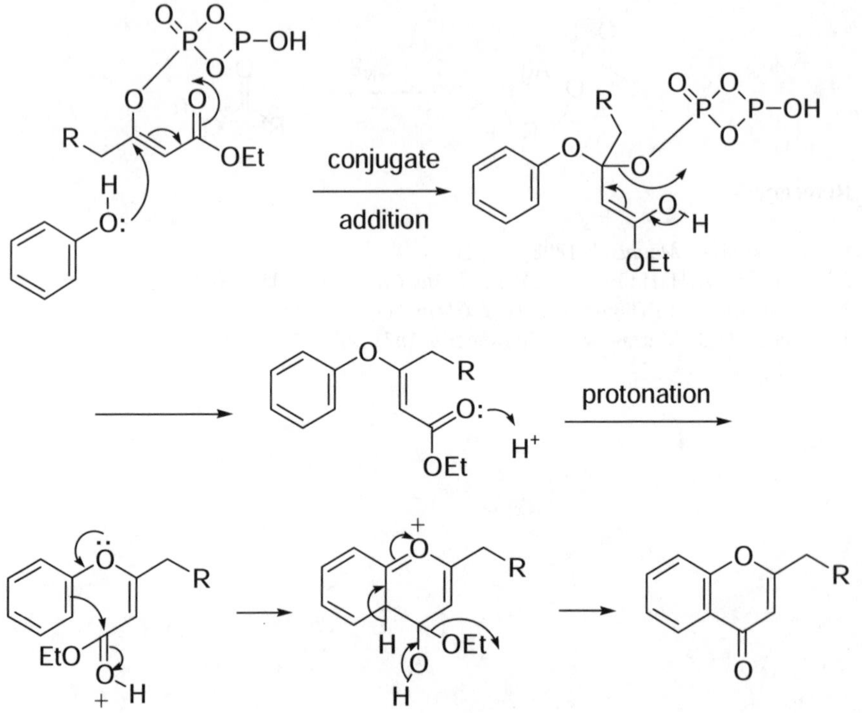

References

1. Petschek, E.; Simonis, H. *Ber.* **1913**, *46*, 2014.
2. Ruwet, A.; Janne, D.; Renson, M. *Bull. Soc. Chim. Belg.* **1970**, *79*, 81.
3. Oyman, U.; Gunaydin, K. *Bull. Soc. Chim. Belg.* **1994**, *103*, 763.

Skraup quinoline synthesis

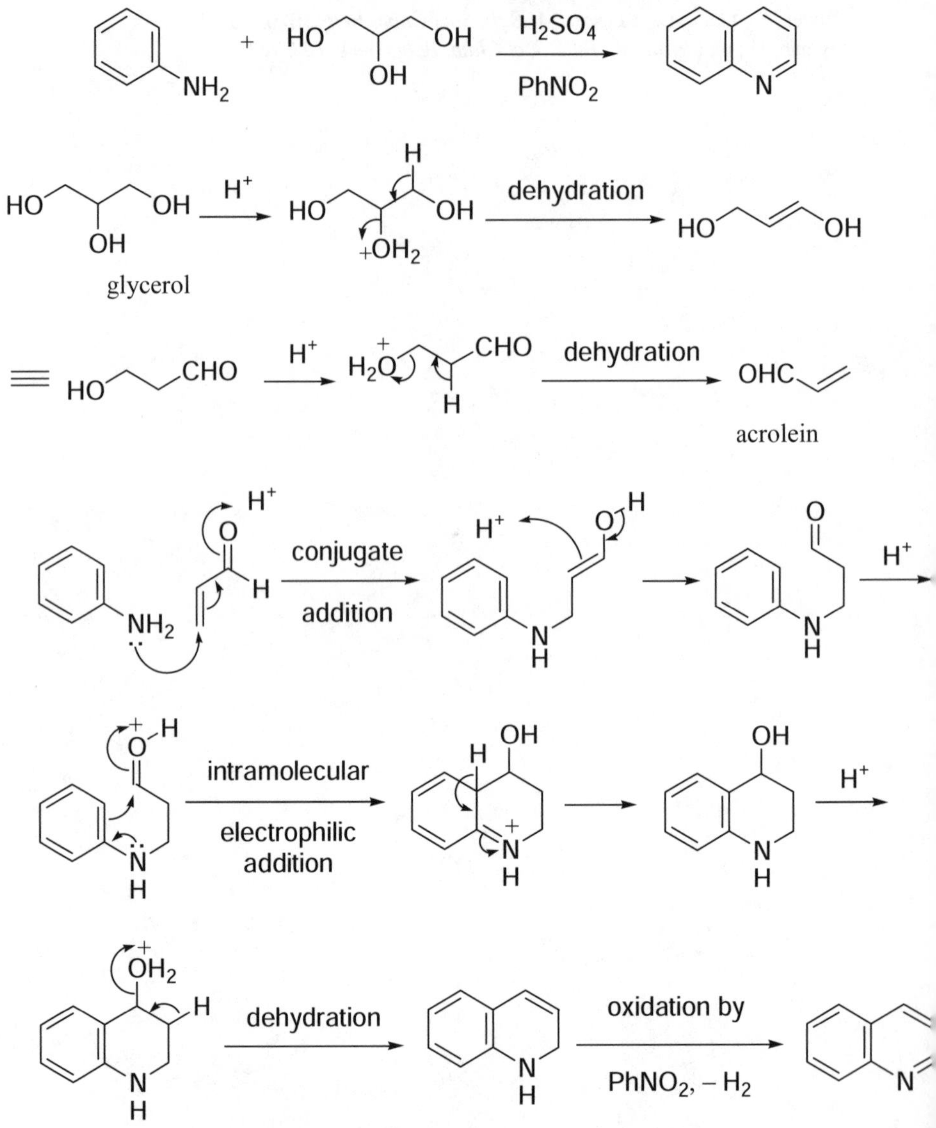

H_2SO_4

$PhNO_2$

H^+

dehydration

glycerol

H^+

dehydration

acrolein

conjugate addition

H^+

intramolecular electrophilic addition

H^+

dehydration

oxidation by $PhNO_2$, $- H_2$

For an alternative mechanism, see that of the Doebner–von Miller reaction (page 104).

References

1. Skraup, Z. H. *Ber.* **1880**, *13*, 2086.
2. Fujiwara, H.; Okabayashi, I. *Chem. Pharm. Bull.* **1994**, *42*, 1322.
3. Fujiwara, H. *Heterocycles* **1997**, *45*, 119.
4. Fujiwara, H.; Kitagawa, K. *ibid.* **2000**, *53*, 409.

Smiles rearrangement

General scheme:

strong base

$X = S, SO, SO_2, O, CO_2$
$YH = OH, NHR, SH, CH_2R, CONHR$
$Z = NO_2, SO_2R$

e.g.

^{-}OH

^{-}OH S_NAr

spirocyclic anion intermediate
(Meisenheimer complex)

References

1. Evans, W. J.; Smiles, S. *J. Chem. Soc.* **1935**, 181.
2. Truce, W. E.; Kreider, E. M.; Brand, W. W. *Org. React.* **1970**, *18*, 99.
3. Gerasimova, T. N.; Kolchina, E. F. *J. Fluorine Chem.* **1994**, *66*, 69.
4. Boschi, D.; Sorba, G.; Bertinaria, M.; Fruttero, R.; Calvino, R.; Gasco, A. *J. Chem. Soc., Perkin Trans. 1* **2001**, 1751.
5. Hirota, T.; Tomita, K.-I.; Sasaki, K.; Okuda, K.; Yoshida, M.; Kashino, S. *Heterocycles* **2001**, *55*, 741.

Sommelet reaction

Transformation of benzyl halides to the corresponding benzaldehydes with the aide of hexamethylenetetramine.

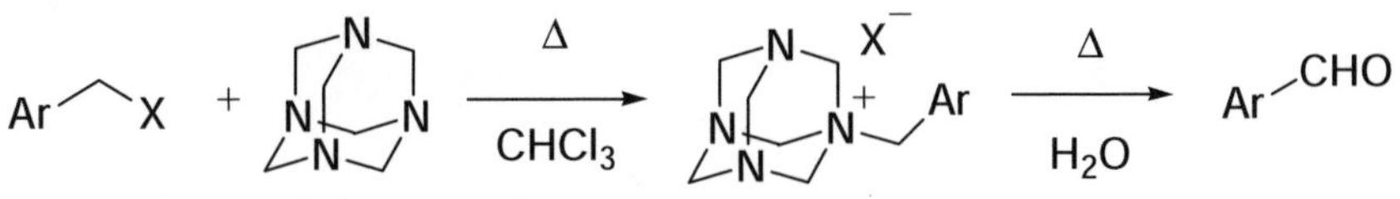

Hexamethylenetetramine

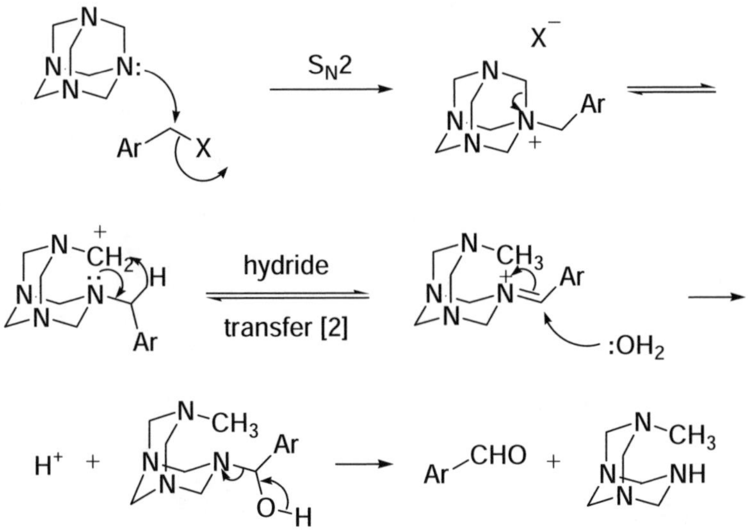

hemiaminal

The hydride transfer and the ring-opening of hexamethylenetetramine may occur in a synchronized fashion:

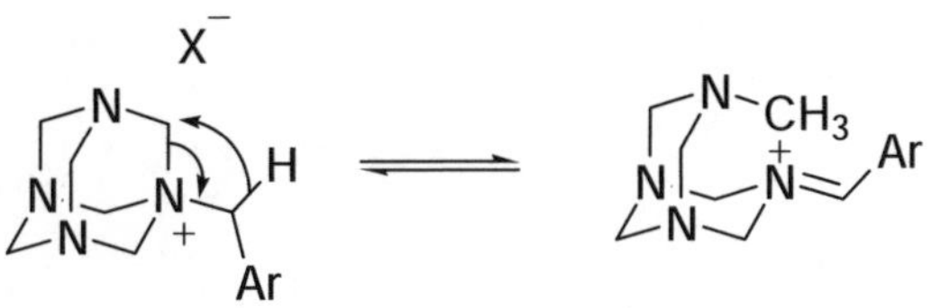

References

1. Sommelet, M. *Compt. Rend.* **1913**, *157*, 852.
2. Le Henaff, P. *Annals Chim. Phys.* **1962**, 367.
3. Zaluski, M. C.; Robba, M.; Bonhomme, M. *Bull. Soc. Chim. Fr.* **1970**, 1445.
4. Smith, W. E. *J. Org. Chem.* **1972**, *37*, 3972.

5. Simiti, I.; Chindris, E. *Arch. Pharm.* **1975**, *308*, 688.
6. Stokker, G. E.; Schultz, E. M. *Synth. Commun.* **1982**, *12*, 847.
7. Armesto, D.; Horspool, W. M.; Martin, J. A. F.; Perez-Ossorio, R. *Tetrahedron Lett.* **1985**, *26*, 5217.
8. Simiti, I.; Oniga, O. *Monatsh. Chem.* **1996**, *127*, 733.

Sommelet–Hauser (ammonium ylide) rearrangement

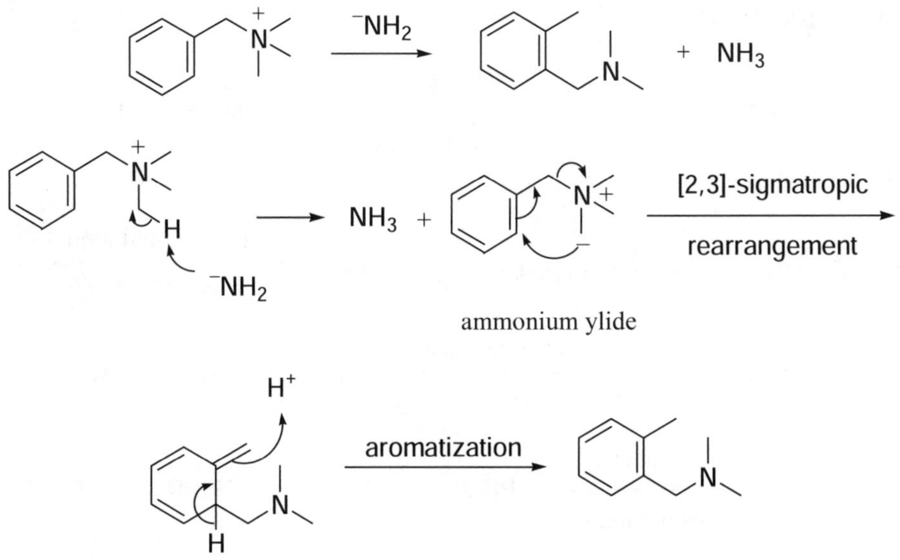

$^{-}NH_2$ + NH_3

$^{-}NH_2$ → NH_3 + [2,3]-sigmatropic rearrangement

ammonium ylide

H^+ aromatization

References

1. Sommelet, M. *Compt. Rend.* **1937**, *205*, 56.
2. Pine, S. H. *Tetrahedron Lett.* **1967**, 3393.
3. Wittig, G. *Bull. Soc. Chim. Fr.* **1971**, 1921.
4. Robert, A.t; Lucas-Thomas, M. T. *J. Chem. Soc., Chem. Commun.* **1980**, 629.
5. Shirai, N.; Sumiya, F.; Sato, Y.; Hori, M. *ibid.* **1988**, 370.
6. Tanaka, T.; Shirai, N.; Sugimori, J.; Sato, Y. *J. Org. Chem.* **1992**, *57*, 5034.
7. Maeda, Y.; Sato, Y. *ibid.* **1996**, *61*, 5188.

Sonogashira reaction

Pd-Cu-catalyzed cross-coupling of organohalides with terminal alkynes. *Cf.* Castro–Stephens reaction.

$$\text{Ar—X} + \equiv\text{—R} \xrightarrow[\text{CuI, Et}_3\text{N, rt}]{\text{PdCl}_2\bullet(\text{PPh}_3)_2} \text{Ar—}\equiv\text{—R}$$

Generation of Pd(0):

$$\equiv\text{—R} \xrightarrow{\text{CuI, Et}_3\text{N}} \text{Et}_3\text{NH}^+\bullet\text{I}^- + \text{Cu—}\equiv\text{—R} \xrightarrow[\text{PdCl}_2\bullet(\text{PPh}_3)_2]{\text{transmetallation}}$$

$$\text{Cl—Pd—}\equiv\text{—R} \xrightarrow[\text{Cu—}\equiv\text{—R}]{\text{transmetallation}} \text{R—}\equiv\text{—Pd—}\equiv\text{—R}$$

$$\xrightarrow[\text{elimination}]{\text{reductive}} \text{Pd(0)} + \text{R—}\equiv\text{——}\equiv\text{—R}$$

Coupling reaction:

$$\text{Ar—X} + \text{Pd(0)} \xrightarrow[\text{addition}]{\text{oxidative}} \text{Ar—Pd—X} \xrightarrow[\text{Cu—}\equiv\text{—R}]{\text{transmetallation}}$$

$$\text{Ar—Pd—}\equiv\text{—R} \xrightarrow[\text{elimination}]{\text{reductive}} \text{Ar—}\equiv\text{—R} + \text{Pd(0)}$$

Note that Et_3N may reduce Pd(II) to Pd(0) as well, where Et_3N is oxidized to iminium ion at the same time [8]:

$$\text{Cl—PdCl} + \text{:NEt}_3 \xrightarrow{\text{coordination}} \text{ClPd(II)—N(Et)}_2\text{—CH(H)CH}_3 \xrightarrow[\text{elimination}]{\beta\text{-hydride}}$$

$$\text{Et}_2\text{N}^+\text{=CHCH}_3 + \text{Cl—Pd—H} \xrightarrow[\text{elimination}]{\text{reductive}} \text{HCl} + \text{Pd(0)}$$

References

1. Sonogashira K.; Tohda, Y.; Hagihara, N. *Tetrahedron Lett.* **1975**, 4467.
2. McCrindle, R.; ferguson, G.; Arsenaut, G. J.; McAlees, A. J.; Stephenson, D. K. *J. Chem. Res. (S)* **1984**, 360.
3. Rossi, R. Carpita, A.; Belina, F. *Org. Prep. Proc. Int.* **1995**, *27*, 129.
4. Campbell, I. B. In *Organocopper Reagents*, Taylor, R. J. K. Ed. Publisher: IRL Press: Oxford, UK, **1994**, 217.
5. Hundermark, T.; Littke, A.; Buchwald, S. L.; Fu, G. C. *Org. Lett.* **2000**, *2*, 1729.
6. Dai, W.-M.; Wu, A. *Tetrahedron Lett.* **2001**, *42*, 81.
7. Alami, M.; Crousse, B.; Ferri, F. *J. Organomet. Chem.* **2001**, *624*, 114.
8. Bates, R. W.; Boonsombat, J. *J. Chem. Soc., Perkin Trans. 1* **2001**, 654.

Staudinger reaction

Reduction of azides to amines by Ph_3P/H_2O.

$$R{-}N_3 \xrightarrow[H_2O]{PPh_3} R{-}NH_2$$

$$R{-}\overset{-}{N}{-}\overset{+}{N}{\equiv}N \xrightarrow{S_N2} N_2\uparrow + R{-}\overset{-}{N}{-}\overset{+}{P}Ph_3 \longleftrightarrow R{-}N{=}PPh_3$$

$$\longrightarrow R{-}NH{-}PPh_3(OH) \longrightarrow R{-}NH_2 + O{=}PPh_3$$

References

1. Staudinger, H.; Meyer, J. *Helv. Chim. Acta* **1919**, *2*, 635.
2. Gololobov, Y. G.; Zhmurova, I. N.; Kasukhin, L. F. *Tetrahedron* **1981**, *37*, 437.
3. Gololobov, Y. G.; Kasukhin, L. F. *ibid.* **1992**, *48*, 1353.
4. Velasco, M. D.; Molina, P.; Fresneda, P. M.; Sanz, M. A. *ibid.* **2000**, *56*, 4079.
5. Bongini, A.; Panunzio, M.; Piersanti, G.; Bandini, E.; Martelli, G.; Spunta, G.; Venturini, A. *Eur. J. Org. Chem.* **2000**, *65*, 2379.
6. Balakrishna, M. S.; Abhyankar, R. M.; Walawalker, M. G. *Tetrahedron Lett.* **2001**, *42*, 2733.

Stetter reaction (Michael–Stetter reaction)

1,4-Dicarbonyl derivatives from aldehydes and α,β-unsaturated ketones. The thiazolium catalyst serves as a safe surrogate for $^{-}$CN. *Cf.* Benzoin condensation.

R–CHO + [α,β-unsaturated ketone] —(thiazolium catalyst, NH_3)→ R–CO–CH$_2$CH$_2$–CO–CH$_3$

deprotonation

Michael addition

tautomerization

References

1. Stetter, H. *Angew. Chem.* **1973**, *85*, 89.
2. Stetter, H. *Angew. Chem., Int. Ed.* **1976**, *15*, 639.
3. Castells, J.; Dunach, E.; Geijo, F.; Lopez-Calahorra, F.; Prats, M.; Sanahuja, O.; Villanova, L. *Tetrahedron Lett.* **1980**, *21*, 2291.
4. Ho, T. L.; Liu, S. H. *Synth. Commun.* **1983**, *13*, 1125.
5. Phillips, R. B.; Herbert, S. A.; Robichaud, A. J. *ibid.* **1986**, *16*, 411.
6. Stetter, H.; Kuhlmann, H.; Haese, W. *Org. Synth.* **1987**, *65*, 26.

7. Powell, P.; Sosabowski, M. H. *J. Chem. Res., (S)* **1995**, 306.
8. Ciganek, E. *Synthesis* **1995**, 1311.
9. Enders, D.; Breuer, K.; Runsink, J.; Teles, J. H. *Helv. Chim. Acta* **1996**, *79*, 1899.
10. Harrington, P. E.; Tius, M. A. *Org. Lett.* **1999**, *1*, 649.

Stevens rearrangement

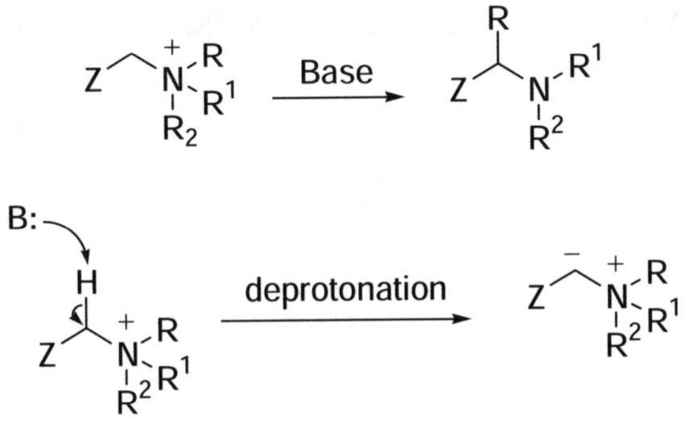

Z N^{+} R R^{1} R$_{2}$ — Base → Z N R R^{1} R^{2}

B: H Z N^{+} R R^{1} R^{2} — deprotonation → Z^{-} N^{+} R R^{1} R^{2}

The contemporary radical mechanism:

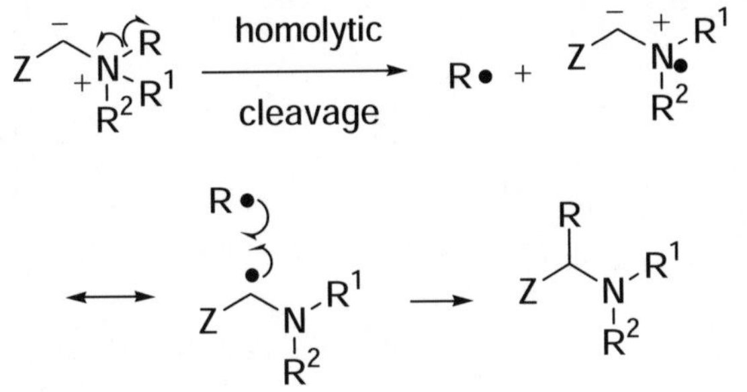

Z^{-} N^{+} R R^{1} R^{2} — homolytic cleavage → R• + Z^{-} N^{+}• R^{1} R^{2}

↔ R• Z N R^{1} R^{2} → Z N R R^{1} R^{2}

The original ionic mechanism:

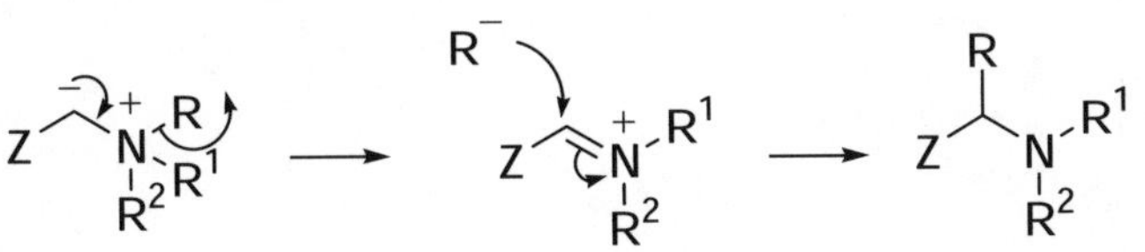

Z^{-} N^{+} R R^{1} R^{2} → R^{-} Z N^{+} R^{1} R^{2} → Z N R R^{1} R^{2}

References

1. Stevens, T. S.; Creighton, E. M.; Gordon, A. B.; MacNicol, M. *J. Chem. Soc.* **1928**, 3193.
2. Schöllkopf, U.; Ludwig, U.; Ostermann, G.; Paysch, M. *Tetrahedron Lett.* **1969**, 3415.
3. Pine, S. H.; Catto, B. A.; Yamagishi, F. G. *J. Org. Chem.* **1970**, *35*, 3663.
4. Lepey, Arthur R.; Giumanini, Angelo G. *Mech. Mol. Migr.* **1971**, *3*, 297.
5. Pant, J.; Joshi, B. C. *Indian J. Chem. Educ.* **1980**, *7*, 11.
6. Doyle, M. P.; Ene, D. G.; Forbes, D. C.; Tedrow, J. S. *Tetrahedron Lett.* **1997**, *38*, 4367.

7. Makita, K.; Koketsu, J.; Ando, F.; Ninomiya, Y.; Koga, N. *J. Am. Chem. Soc.* **1998**, *120*, 5764.
8. Feldman, K. S.; Wrobleski, M. L. *J. Org. Chem.* **2000**, *65*, 8659.
9. Kitagaki, S.; Yanamoto, Y.; Tsutsui, H.; Anada, M.; Nakajima, M.; Hashimoto, S. *Tetrahedron Lett.* **2001**, *42*, 6361.

Stieglitz rearrangement

$Ph_3C{-}NH{-}OH \xrightarrow{PCl_5} Ph_2C{=}N^+H{-}Ph\ Cl^- \xrightarrow{H_2O} Ph_2C{=}O + Ph{-}NH_3^+\ Cl^-$

phenyl migration

$O{=}PCl_3$ +

S_N2

hydrolysis

References

1. Stieglitz, J.; Leech, P. N. *Ber.* **1913**, *46*, 2147.
2. Koga, N.; Anselme, J. P. *Tetrahedron Lett.* **1969**, 4773.
3. Sisti, A. J.; Milstein, S. R. *J. Org. Chem.* **1974**, *39*, 3932.
4. Hoffman, R. V.; Poelker, D. J. *ibid.* **1979**, *44*, 2364.
5. Renslo, A. R.; Danheiser, R. L. *ibid.* **1998**, *63*, 7840.

Still–Gennari phosphonate reaction

Horner–Emmons reaction using bis(trifluoroethyl)phosphonate to give *Z*-olefins.

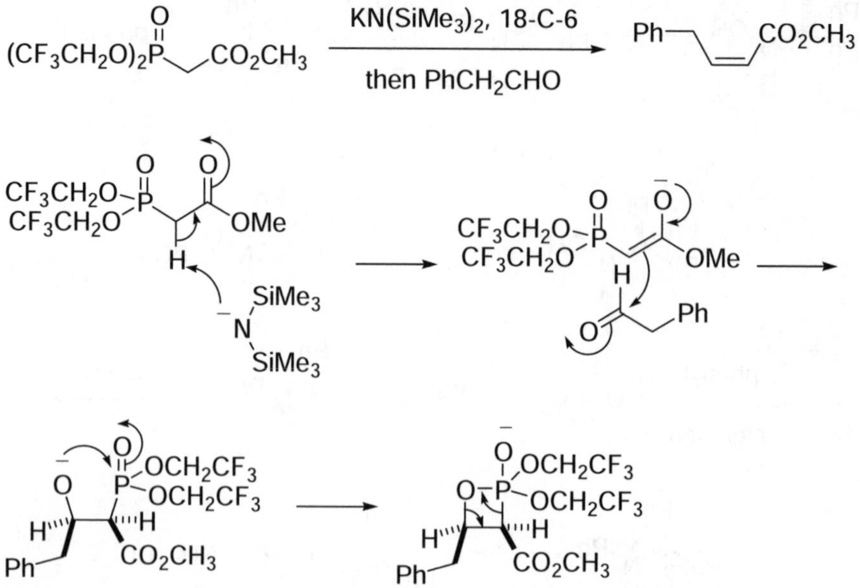

$$(CF_3CH_2O)_2P(O)CH_2CO_2CH_3 \xrightarrow[\text{then } PhCH_2CHO]{KN(SiMe_3)_2,\ 18\text{-C-6}} PhCH_2CH{=}CHCO_2CH_3$$

erythro isomer, kinetic adduct

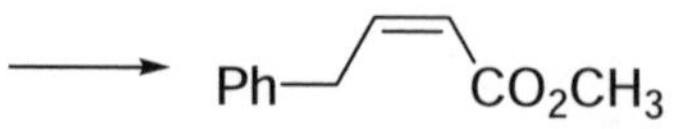

References

1. Still, W. C.; Gennari, C. *Tetrahedron Lett.* **1983**, *24,* 4405.
2. Ralph, J.; Zhang, Y. *Tetrahedron* **1998**, *54*, 1349.
3. Mulzer, J.; Mantoulidis, A.; Ohler, E. *Tetrahedron Lett.* **1998**, *39*, 8633.
4. Jung, M. E.; Marquez, R. *Org. Lett.* **2000**, *2,* 1669.

Stille coupling

Palladium-catalyzed cross-coupling reaction of organostannanes with organic halides, triflates, *etc.* For the catalytic cycle, see Kumada coupling on page 208.

$$\mathrm{R{-}X} + \mathrm{R^1{-}Sn(R^2)_3} \xrightarrow{\mathrm{Pd(0)}} \mathrm{R{-}R^1} + \mathrm{X{-}Sn(R^2)_3}$$

$$\mathrm{R{-}X} + \mathrm{L_2Pd(0)} \xrightarrow[\text{addition}]{\text{oxidative}} \mathrm{R(L)_2Pd{-}X} \xrightarrow[\text{transmetallation, isomerization}]{\mathrm{R_1{-}Sn(R^2)_3}}$$

$$\mathrm{X{-}Sn(R^2)_3} + \mathrm{L_2Pd(R)(R^1)} \xrightarrow[\text{elimination}]{\text{reductive}} \mathrm{R{-}R^1} + \mathrm{L_2Pd(0)}$$

References

1. Milstein, D.; Stille, J. K. *J. Am. Chem. Soc.* **1978**, *100*, 3636.
2. Milstein, D.; Stille, J. K. *ibid.* **1979**, *101*, 4992.
3. Stille, J. K. *Angew. Chem., Int. Ed. Engl.* **1986**, *25*, 508.
4. Farina, V.; Krishnamurphy, V.; Scott, W. J. *Organic Reactions* **1997**, *50*, 1–652.
5. For an excellent review on the intramolecular Stille reaction, see, Duncton, M. A. J.; Pattenden, G. *J. Chem. Soc., Perkin Trans. 1* **1999**, 1235.
6. Nakamura, H.; Bao, M.; Yamamoto, Y. *Angew. Chem., Int. Ed.* **2001**, *40*, 3208.

Stille–Kelly reaction

Palladium-catalyzed intramolecular cross-coupling reaction of bis-aryl halides using ditin reagents.

$Me_3Sn{-}SnMe_3$

$Pd(Ph_3P)_4$

Pd(0)

oxidative addition

$Me_3Sn{-}SnMe_3$

transmetallation

reductive elimination

Pd(0) +

Pd(0)

oxidative addition

transmetallation

reductive elimination

+ Pd(0)

References

1. Kelly, T. R.; Li, Q.; Bhushan, V. *Tetrahedron Lett.* **1990**, *31*, 161.
2. Grigg, R.; Teasdale, A.; Sridharan, V. *ibid.* **1991**, *32*, 3859.
3. Sakamoto, T.; Yasuhara, A.; Kondo, Y.; Yamanaka, H. *Heterocycles* **1993**, *36*, 2597.

4. Iyoda, M.; Miura, M.i; Sasaki, S.; Kabir, S. M. H.; Kuwatani, Y.; Yoshida, M. *ibid.* **1997**, *38*, 4581.
5. Fukuyama, Y.; Yaso, H.; Nakamura, K.; Kodama, M. *Tetrahedron Lett.* **1999**, *40*, 105.
6. Iwaki, T.; Yasuhara, A.; Sakamoto, T. *J. Chem. Soc., Perkin Trans. 1* **1999**, 1505.
7. Fukuyama, Y.; Yaso, H.; Mori, T.; Takahashi, H.; Minami, H.; Kodama, M. *Heterocycles* **2001**, *54*, 259.

Stobbe condensation

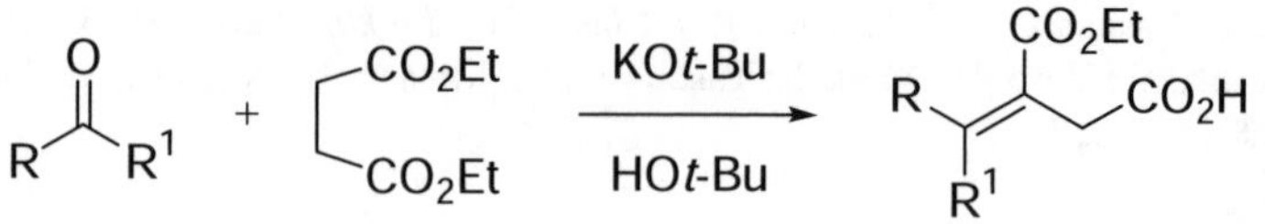

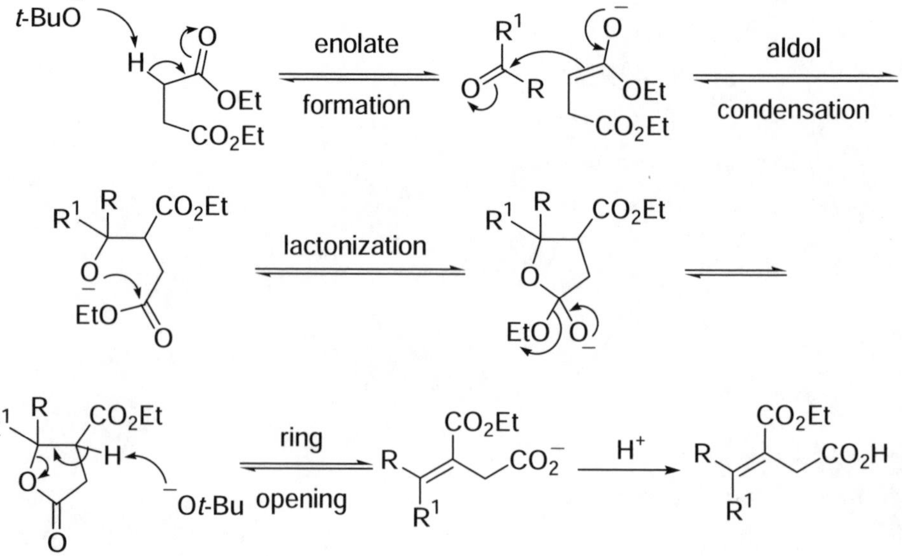

References

1. Stobbe, H. *Ber.* **1893**, *26*, 2312.
2. El-Rayyes, N. R.; Al-Salman, Mrs. N. A. *J. Heterocycl. Chem.* **1976**, *13*, 285.
3. Baghos, V. B.; Nasr, F. H.; Gindy, M. *Helv. Chim. Acta* **1979**, *62*, 90.
4. Baghos, V. B.; Doss, S. H.; Eskander, E. F. *Org. Prep. Proced. Int.* **1993**, *25*, 301.
5. Moldvai, I.; Temesvari-Major, E.; Balazs, M.; Gacs-Baitz, E.; Egyed, O.; Szantay, C. *J. Chem. Res., (S)* **1999**, 3018.
6. Moldvai, I.; Temesvari-Major, E.; Gacs-Baitz, E.; Egyed, O.; Gomory, A.; Nyulaszi, L.; Szantay, C. *Heterocycles* **2001**, *53*, 759.

Stollé synthesis

Acid-catalyzed indolinone formation from aniline and α-chlorocarboxylic acid chloride.

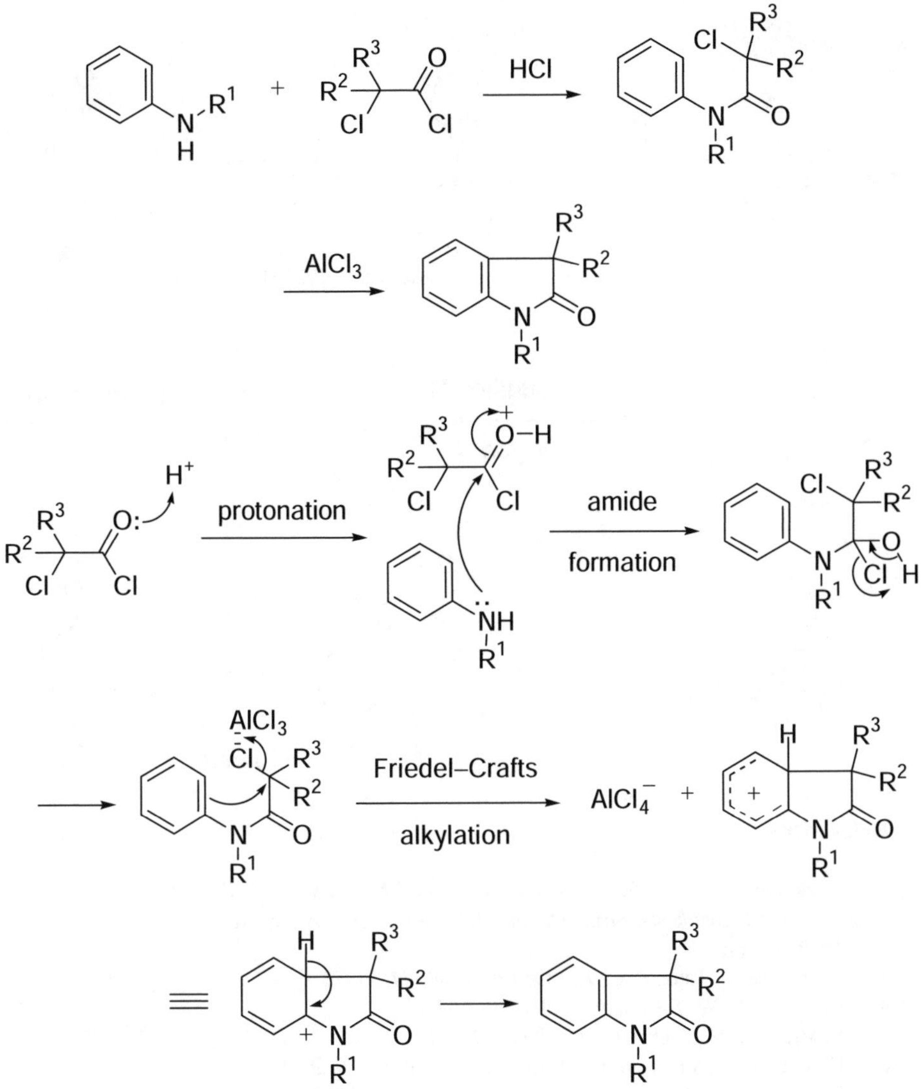

References

1. Stollé, R. *Ber.* **1913**, *46*, 3915.
2. Stollé, R. *ibid.* **1914**, *47*, 2120.
3. Przheval'skii, N. M.; Grandberg, I. I. *Khim. Geterotsikl. Soedin.* **1982**, 940.

Stork enamine reaction

A variant of the Robinson annulation, where bulky amines such as pyrrolidine are used, making the conjugate addition to MVK take place at the less hindered side of two possible enamines.

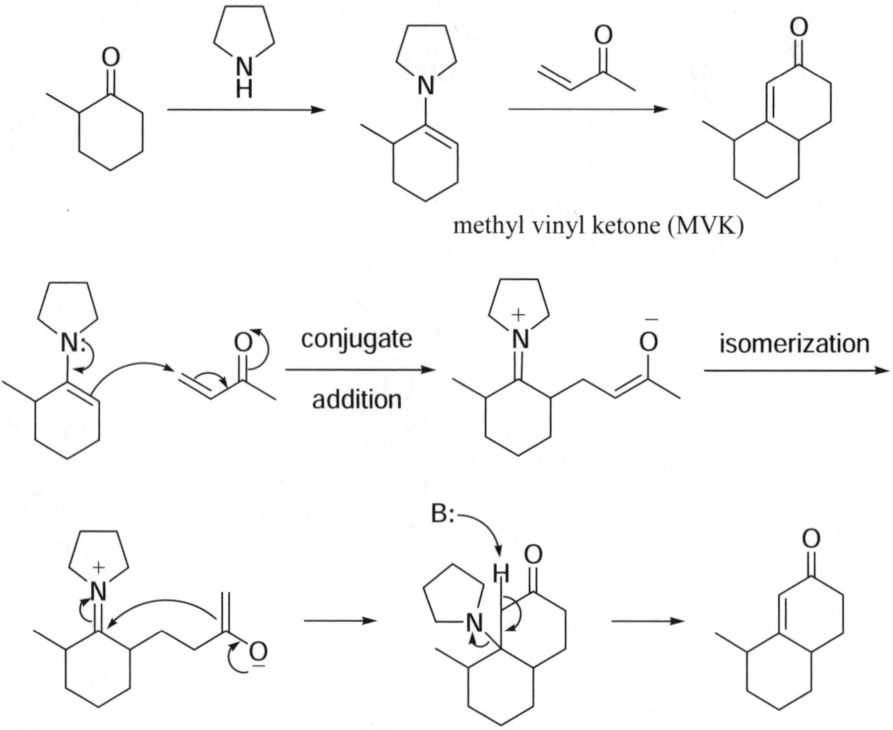

methyl vinyl ketone (MVK)

conjugate addition

isomerization

References

1. Stork, G.; Terrell, R.; Szmuszkovicz, J. *J. Am. Chem. Soc.* **1954**, *76*, 2029.
2. *Enamines: Synthesis, Structure, and Reactions* Cook, A. G; Ed. Dekker: New York, **1969**, 514 pp.
3. Autrey, R. L.; Tahk, F. C. *Tetrahedron* **1968**, *24*, 3337.
4. Hickmott, P. W. *Tetrahedron* **1982**, *38*, 1975.
5. Szablewski, M. *J. Org. Chem.* **1994**, *59*, 954.
6. Hammadi, M.; Villemin, D. *Synth. Commun.* **1996**, *26*, 2901.
7. Bridge, C. F.; O'Hagan, D. *J. Fluorine Chem.* **1997**, *82*, 21.
8. Li, J. J.; Trivedi, B. K.; Rubin, J. R.; Roth, B. D. *Tetrahedron Lett.* **1998**, *39*, 6111.

Strecker amino acid synthesis

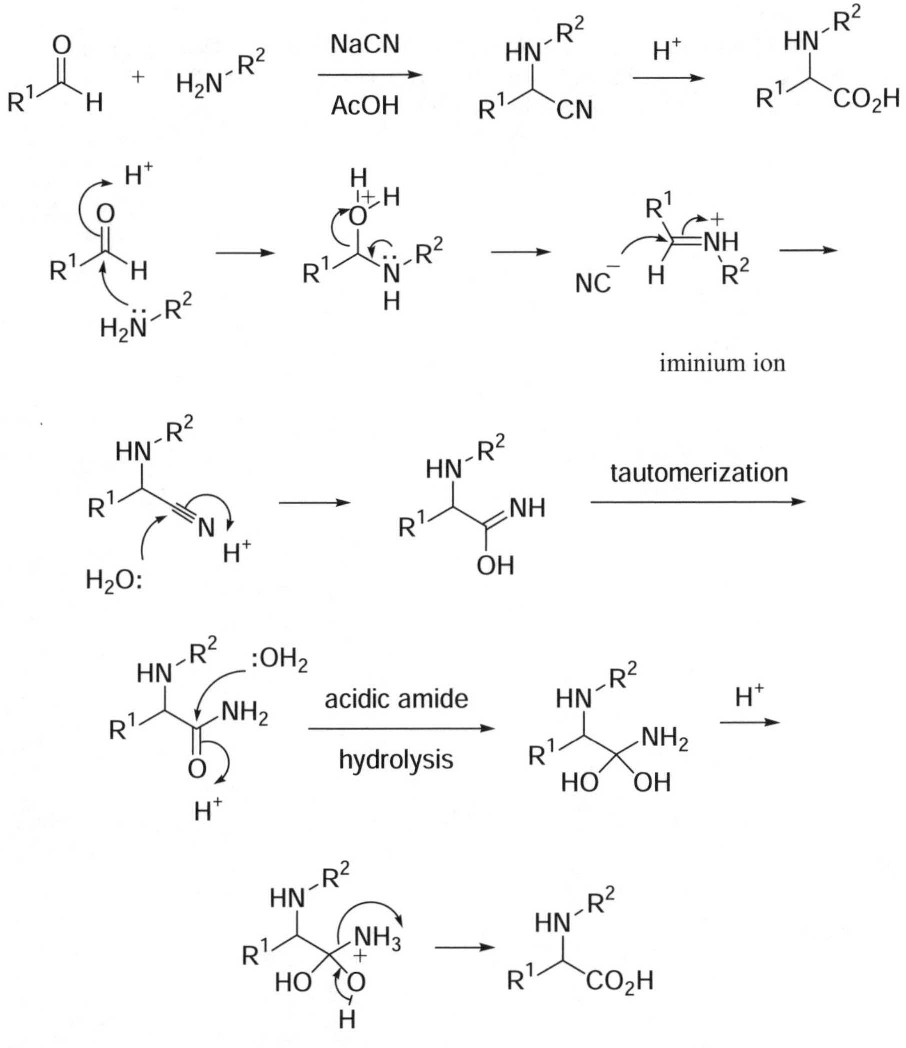

NaCN
AcOH
H^+

H^+

iminium ion

tautomerization

acidic amide
hydrolysis

H^+

References

1. Strecker, A. *Liebigs Ann. Chem.* **1850**, *75*, 27.
2. Chakraborty, T. K.; Hussain, K. A; Reddy, G. V. *Tetrahedron* **1995**, *51*, 9179.
3. Iyer, M. S.; Gigstad, K. M.; Namdev, N. D.; Lipton, M. *J. Am. Chem. Soc.* **1996**, *118*, 4910.
4. Iyer, M. S.; Gigstad, K. M.; Namdev, N. D.; Lipton, M. *Amino Acids* **1996**, *11*, 259.

5. Mori, A.; Inoue, S. *Compr. Asymmetric Catal. I-III* **1999**, *2*, 983.
6. Ishitani, H.; Komiyama, S.; Hasegawa, Y.; Kobayashi, S. *J. Am. Chem. Soc.* **2000**, *122*, 762.
7. Wede, J.; Volk, Franz-J.; Frahm, A. W. *Tetrahedron: Asymmetry* **2000**, *11*, 3231.
8. Davis, F. A.; Lee, S.; Zhang, H.; Fanelli, D. L. *J. Org. Chem.* **2000**, *65*, 8704.
9. Ding, K.; Ma, D. *Tetrahedron* **2001**, *57*, 6361.

Suzuki coupling

Palladium-catalyzed cross-coupling reaction of organoboranes with organic halides, triflates, *etc.* in the presence of a base (transmetallation is reluctant to occur without the activating effect of a base). For the catalytic cycle, see Kumada coupling on page 208.

$$R{-}X + R^1{-}B(R_2)_2 \xrightarrow[\text{NaOR}^3]{L_2Pd(0)} R{-}R^1$$

$$R{-}X + L_2Pd(0) \xrightarrow[\text{addition}]{\text{oxidative}} RPd(L)_2X$$

$$R^1{-}B(R_2)_2 \xrightarrow[\text{addition of base}]{\text{NaOR}^3} R^1{-}B^{-}(OR_3)(R^2)_2$$

$$RPd(L)_2X + R^1{-}B^{-}(OR^3)(R^2)_2 \xrightarrow[\text{isomerization}]{\text{transmetallation}}$$

$$R^3O{-}B(R^2)_2 + L_2Pd(R)R^1 \xrightarrow[\text{elimination}]{\text{reductive}} R{-}R^1 + L_2Pd(0)$$

References

1. Miyaura, N.; Suzuki, A. *Chem. Rev.* **1995**, *95*, 2457.
2. Suzuki, A. In *Metal-catalyzed Cross-coupling Reactions*; Diederich, F.; Stang, P. J. Eds. Wiley-VCH: Weinhein, Germany, **1998**, 49–97.
3. Stanforth, S. P. *Tetrahedron* **1998**, *54*, 263.
4. Li, J. J. *Alkaloids: Chem. Biol. Perspect.* **1999**, *14*, 437.
5. Groger, H. *J. Prakt. Chem.* **2000**, *342*, 334.
6. Franzen, R. *Can. J. Chem.* **2000**, *78*, 957.
7. LeBlond, C. R.; Andrews, A. T.; Sun, Y.; and Sowa, J. R., Jr. *Org. Lett.* **2001**, *3*, 1557.

Swern oxidation

Oxidation of alcohols to the corresponding carbonyl compounds using $(COCl)_2$, DMSO, and quenching with Et_3N. Not applicable to allylic and benzylic alcohols.

Alternatively:

$$\xrightarrow{\text{Ei}} R^1C(=O)R^2 + (CH_3)_2S\uparrow$$

References

1. Huang, S. L.; Omura, K.; Swern, D. *J. Org. Chem.* **1976**, *41*, 3329.
2. Huang, S. L.; Omura, K.; Swern, D. *ibid.* **1978**, *43*, 297.
3. Mancuso, A. J.; Huang, S.-L.; Swern, D. *ibid.* **1978**, *43*, 2489.
4. Tidwell, T. T. *Org. React.* **1990**, *39*, 297.
5. Nakajima, N.; Ubukata, M. *Tetrahedron Lett.* **1997**, *38*, 2099.
6. Harris, J. M.; Liu, Y.; Chai, S.; Andrews, M. D.; Vederas, J. C. *J. Org. Chem.* **1998**, *63*, 2407.
7. Bailey, P. D.; Cochrane, P. J.; Irvine, F.; Morgan, K. M.; Pearson, D. P. J.; Veal, K. T. *Tetrahedron Lett.* **1998**, *40*, 4593.
8. Rodriguez, A.; Nomen, M.; Spur, B. W.; Godfroid, J. J. *ibid.* **1999**, *40*, 5161.
9. Dupont, J.; Bemish, R. J.; McCarthy, K. E.; Payne, E. R.; Pollard, E. B.; Ripin, D. H. B.; Vanderplas, B. C.; Watrous, R. M. *Tetrahedron Lett.* **2001**, *42*, 1453.

Tamao–Kumada oxidation

Oxidation of alkyl fluorosilanes to the corresponding alcohols.
Cf. Fleming oxidation.

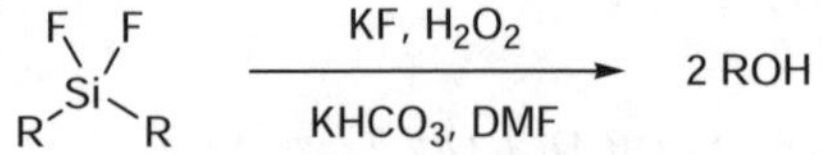

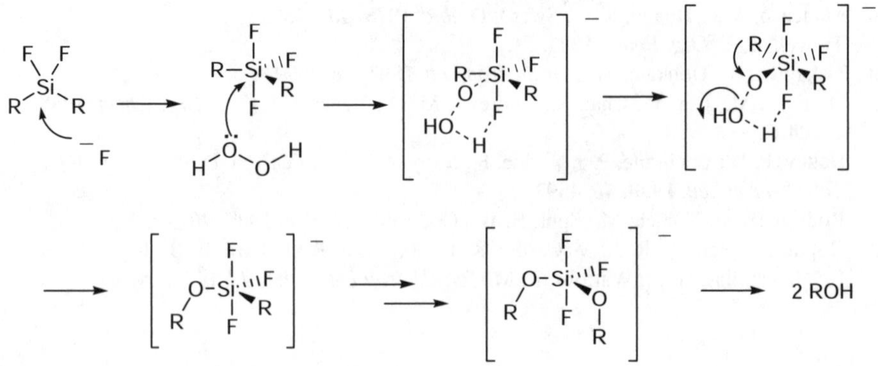

$$R_2SiF_2 \xrightarrow[KHCO_3,\ DMF]{KF,\ H_2O_2} 2\ ROH$$

References

1. Tamao, K.; Ishida, N.; Kumada, M. *J. Org. Chem.* **1983**, *48*, 2120.
2. Kim, S.; Emeric, G.; Fuchs, P. L. *ibid.* **1992**, *57*, 7362.
3. Jones, G. R.; Landais, Y. *Tetrahedron* **1996**, *52*, 7599.
4. Hunt, J. A.; Roush, W. R. *J. Org. Chem.* **1997**, *62*, 1112.
5. Knölker, H.-J.; Jones, P. G.; Wanzl, G. *Synlett* **1997**, 613.
6. Studer, A.; Steen, H. *Chem.--Eur. J.* **1999**, *5*, 759.
7. Barrett, A. G. M.; Head, J.; Smith, M. L.; Stock, N. S.; White, A. J. P.; Williams, D. J. *J. Org. Chem.* **1999**, *64*, 6005.

Tebbe olefination (Petasis alkenylation)

$Cp_2Ti(\mu\text{-}CH_2)(\mu\text{-}Cl)Al(CH_3)_2$ + $RC(O)R^1$ ⟶ $RC(=CH_2)R^1$ + $Cp_2Ti{=}O$

Tebbe's reagent

$Cp_2TiCl_2 + 2\ Al(CH_3)_3$ —transmetallation⟶ $Cp_2Ti(CH_3)_2$ —α-hydride abstraction⟶

$CH_4\uparrow + Cp_2Ti{=}CH_2$ —$Cl{-}Al(CH_3)_2$, coordination⟶ $Cp_2Ti(\mu\text{-}CH_2)(\mu\text{-}Cl)Al(CH_3)_2$

Tebbe's reagent

$Cp_2Ti(\mu\text{-}CH_2)(\mu\text{-}Cl)Al(CH_3)_2$ ⇌ (dissociation) $Cl{-}Al(CH_3)_2$ + $Cp_2Ti{=}CH_2$ / $O{=}CRR^1$

—[2 + 2] cycloaddition⟶ $Cp_2Ti{-}CH_2{-}CRR^1{-}O$ (oxatitanacyclobutane) —retro-[2 + 2] cycloaddition⟶ $RC(=CH_2)R^1$ + $Cp_2Ti{=}O$

The Petasis reagent (Me_2TiCp_2, dimethyltitanocene) undergoes similar olefination reactions with ketones and aldehydes [5]. However, the mechanism is very different.

References

1. Tebbe, F. N.; Parshall, G. W.; Reddy, G. S. *J. Am. Chem. Soc.* **1978**, *100*, 3611.
2. Chou, T. S.; Huang, S. B. *Tetrahedron Lett.* **1983**, *24*, 2169.
3. Petasis, N. A.; Bzowej, E. I. *J. Am. Chem. Soc.* **1990**, *112*, 6392.
4. Schioett, B.; Joergensen, K. A. *J. Chem. Soc., Dalton Trans.* **1993**, 337.
5. Nicolaou, K. C.; Postema, M. H. D.; Claiborne, C. F. *J. Am. Chem. Soc.* **1996**, *118*, 1565.
6. Godage, H. Y.; Fairbanks, A. J. *Tetrahedron Lett.* **2000**, *41*, 7589.

Thorpe–Ziegler reaction

The intramolecular version of the Thorpe reaction.

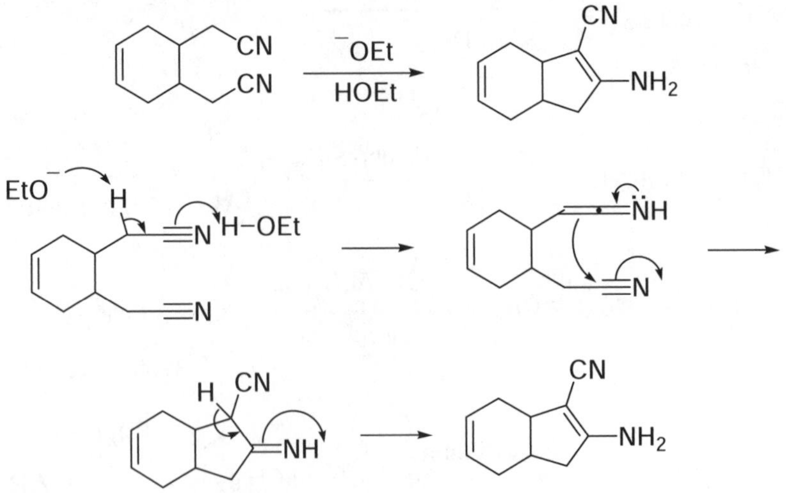

References

1. Baron, H.; Remfry, F. G. P.; Thorpe, Y. F. *J. Chem. Soc.* **1904**, *85*, 1726.
2. Yakovlev, M. Yu.; Kadushkin, A. V.; Solov'eva, N. P.; Granik, V. G. *Heterocycl. Commun.* **1998**, *4*, 245.
3. Curran, D. P.; Liu, W. *Synlett* **1999**, 117.
4. Kovacs, L. *Molecules* **2000**, *5*, 127.
5. Gutschow, M.; Powers, J. C. *J. Heterocycl. Chem.* **2001**, *38*, 419.

Tiemann rearrangement

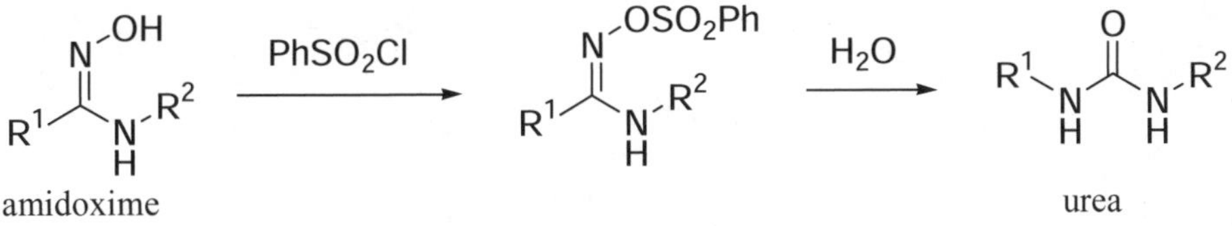

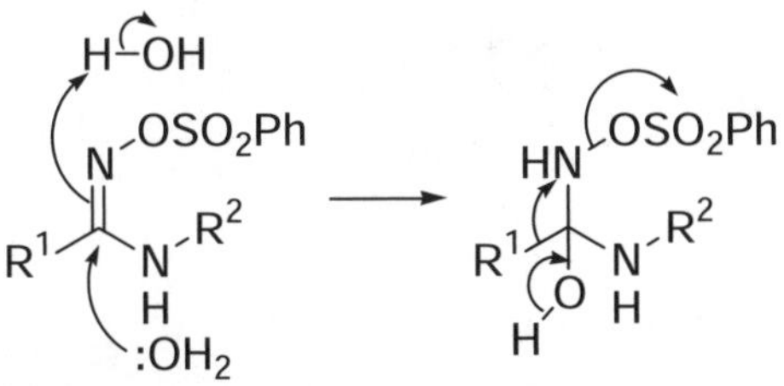

amidoxime

$PhSO_2Cl$

H_2O

urea

The substituent *anti* to the leaving group ($^{-}OSO_2Ph$) migrates.

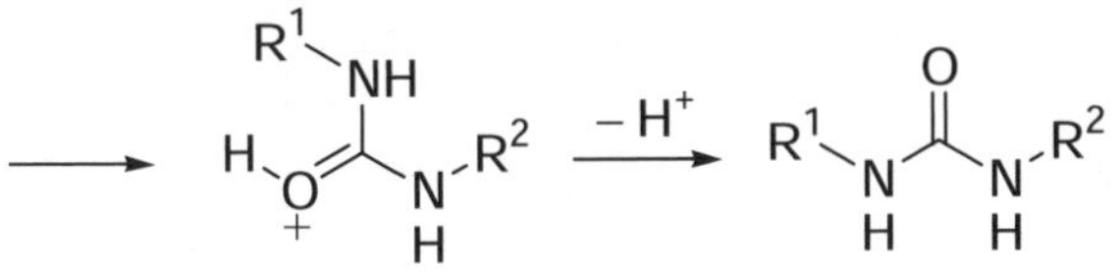

$-H^+$

References

1. Tiemann, F. *Ber.* **1891**, *24*, 4162.
2. Garapon, J.; Sillion, B.; Bonnier, J. M. *Tetrahedron Lett.* **1970**, 4905.
3. Adams, G. W.; Bowie, J. H.; Hayes, R. N.; Gross, M. L. *J. Chem. Soc., Perkin Trans. 2* **1992**, 897.
4. Bakunov, S. A.; Rukavishnikov, A. V.; Tkachev, A. V. *Synthesis* **2000**, 1148.

Tiffeneau–Demjanov rearrangement

OH
$NaNO_2$
NH_2
H^+
O

Step 1, Generation of N_2O_3

HO N O
H^+
H O N O
H_2O^+ N O
$- H_2O$
O N O N O

Step 2, Transformation of amine to diazonium salt

OH
O N O N O
NH_2
$- NO_2^-$
OH
N–N=O
H
H^+
OH
NO_2^-

H^+
OH
$N=N-OH_2^+$
$- H_2O$
OH
N^+
N

Step 3, Ring-expansion *via* rearrangement

OH $^+N \equiv N$
$- N_2$
+
O–H
$- H^+$
O

References

1. Tiffeneau, M.; Weil, P.; Tehoubar, B. *Compt. Rend* **1937**, *205*, 54.
2. Smith, P. A. S.; Baer, D. R. *Org. React.* **1960**, *11*, 157.
3. Fattori, D.; Henry, S.; Vogel, P. *Tetrahedron* **1993**, *49*, 1649.
4. Houdai, T.; Matsuoka, S.; Murata, M.; Satake, M.; Ota, S.; Oshima, Y.; Rhodes, L. L. *Tetrahedron* **2001**, *57*, 5551.

Tishchenko reaction

Esters from the corresponding aldehydes and $Al(OEt)_3$.

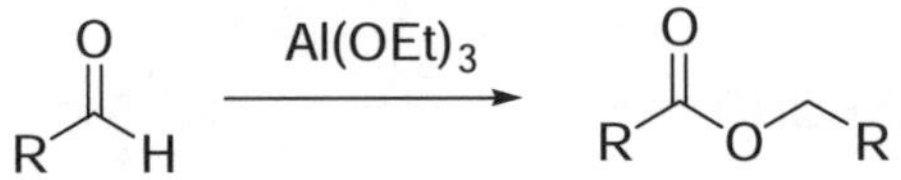

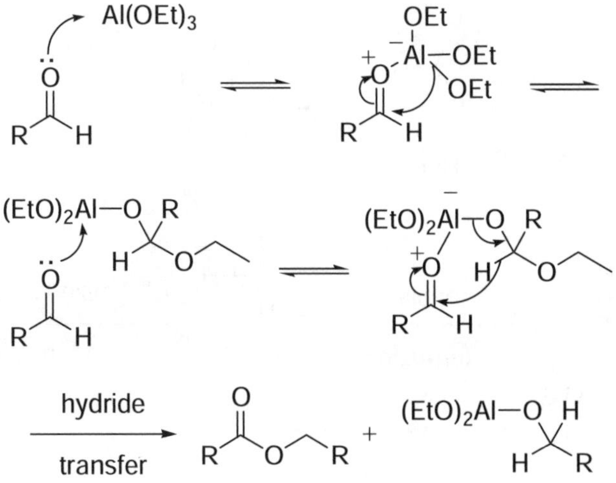

References

1. Tishchenko, V. *J. Russ. Phys. Chem. Soc.* **1906**, *38*, 355.
2. Saegusa, T.; Ueshima, T.; Kitagawa, S. *Bull. Chem. Soc. Jpn.* **1969**, *42*, 248.
3. Ugata, Y.; Kishi, I. *Tetrahedron* **1969**, *25*, 929.
4. Berberich, H.; Roesky, P. W. *Angew. Chem., Int. Ed.* **1998**, *37*, 1569.
5. Lu, L.; Chang, H.-Y.; Fang, J.-M. *J. Org. Chem.* **1999**, *64*, 843.
6. Mascarenhas, C.; Duffey, M. O.; Liu, S.-Y.; Morken, J. P. *Org. Lett.* **1999**, *1*, 1427.
7. Bideau, F. L.; Coradin, T.; Gourier, D.; Hénique, J.; Samuel, E. *Tetrahedron Lett.* **2000**, *41*, 5215.
8. Toermaekangas, O. P.; Koskinen, A. M. P. *Org. Process Res. Dev.* **2001**, *5*, 421.

Tollens reaction

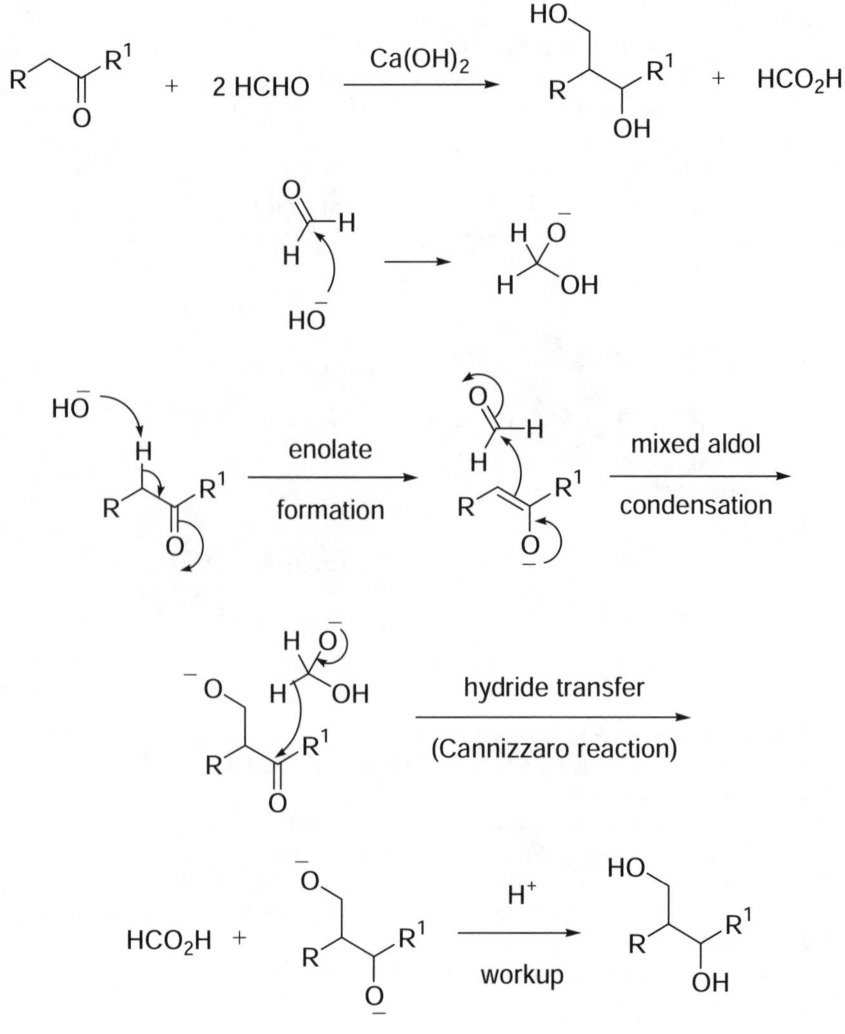

$$R-CH_2-C(=O)-R^1 + 2\,HCHO \xrightarrow{Ca(OH)_2} R-CH(CH_2OH)-CH(OH)-R^1 + HCO_2H$$

$$H_2C=O + HO^- \longrightarrow H_2C(O^-)(OH)$$

enolate formation

mixed aldol condensation

hydride transfer (Cannizzaro reaction)

$$HCO_2H + {}^-OCH_2-CH(R)-CH(O^-)-R^1 \xrightarrow[\text{workup}]{H^+} HOCH_2-CH(R)-CH(OH)-R^1$$

References

1. Parry-Jones, R.; Kumar, J. *Educ. Chem.* **1985**, *22*, 114.
2. Jenkins, I. D. *J. Chem. Educ.* **1987**, *64*, 164.
3. Munoz, S.; Gokel, G. W. *J. Am. Chem. Soc.* **1993**, *115*, 4899.

Tsuji–Trost reaction

BnO OPh — $Pd_2(dba)_3$, dppb / THF, 60–70 °C → imidazole (N, NH), 72% →

Pd(0)L_n coordination

Pd(II)L_n

π-allyl complex

S_N2 → + Pd(0)L_n

References

1. Tsuji, J.; Takahashi, H.; Morikawa, M. *Tetrahedron Lett.* **1965**, 4387.
2. Tsuji, J. *Acc. Chem. Res.* **1969**, *2*, 144.
3. Godleski, S. A. In *Comprehensive Organic Synthesis* Trost, B. M.; and Fleming, I.; eds., *vol. 4.* Chapter 3.3. Pergamon: Oxford, **1991**.
4. Bolitt, V.; Chaguir, B.; Sinou, D. *Tetrahedron Lett.* **1992**, *33*, 2481.
5. Moreno-Mañas, M.; Pleixats, R. In *Advances in Heterocyclic Chemistry* A.R. Katritzky, ed.; Academic Press: San Diego, **1996**, *66*, 73.
6. Tietze, L. F.; Nordmann, G. *Eur. J. Org. Chem.* **2001**, 3247.

Ueno–Stork cyclization

OEt

NBS

Br

OH

O

OEt

Bu_3SnH, AIBN

H

H

O

OEt

NC—N=N—CN

Δ

N_2↑ + 2 •—CN

2,2'-azobisisobutyronitrile (AIBN)

n-Bu_3Sn—H •—CN ⟶ n-Bu_3Sn• + H—CN

n-Bu_3Sn•

Br

O OEt

O OEt

5-*exo-trig*

cyclization

n-Bu_3Sn—H

H

H O OEt

hydrogen atom

abstraction

H

H O OEt

References

1. Ueno, Y.; Chino, K.; Watanabe, M.; Moriya, O.; Okawara, M. *J. Am. Chem. Soc.* **1982**, *104,* 5564.
2. Stork, G.; Mook, R.; Biller, S. A.; Rychnovsky, S. D. *ibid.* **1983**, *105,* 3741.
3. Villar, F.; Renaud, P. *Tetrahedron Lett.* **1998**, *39,* 8655.
4. Villar, F.; Andrey, O.; Renaud, P. *ibid.* **1999**, *40,* 3375.
5. Villar, F.; Equey, O.; Renaud, P. *Org. Lett.* **2000**, *2,* 1061.

Ugi reaction

Four-component condensation (4CC) of carboxylic acids, *C*-isocyanides, amines, and oxo compounds to afford peptides. *Cf.* Passerini reaction.

$$R{-}CO_2H + R^1{-}NH_2 + R^2{-}CHO + R^3{-}\overset{+}{N}{\equiv}\overset{-}{C} \longrightarrow$$

isocyanide

imine

R−CO$_2$H

References

1. Ugi, I. *Angew. Chem., Int. Ed. Engl.* **1962**, *1*, 8.
2. Ugi, I.; Lohberger, S.; Karl, R. In *Comprehensive Organic Synthesis*, Trost, B. M.; Fleming, I. Eds, Pergamon: Oxford, **1991**, *Vol. 2*, 1083,
3. Dömling, A.; Ugi, I. *Angew. Chem., Int. Ed.* **2000**, *39*, 3168.
4. Zimmer, R.; Ziemer, A.; Grunner, M.; Brüdgam, I.; Hartl, H.; Reissig, H.-U. *Synthesis* **2001**, 1649.

Ullmann reaction

Homocoupling of aryl iodide in the presence of Cu.

$$2\ PhI \xrightarrow{Cu} CuI_2 + Ph{-}Ph$$

$$PhI \xrightarrow[\text{electron transfer}]{\text{Cu, single}} CuI + Ph\cdot \xrightarrow{CuI}$$

$$PhCuI \xrightarrow[\text{SET}]{PhI} CuI_2 + Ph{-}Ph$$

The overall transformation of PhI to PhCuI is an oxidative addition process.

References

1. Ullmann, F. *Liebigs Ann. Chem.* **1904**, *332,* 38.
2. Fanta, P. E. *Synthesis* **1974**, 9.
3. Stark, L. M.; Lin, X.-F.; Flippin, L. A. *J. Org. Chem.* **2000**, *65,* 3227.
4. Belfield, K. D.; Schafer, K. J.; Mourad, W.; Reinhardt, B. A. *ibid.* **2000**, *65,* 4475.
5. Venkatraman, S.; Li, C.-J. *Tetrahedron Lett.* **2000**, *41,* 4831.
6. Farrar, J. M.; Sienkowska, M.; Kaszynski, P. *Synth. Commun.* **2000**, *30,* 4039.
7. Ma, D.; Xia, C. *Org. Lett.* **2001**, *3,* 2583.

Vilsmeier–Haack reaction

O

DMF, $POCl_3$

100 °C, 24 h

O

CHO

34%

+

O

CHO

4%

O

Cl–P–Cl

Cl

N

O

H

S_N2

O

+N

O–PCl_2

H

$^-$Cl

N

O

O–PCl_2

H

Cl

+N

H

Cl

O

$^-$O –PCl_2

Vilsmeier–Haack reagent

O:

+N

H

Cl

O+

H

Cl–H

N

O

Cl–H

N:

O

H

O:

H

H

N+

aqueous

workup

O

HO–H

N

O

CHO

References

1. Vilsmeier, A.; Haack, A. *Ber.* **1927**, *60*, 119.
2. Marson, C. M.; Giles, P. R. *Synthesis Using Vilsmeier Reagents* CRC Press, **1994**.
3. Jones, G.; Stanforth, S. P. *Org. React.* **1997**, *49*, 1.
4. Ali, M. M.; Tasneem; Rajanna, K. C.; Sai Prakash, P. K. *Synlett* **2001**, 251.

von Braun reaction

Treatment of tertiary amines with cyanogen bromide, resulting in cyanamide and alkyl halides.

$$R^1R_2NR \xrightarrow{Br-CN} R^1R_2N-CN + Br-R$$

Cyanogen bromide (BrCN) is a *counterattack reagent.*

$$R^1R_2N\!:\!R + NC-Br \xrightarrow{S_N2} Br^- + R^1R_2N^+(R)-CN \xrightarrow{S_N2} R^1R_2N-CN + Br-R$$

References

1. von Braun, J. *Ber.* **1907**, *40*, 3914.
2. Hageman, H. A. *Org. React.* **1953**, *7*, 198.
3. Nakahara, Y.; Niwaguchi, T.; Ishii, H. *Tetrahedron* **1977**, *33*, 1591.
4. Fodor, G.; Nagubandi, S. *Tetrahedron* **1980**, *36*, 1279.
5. Perni, R. B.; Gribble, G. W. *Org. Prep. Proced. Int.* **1980**, *15*, 297.
6. McLean, S.; Reynolds, W. F.; Zhu, X. *Can. J. Chem.* **1987**, *65*, 200.
7. Cooley, J. H.; Evain, E. J. *synthesis* **1989**, 1.
8. Aguirre, J. M.; Alesso, E. N.; Ibanez, A. F.; Tombari, D. G.; Moltrasio Iglesias, G. Y. *J. Heterocycl. Chem.* **1989**, *26*, 25.
9. Laabs, S.; Scherrmann, A.; Sudau, A.; Diederich, M.; Kierig, C.; Nubbemeyer, U. *Synlett* **1999**, 25.
10. Ouyang, A.; Ghoshal, M.; Sigler, G. *Abstracts of Papers, 222nd ACS National Meeting*, Chicago, IL, United States, August 26–30, **2001**, ORGN-378.

von Richter reaction

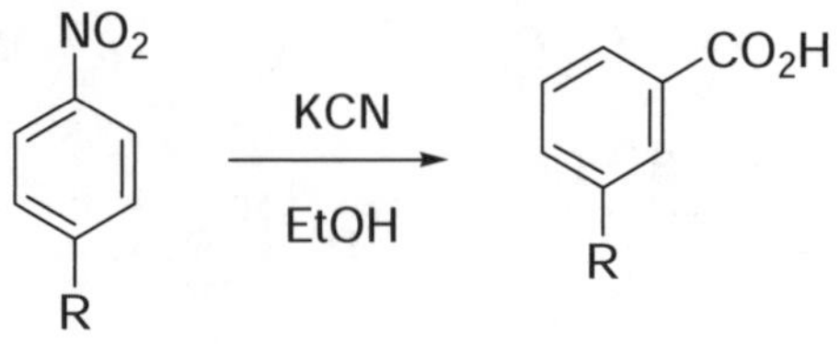

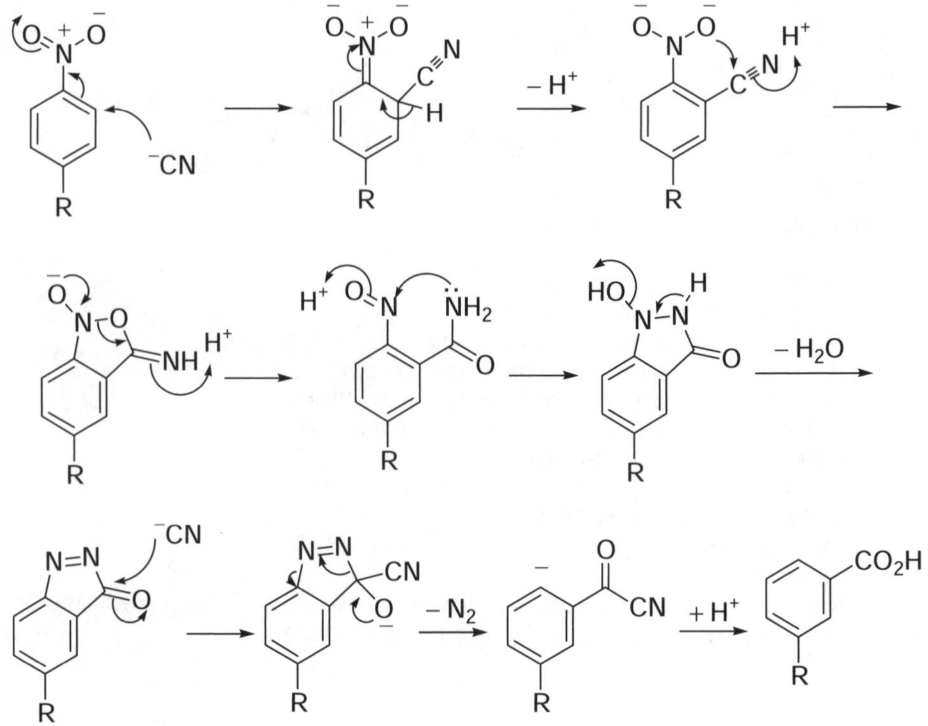

KCN

EtOH

NO_2 R CO_2H R

^{-}CN $-H^+$ H^+ $-H_2O$ NH_2 HO $-N_2$ $+H^+$ CO_2H

pyrazolone intermediate

References

1. von Richter, V. *Ber.* **1871**, *4*, 21, 459, 553.
2. Tretyakov, E. V.; Knight, D. W.; Vasilevsky, S. F. *Heterocycl. Commun.* **1998**, *4*, 519.
3. Tretyakov, E. V.; Knight, D. W.; Vasilevsky, S. F. *J. Chem. Soc., Perkin Trans. 1* **1999**, 3721.
4. Brase, S.; Dahmen, S.; Heuts, J. *Tetrahedron Lett.* **1999**, *40*, 6201.

Wacker oxidation

$$R\text{-}CH{=}CH_2 \xrightarrow[CuCl_2,\ O_2]{PdCl_2,\ H_2O} R\text{-}C(=O)CH_3$$

$$R\text{-}CH{=}CH_2 + PdCl_2 \xrightarrow{\text{palladation}}$$

$$\xrightarrow[\text{nucleophilic attack}]{H_2O} R\text{-}CH(OH)\text{-}CH_2\text{-}PdCl \xrightarrow[\text{elimination}]{\beta\text{-hydride}}$$

$$H\text{-}Pd\text{-}Cl + R\text{-}C(OH){=}CH_2 \xrightarrow{\text{tautomerization}} R\text{-}C(=O)CH_3$$

$$H\text{-}Pd\text{-}Cl \longrightarrow Pd(0) + HCl$$

Regeneration of Pd(II):

$$Pd(0) + 2\ CuCl_2 \longrightarrow PdCl_2 + 2\ CuCl$$

Regeneration of Cu(II):

$$CuCl + O_2 \longrightarrow CuCl_2 + H_2O$$

References

1. Tsuji, J. *Synthesis* **1984**, 369.
2. Hegedus, L. S. In *Comp. Org. Syn.* Trost, B. M.; Fleming, I., Eds, Pergamon, **1991**, *Vol. 4*, 552.
3. Tsuji, J. *ibid.* **1991**, *7*, 449.
4. Feringa, B. L. *Transition Met. Org. Synth.* **1998**, *2*, 307.
5. Gaunt, M. J.; Yu, J.; Spencer, J. B. *Chem. Commun.* **2001**, 1844.

Wagner–Meerwein rearrangement

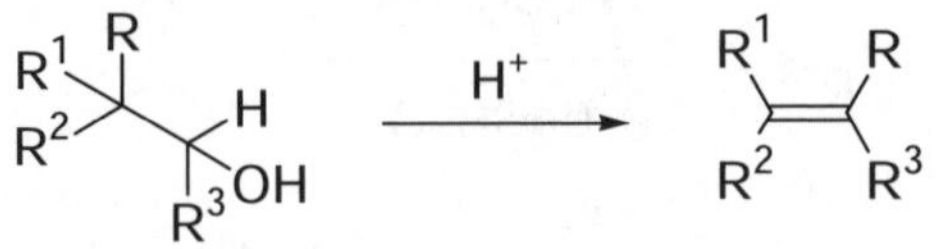

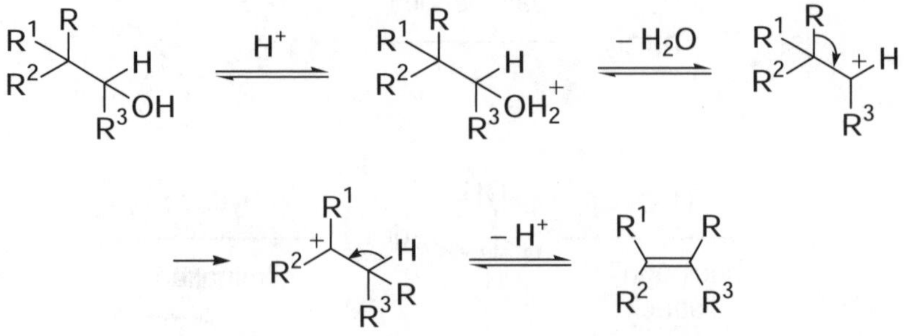

References

1. Wagner, G. *J. Russ. Phys. Chem. Soc.* **1899**, *31,* 690.
2. Hogeveen, H.; Van Kruchten, E. M. G. A. *Top. Curr. Chem.* **1979**, *80,* 89.
3. Martinez, A. G.; Vilar, E. T.; Fraile, A. G.; Fernandez, A. H.; De La Moya Cerero, S.; Jimenez, F. M. *Tetrahedron* **1998**, *54,* 4607.
4. Birladeanu, L. *J. Chem. Educ.* **2000**, *77,* 858.
5. Kobayashi, T.; Uchiyama, Y. *Perkin 1* **2000**, 2731.
6. Trost, B. M.; Yasukata, T. *J. Am. Chem. Soc.* **2001**, *123,* 7162.

Wallach rearrangement

Ar
N=N
O⁻
+
Δ
H+
Ar
N=N
OH

Ar
N=N+
O–
H+
Ar
N=N+
OH
Ar—N≡N
+ +
:OH2

Ar
N=N
+
OH
H
Ar
N=N
OH

References

1. Wallach, O.; Belli, L. *Ber.* **1880**, *13,* 525.
2. Cichon, L. *Wiad. Chem.* **1966**, *20,* 641.
3. Buncel, E.; Keum, S. R.; *J. Chem. Soc., Chem. Commun.* **1983**, 578.
4. Shine, H. J.; Subotkowski, W.; Gruszecka, E. *Can. J. Chem.* **1986**, *64*, 1108.
5. Okano, T. *Kikan Kagaku Sosetsu* **1998**, *37,* 130.
6. Hattori, H. *Kikan Kagaku Sosetsu* **1999**, *41*, 46.
7. Lalitha, A.; Pitchumani, K.; Srinivasan, C. *J. Mol. Catal. A: Chem.* **2000**, *162*, 429.

Weinreb amide

$$\mathrm{RCO_2R^1} \xrightarrow[\mathrm{MeONHMe{\bullet}HCl}]{\mathrm{AlMe_3}} \mathrm{RC(O)N(OMe)Me}$$

$$\xrightarrow{\mathrm{R^2M}} \left[\ \text{chelated tetrahedral intermediate}\ \right] \xrightarrow[\text{workup}]{\mathrm{H^+}} \mathrm{RC(O)R^2}$$

References

1. Nahm, S.; Weinreb, S. M. *Tetrahedron Lett.* **1981**, *22,* 3815.
2. Sibi, M. P. *Org. Prep. Proc. Int.* **1993**, *25,* 15.
3. Mentzel, M.; Hoffmann, H. M. R. *J. Prakt. Chem.* **1997**, *339,* 517.
4. Singh, J.; Satyamurthi, N.; Aidhen, I. S. *ibid.* **2000**, *342,* 340.
5. McNulty, J.; Grunner, V.; Mao, J. *Tetrahedron Lett.* **2001**, *42,* 5609.

Weiss reaction

Synthesis of *cis*-bicyclo[3.3.0]octane-3,7-dione.

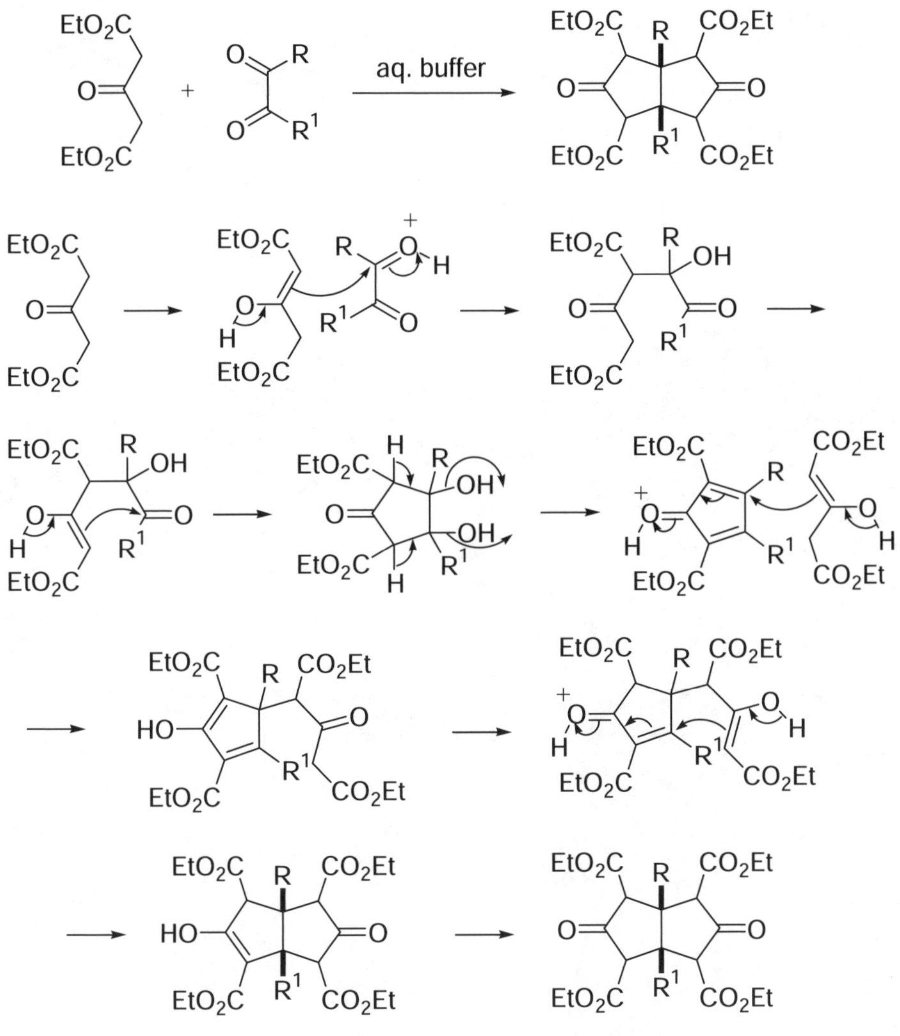

References

1. Weiss, U.; Edwards, J. M. *Tetrahedron Lett.* **1968**, 4885.
2. Gupta, A. K.; Fu, X.; Snyder, J. P.; Cook, J. M. *Tetrahedron* **1991**, *47*, 3665.
3. Reissig, H. U. *Org. Synth. Highlights* **1991**, 121.

4. Fu, X.; Cook, J. M. *Aldrichimica Acta* **1992**, *25*, 43.
5. Fu, X.; Kubiak, G.; Zhang, W.; Han, W.; Gupta, A. K.; Cook, J. M. *Tetrahedron* **1993**, *49*, 1511.
6. Van Ornum, S. G.; Li, J.; Kubiak, G. G.; Cook, J. M. *J. Chem. Soc., Perkin Trans. 1* **1997**, 3471.

Wharton oxygen transposition reaction

Reduction of α,β-epoxy ketones by hydrazine to allylic alcohols.

NH_2NH_2, Δ; + N_2↑ + H_2O

:NH_2NH_2; OH; NH–NH_2

H_2O + ; tautomerization; :OH_2

N≡N↑ + ; + H_3O^+; OH

References

1. Wharton, P. S.; Bohlen, D. H. *J. Org. Chem.* **1961**, *26*, 3615.
2. Wharton, P. S. *ibid.* **1961**, *26*, 4781.
3. Caine, D. *Org. Prep. Proced. Int.* **1988**, *20*, 1.
4. Dupuy, C.; Luche, J. L. *Tetrahedron* **1989**, *45*, 3437.
5. Di Filippo, M.; Fezza, F.; Izzo, I.; De Riccardis, F.; Sodano, G. *Eur. J. Org. Chem.* **2000**, 3247.

Willgerodt–Kindler reaction

Conversion of ketones to the corresponding thioamide and/or ammonium salt.

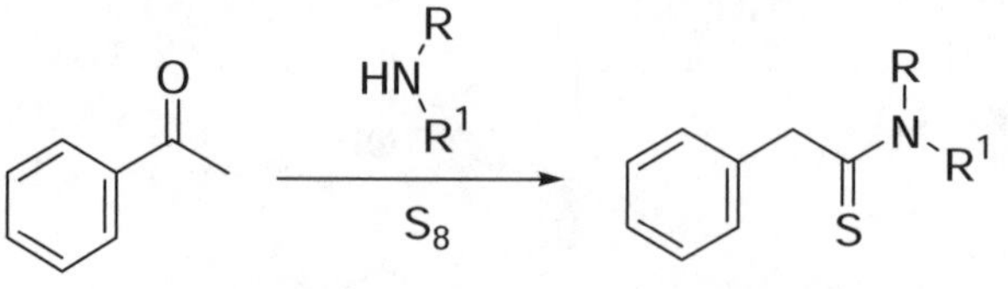

thioamide

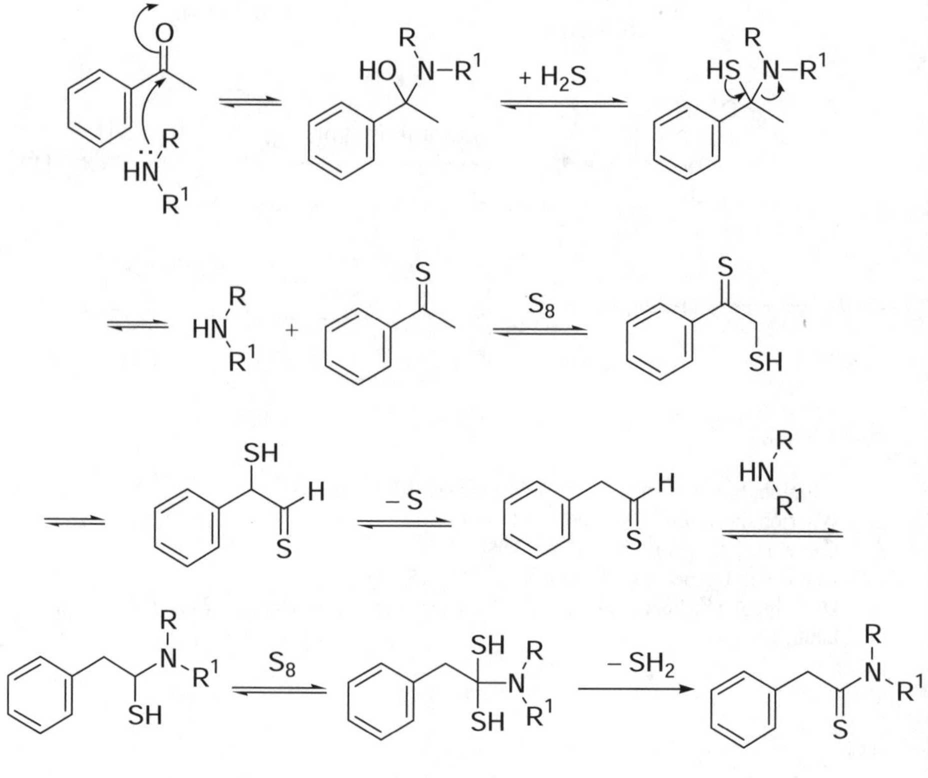

A slightly different mechanism has also been proposed:

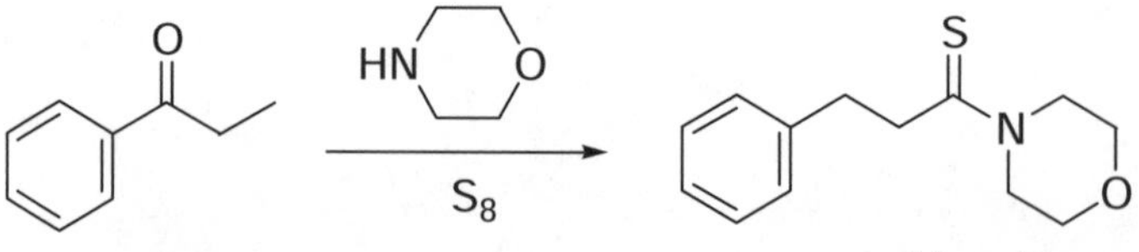

thioamide

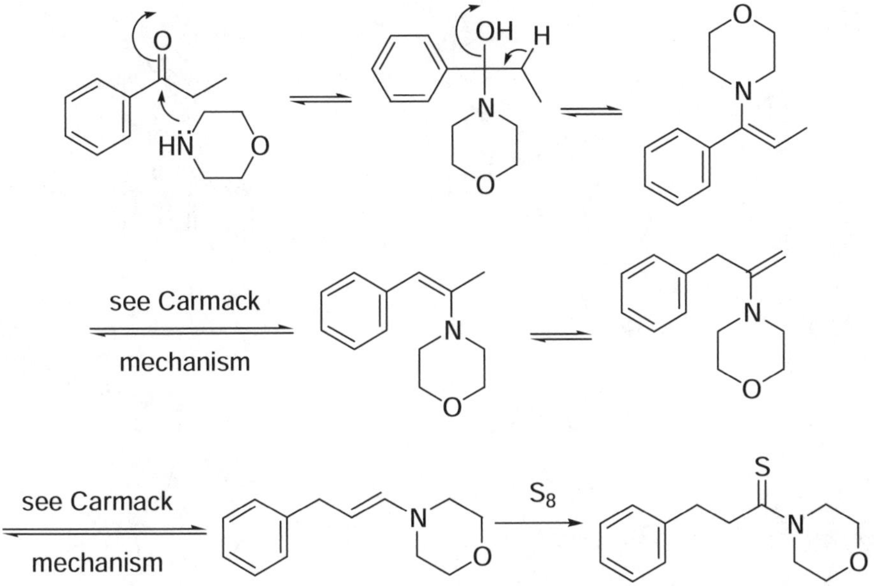

see Carmack mechanism

see Carmack mechanism

S_8

In Carmack's mechanism [5], the most unusual movement of a carbonyl group from methylene carbon to methylene carbon was proposed to go through an intricate pathway *via* a highly reactive intermediate with a sulfur-containing heterocyclic ring. The sulfenamide serves as the isomerization catalyst:

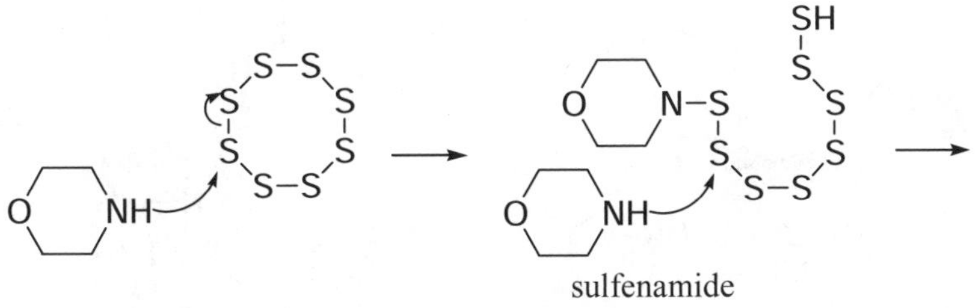

sulfenamide

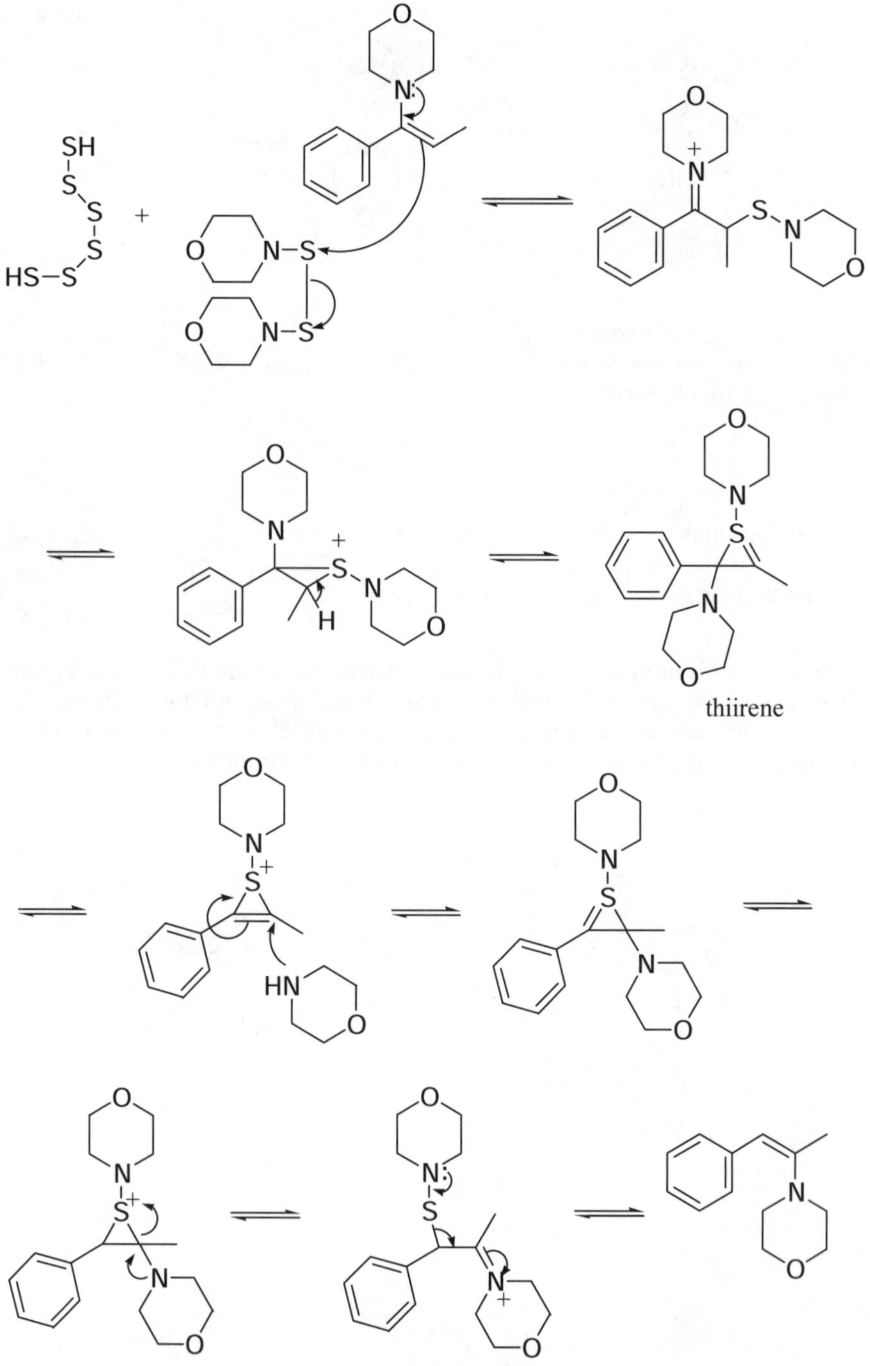

thiirene

References

1. Willgerodt, C. *Ber.* **1887**, *20*, 2467.
2. Schneller, S. W. *Int. J. Sulfur Chem. B* **1972**, *7*, 155.
3. Schneller, S. W. *Int. J. Sulfur Chem.* **1973**, *8*, 485.
4. Schneller, S. W. *ibid,* **1976**, *8*, 579.
5. Carmack, M. *J. Heterocycl. Chem.* **1989**, *26*, 1319.
6. You, Q.; Zhou, H.; Wang, Q.; Lei, X. *Org. Prep. Proced. Int.* **1991**, *23*, 435.
7. Chatterjea, J. N.; Singh, R. P.; Ojha, N.; Prasad, R. *J. Inst. Chem. (India)* **1998**, *70*, 108.
8. Moghaddam, F. M.; Ghaffarzadeh, M.; Dakamin, M. G. *J. Chem. Res., (S)* **2000**, 228.
9. Poupaert, J. H.; Bouinidane, K.; Renard, M.; Lambert, D. M.; Isa, M. *Org. Prep. Proced. Int.* **2001**, *33*, 335.

Wittig reaction

Olefination of carbonyls using phosphorus ylides.

Ph_3P + R^1 R X ⟶ X^- $Ph_3\overset{+}{P}$–C(R^1)(R)–H —base⟶

Ph_3P=C(R^1)(R) —[R_3, R^2, O]⟶ $Ph_3P=O$ + R_3 R^2 R^1 R

Ph_3P: R^1 R X —S_N2⟶ X^- $Ph_3\overset{+}{P}$ R^1 H R :B ⟶

Ph_3P=C(R^1)(R) ⟷ + Ph_3P O R^2 R^1 – R^3 R —[2 + 2] cycloaddition⟶

“puckered” transition state, irreversible and concerted

Ph_3P–O R R^3 R^1 R^2 ⟶ $Ph_3P=O$ + R^3 R^2 R^1 R

oxaphosphetane intermediate

References

1. Wittig, G.; Schöllkopf, U. *Ber.* **1954**, *87,* 1318.
2. Vedejs, E.; Peterson, M. J. *Top. Stereochem.* **1994**, *21,* 1.
3. Heron, B. M. *Heterocycles* **1995**, *41,* 2357.
4. Rein, T.; Reiser, O. *Acta Chem. Scand.* **1996**, *50,* 369.
5. Murphy, P. J.; Lee, S. E. *J. Chem. Soc., Perkin Trans. 1* **1999**, 3049.
6. Frattini, S.; Quai, M.; Cereda, E. *Tetrahedron Lett.* **2001**, *42,* 6827.

[1,2]-Wittig rearrangement

Treatment of ethers with alkyl lithium results in alcohols.

R^2Li

deprotonation

[1,2]-sigmatropic rearrangement

workup

The radical mechanism is also possible as radical intermediates have been identified.

References

1. Wittig, G.; Löhmann *Ann.* **1942**, *550*, 260.
2. Hoffmann, R. W. *Angew. Chem.* **1979**, *91*, 625.
3. Tomooka, K.; Yamamoto, H.; Nakai, T. *Liebigs Ann.* **1997**, 1275.
4. Maleczka, R. E., Jr.; Geng, F. J. *Am. Chem. Soc.* **1998**, *120*, 8551.
5. Tomooka, K.; Kikuchi, M.; Igawa, K.; Suzuki, M.; Keong, P.-H.; Nakai, T. *Angew. Chem., Int. Ed.* **2000**, *39*, 4502.
6. Katritzky, A. R.; Fang, Y. *Heterocycles* **2000**, *53*, 1783.
7. Kitagawa, O.; Momose, S.-i.; Yamada, Y.; Shiro, M.; Taguchi, T. *Tetrahedron Lett.* **2001**, *42*, 4865.

[2,3]-Wittig rearrangement

Transformation of allyl ethers into homoallylic alcohols by treatment with base. Also known as Still–Wittig rearrangement.

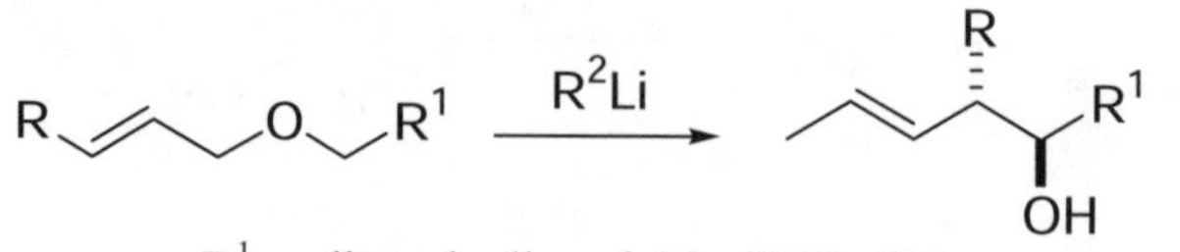

R^1 = alkynyl, alkenyl, Ph, COR, CN.

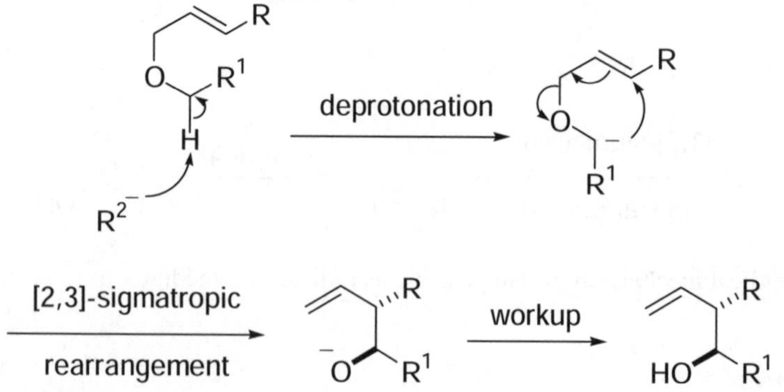

References

1. Cast, J.; Stevens, T. S.; Holmes, J. *J. Chem. Soc.* **1960**, 3521.
2. Nakai, T.; Mikami, K. *Org. React.* **1994**, *46*, 105.
3. Bertrand, P.; Gesson, J.-P.; Renoux, B.; Tranoy, I. *Tetrahedron Lett.* **1995**, *36*, 4073.
4. Maleczka, R. E., Jr.; Geng, F. *Org. Lett.* **1999**, *1*, 1111.
5. Tsubuki, M.; Kamata, T.; Nakatani, M.; Yamazaki, K.; Matsui, T.; Honda, T. *Tetrahedron: Asymmetry* **2000**, *11*, 4725.
6. Itoh, T.; Kudo, K. *Tetrahedron Lett.* **2001**, *42*, 1317.
7. Pévet, I.; Meyer, C.; Cossy, J. *Tetrahedron Lett.* **2001**, *42*, 5215.

Wohl–Ziegler reaction

Allylic bromination.

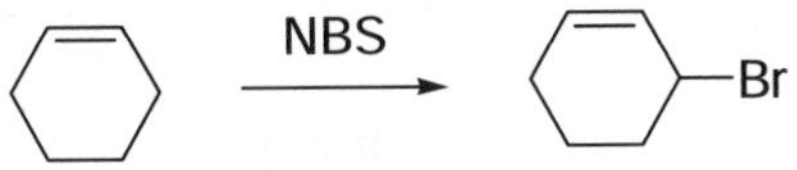

NBS

Initiation:

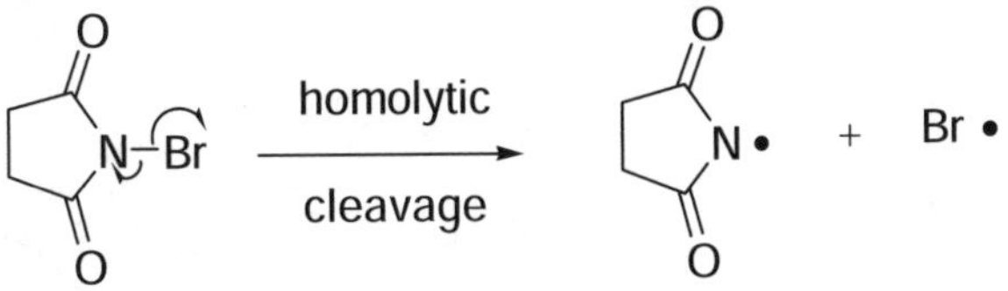

homolytic cleavage

Propagation:

H abstraction

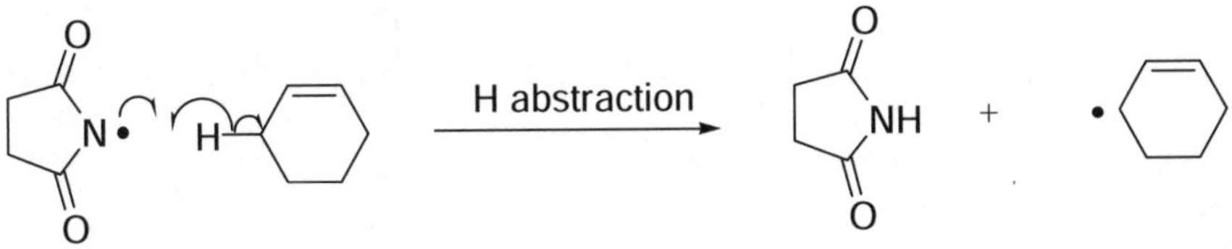

Termination:

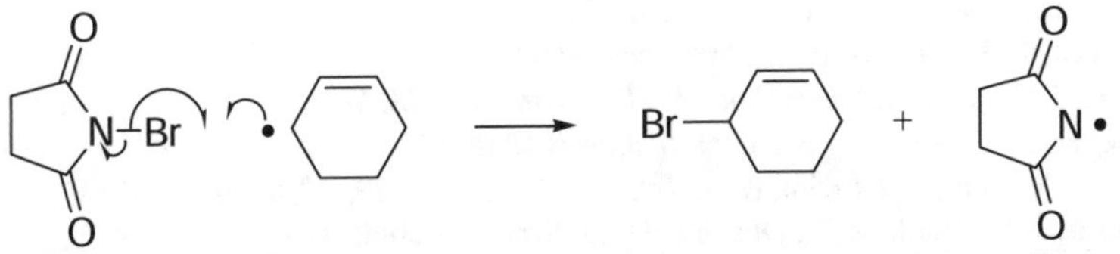

The succinimidyl radical now is available for the next cycle of the radical chain reaction

References

1. Wohl, A. *Ber.* **1919**, *52*, 51.
2. Wolfe, S.; Awang, D. V. C. *Can. J. Chem.* **1971**, *49*, 1384.
3. Strunz, G. M.; Court, A. S. *Experientia* **1970**, *26*, 1054.
4. Ito, I.; Ueda, T. *Chem. Pharm. Bull.* **1975**, *23*, 1646.
5. Pennanen, S. I. *Heterocycles* **1978**, *9*, 1047.
6. Rose, U. *J. Heterocycl. Chem.* **1991**, *28*, 2005.
7. Gavriliu, D.; Draghici, C.; Maior, O. *An. Univ. Bucuresti, Chim.* **1997**, *6*, 93.

Wolff rearrangement

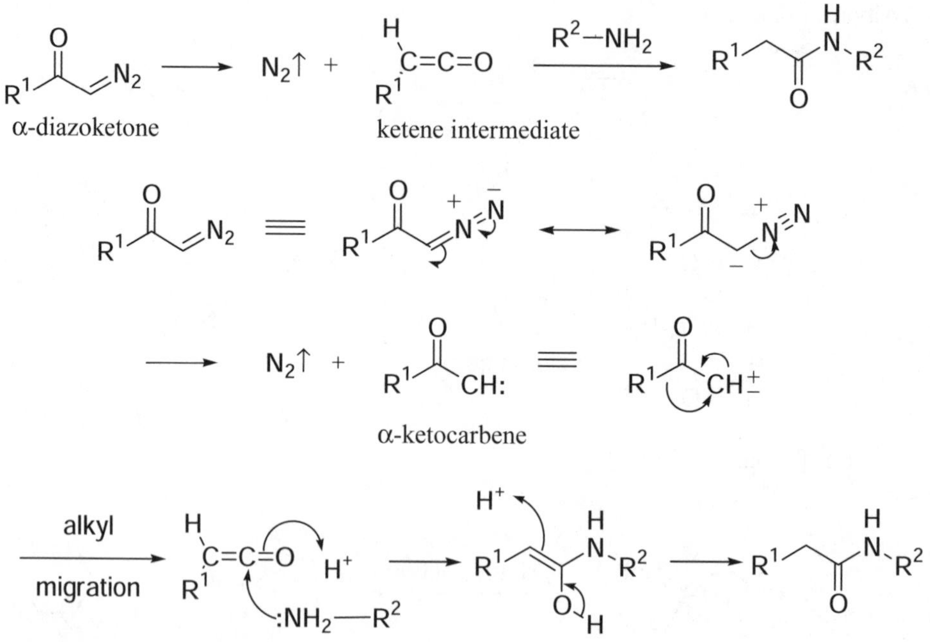

α-diazoketone

ketene intermediate

α-ketocarbene

alkyl migration

References

1. Wolff, L. *Ann.* **1912**, *394*, 25.
2. Meier, H.; Zeller, K. P. *Angew. Chem.* **1975**, *87*, 52.
3. Podlech, J.; Linder, M. R. *J. Org. Chem.* **1997**, *62,* 5873.
4. Wang, J.; Hou, Y. *J. Chem. Soc., Perkin Trans. 1* **1998**, 1919.
5. Mueller, A.; Vogt, C.; Sewald, N. *Synthesis* **1998**, 837.
6. Lee, Y. R.; Suk, J. Y.; Kim, B. S. *Tetrahedron Lett.* **1999**, *40,* 8219.
7. Tilekar, J. N.; Patil, N. T.; Dhavale, D. D. *Synthesis* **2000**, 395.
8. Yang, H.; Foster, K.; Stephenson, C. R. J.; Brown, W.; Roberts, E. *Org. Lett.* **2000**, *2,* 2177.
9. Xu, J.; Zhang, Q.; Chen, L.; Chen, H. *J. Chem. Soc., Perkin Trans. 1* **2001**, 2256.

Wolff–Kishner reduction

Carbonyl reduction to methylene using basic hydrazine.

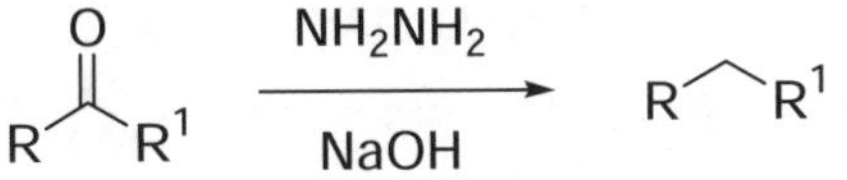

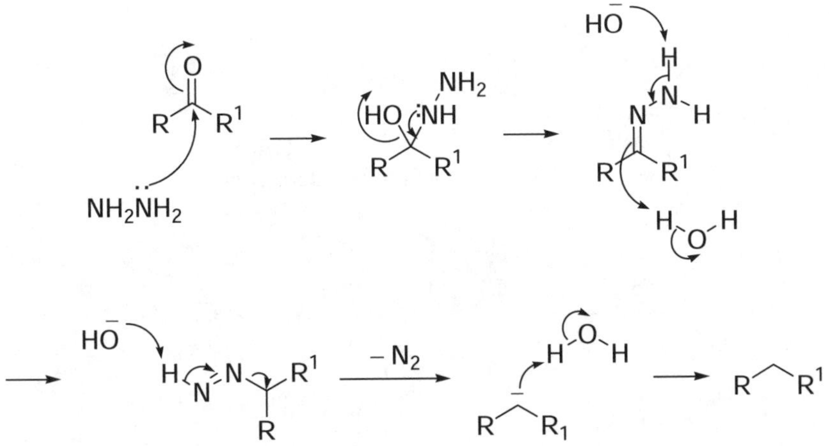

References

1. Kishner, N. *J. Russ. Phys. Chem. Soc.* **1911**, *43,* 582.
2. Szmant, H. H. *Angew. Chem., Int. Ed. Engl.* **1969**, *7,* 120.
3. Murray, R. K., Jr.; Babiak, K. A. *J. Org. Chem.* **1973**, *38,* 2556.
4. Akhila, A.; Banthorpe, D. V. *Indian J. Chem.* **1980**, *19B,* 998.
5. Bosch, J.; Moral, M.; Rubiralta, M. *Heterocycles* **1983**, *20,* 509.
6. Taber, D. F.; Stachel, S. J. *Tetrahedron Lett.* **1992**, *33,* 903.
7. Gadhwal, S.; Baruah, M.; Sandhu, J. S. *Synlett* **1999**, 1573.
8. Eisenbraun, E. J.; Payne, K. W.; Bymaster, J. S. *Ind. Eng. Chem. Res.* **2000**, *39,* 1119.

Woodward *cis*-dihydroxylation

Cf. Prévost *trans*-dihydroxylation

1. I_2, AgOAc

2. H_2O, H^+

cyclic iodonium ion intermediate

neighboring group assistance

References

1. Woodward, R. B. *J. Am. Chem. Soc.* **1958**, *80*, 209.
2. Mangoni, L.; Dovinola, V. *Tetrahedron Lett.* **1969**, 5235.
3. Kamano, Y.; Pettit, G. R.; Tozawa, M.; Komeichi, Y.; Inoue, M. *J. Org. Chem.* **1975**, *40*, 2136.
4. Brimble, M. A.; Nairn, M. R. *J. Org. Chem.* **1996**, *61*, 4801.
5. Hamm, S.; Hennig, L.; Findeisen, M.; Muller, D.; Welzel, P. *Tetrahedron* **2000**, *56*, 134.

Yamada coupling reagent

The use of diethyl phosphoryl cyanide (diethyl cyanophosphonate) for the activation of carboxylic acids.

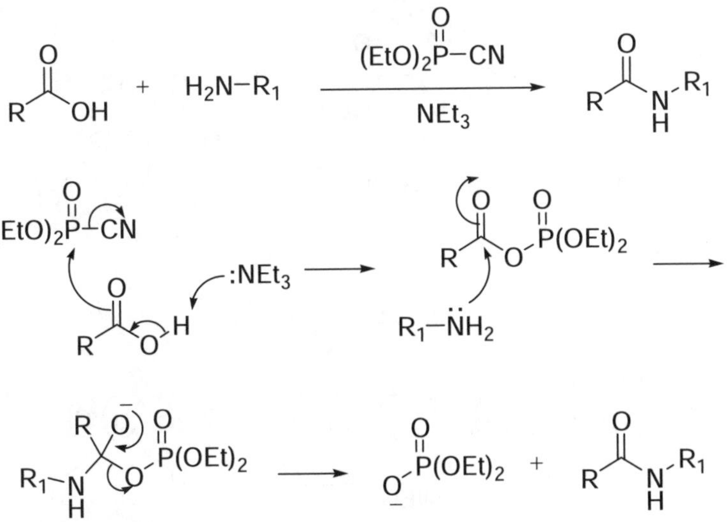

$RCOOH + H_2N{-}R_1 \xrightarrow[NEt_3]{(EtO)_2P(O){-}CN} RC(O)NHR_1$

(EtO)$_2$P(O)–CN

:NEt$_3$

R–C(O)–O–P(O)(OEt)$_2$

$R_1{-}\ddot{N}H_2$

R–C(O$^-$)(NHR$_1$)–O–P(O)(OEt)$_2$ ⟶ $^-$O–P(O)(OEt)$_2$ + RC(O)NHR$_1$

References

1. Yamada, S. *Tetrahedron Lett.* **1971**, 3595.
2. Yamada, S.-i.; Kasai, Y.; Shioiri, T. *ibid.* **1973**, 1595.
3. Yokoyama, Y.; Shioiri, T.; Yamada, S. *Chem. Pharm. Bull.* **1977**, *25*, 2423.
4. Shioiri, T.; Hamada, Y. *J. Org. Chem.* **1978**, *43*, 3631.
5. Kato, N.; Hamada, Y.; Shioiri, T. *Chem. Pharm. Bull.* **1984**, *32*, 3323.
6. Hamada, Y.; Mizuno, A.; Ohno, T.; Shioiri, T. *ibid.* **1984**, *32*, 3683.

Yamaguchi esterification

Esterification using 2,4,6-trichlorobenzoyl chloride (Yamaguchi reagent).

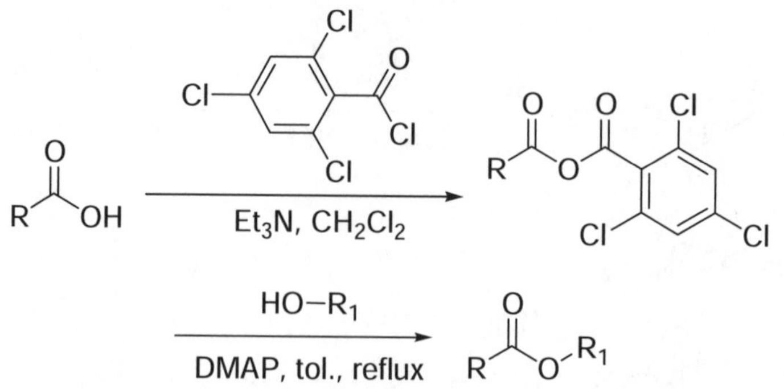

Et_3N, CH_2Cl_2

$HO{-}R_1$

DMAP, tol., reflux

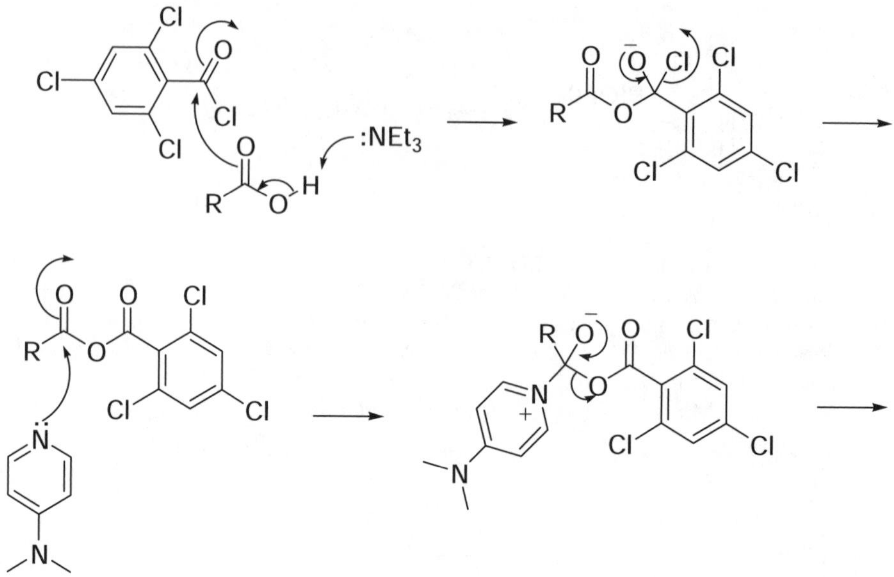

:NEt_3

DMAP (Dimethylaminopyridine)

Steric hindrance of the chloro substituents blocks attack of the other carbonyl.

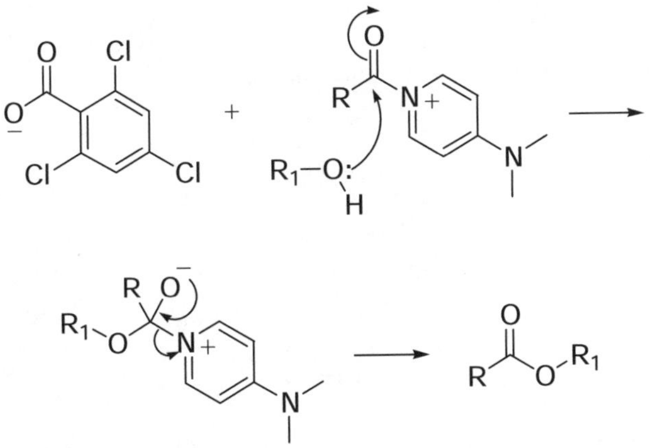

References

1. Yamaguchi, M. *Bull. Chem. Soc. Jpn.* **1979**, *52*, 1989.
2. Yamaguchi, M. *J. Org. Chem.* **1990**, *55*, 7.
3. Richardson, T.; Rychnovsky, S. D. *Tetrahedron* **1999**, *55*, 8977.
4. Berger, M.; Mulzer, J. *J. Am. Chem. Soc.* **1999**, *121*, 8393.
5. Paterson, I.; Chen, D. Y.-K.; Acena, J. L.; Franklin, A. S. *Org. Lett.* **2000**, *2*, 1513.

Zaitsev elimination

E2 elimination to give the more substituted olefin.

Br H $^{-}$OEt KOEt E2 + major minor

Hofmann elimination, on the other hand, furnishes the least highly substituted olefins.

B: H $\overset{+}{N}Me_3$ Base E2

References

Zaitsev elimination

1. Brown, H. C.; Wheeler, O. H. *J. Am. Chem. Soc.* **1956**, *78*, 2199.
2. Elrod, D. W.; Maggiora, G. M.; Trenary, R. G*Tetrahedron Comput. Methodol.* **1990**, *3*, 163.
3. Reinecke, M. G.; Smith, W. B. *J. Chem. Educ.* **1995**, *72*, 541.
4. Lewis, D. E. *Book of Abstracts, 214th ACS National Meeting*, Las Vegas, NV, September 7-11, (1997).

Hofmann elimination

1. Eubanks, J. R. I.; Sims, L. B.; Fry, A. *J. Am. Chem. Soc.* **1991**, *113*, 8821.
2. Bach, R. D.; Braden, M. L. *J. Org. Chem.* **1991**, *56*, 7194.
3. Lai, Y. H.; Eu, H. L. *J. Chem. Soc., Perkin Trans. 1* **1993**, 233.
4. Sepulveda-Arques, J.; Rosende, E. Go.; Marmol, D. P.; Garcia, E. Z.; Yruretagoyena, B.; Ezquerra, J. *Monatsh. Chem.* **1993**, *124*, 323.
5. Woolhouse, A. D.; Gainsford, G. J.; Crump, D. R. *J. Heterocycl. Chem.* **1993**, *30*, 873.
6. Bhonsle, J. B. *Synth. Commun.* **1995**, *25*, 289.
7. Berkes, D.; Netchitailo, P.; Morel, J.; Decroix, B. *ibid.* **1998**, *28*, 949.

Zinin benzidine rearrangement (semidine rearrangement)

cat. H^+

70%

30%

H^+

[5,5]-sigmatropic rearrangement

$-H^+$

[5,3]-sigmatropic rearrangement

References

1. Zinin, N. *J. Prakt. Chem.* **1845**, *36*, 93.
2. Hofmann, A. W. *Proc. Roy. Soc., London* **1863**, *12*, 576.
3. Banthorpe, D. V.; O'Sullivan, M. *J. Chem. Soc., Perkin Trans. 2* **1973**, 551.
4. Shine, H. J. *J. Phys. Org. Chem.* **1989**, *2*, 491.
5. Shine, H. J. *J. Chem. Educ.* **1989**, *66*, 793.
6. Davies, C. J.; heaton, B. T.; Jacob, C. *J. Chem. Soc., Chem. Commun.* **1995**, 1177.

Subject Index